Contributions to Economics

More information about this series at http://www.springer.com/series/1262

Tiago Sequeira • Liliana Reis
Editors

# Climate Change and Global Development

## Market, Global Players and Empirical Evidence

Springer

*Editors*
Tiago Sequeira
Departamento de Gestao e Economia
Universidade da Beira Interior
Covilha, Portugal

Liliana Reis
Departamento de Gestao e Economia
Universidade da Beira Interior
Covilha, Portugal

ISSN 1431-1933 ISSN 2197-7178 (electronic)
Contributions to Economics
ISBN 978-3-030-02661-5 ISBN 978-3-030-02662-2 (eBook)
https://doi.org/10.1007/978-3-030-02662-2



This Springer imprint is published by the registered company Springer Nature Switzerland AG.
The registered company address is: Gewerbestrasse 11, 6330 Cham, Switzerland

# Contents

**Introduction** . . . . . . . . . . . . . . . . . . . . . . . . . . . . . . . . . . . . . . . 1
Tiago Sequeira and Liliana Reis

**Part I A Market Approach**

**Global Development and Climate Change: A Game Theory Approach** . . . . . . . . . . . . . . . . . . . . . . . . . . . . . . . . . . . . 17
António Bento Caleiro, Miguel Rocha de Sousa, and Ingo Andrade de Oliveira

**On the Edge of Climate Change: In a Search of an Adequate Agent-Based Methodology to Model Environmental Dynamics** . . . . . . . . 37
Mariya Gubareva and Orlando Gomes

**Greenhouse Gas, Carbon Dioxide Emissions and Economic Growth: Empirical Evidence from Threshold Effect** . . . . . . . . . . . . . . . . . . . . . . . . 59
Qaiser Munir and Sook Ching Kok

**Effects of Climate Change to Industrial Outputs and Employment in Asian Emerging Economies** . . . . . . . . . . . . . . . . . . . . . . . . . . . 83
Sook Ching Kok and Qaiser Munir

**The Role of Carbon Markets in the Paris Agreement: Mitigation and Development** . . . . . . . . . . . . . . . . . . . . . . . . . . . . . . . . . . . . 109
Henrique Schneider

**Part II The Role of Global Players**

**The European Union as a Leading Environmental Player? A Critical Analysis on the Policy and Commitments Towards Global Development and Climate Change** . . . . . . . . . . . . . . . . . . . . . . 135
Ana Isabel Xavier

**The European Union Accession and Climate Change Policies in the Western Balkan Countries** . . . . . 153
Teresa Maria Resende Cierco Gomes

**The BRICS Commitment on Climate Change: Process Towards an Effective Approach in the Path of Sustainable Development** . . . . . 175
Marco António Baptista Martins

**Reducing Catastrophic Climate Risk by Revolutionizing the Amazon: Novel Pathways for Brazilian Diplomacy** . . . . . 189
Joana Castro Pereira

**Climate Change Mitigation and Adaptation in Africa: Strategies, Synergies, and Constraints** . . . . . 219
Linus M. Nyiwul

**Part III Climate Change Mitigation: Micro Evidence**

**Climate Change Mitigation Through the Application of LCA Methodology on the Environmental Performance of Two Vehicles with Distinct Engines** . . . . . 245
Bruno Pereira, Radu Godina, João C. O. Matias, and Susana Garrido Azevedo

**The Role of Marketing in Reducing Climate Change: An Approach to the Sustainable Marketing Orientation** . . . . . 261
Cristina Calvo-Porral

**Madeira Island: Tourism, Natural Disasters and Destination Image** . . . . . 285
António Manuel Martins de Almeida and Luiz Pinto Machado

**Conclusion** . . . . . 303
Tiago Sequeira and Liliana Reis

# Introduction

**Tiago Sequeira and Liliana Reis**

**Contents**

Bibliography .................................................................................. 13

**Abstract** Nowadays, one of the most important issues crossing international relations with economics is climate change. "Environmental issues cut across a range of topics, namely security and economics, two areas of major importance to the state" (Pereira, Revista Brasileira de Política Internacional, 58(1):194, 2015). Indeed, the editors of this book realized that this topic is still receiving too little attention from the two scientific areas.

The aim of this book from the outset is therefore to understand the implications of climate change on several questions pertaining to economics and international relations fields and how climate change could jeopardize the overall development of the design of the international system, as we have known until the present. Truly, the main subject of concern to international relations and economics is governance, namely, good governance, and how it affects power to maximize state interests and resources in order to obtain quality of life for individuals. The rational choices taken by sovereign states concerning environmental issues and policy decisions on climate change lead to good or bad outcomes for all human beings, unlike decisions on internal policies that affect only the state's citizens themselves.

T. Sequeira (✉)
Universidade da Beira Interior and CEFAGE-UBI, Covilhã, Portugal
e-mail: sequeira@ubi.pt

L. Reis
Universidade da Beira Interior, Covilhã, Portugal

Instituto Português de Relações Internacionais (IPRI-NOVA), Lisboa, Portugal



T. Sequeira, L. Reis (eds.), *Climate Change and Global Development*,
Contributions to Economics, https://doi.org/10.1007/978-3-030-02662-2_1

Nowadays, one of the most important issues crossing international relations with economics is *climate change*. In fact, "Environmental issues cut across a range of topics, namely security and economics, two areas of major importance to the state" (Pereira 2015, p. 194). Indeed, the editors of this book realized that this topic is still receiving too little attention from the two scientific areas.

The aim of this book from the outset is therefore to understand the implications of climate change on several questions pertaining to economics and international relations fields and how climate change could jeopardize the overall development of the design of the international system, as we have known until the present.

Truly, the main subject of concern to international relations and economics is governance, namely, good governance and how it affects power to maximize state interests and resources in order to obtain quality of life for individuals. The rational choices taken by sovereign states concerning environmental issues and policy decisions on climate change lead to good or bad outcomes for all human beings, unlike decisions on internal policies that affect only the state's citizens themselves.

This book on climate change and global development brings together a number of texts from academics from various countries (and continents) and from the two scientific areas, with a view to enlightening what is at stake in relation to climate change and the future of global development. For this, the overall aim of this book is not to provide a guide to the complexity that covers all areas of climate change to IR and economics but to realize from an empirical basis and epistemological framework, with the use of several specific case studies regarding the experience of climate change to global development, the commitment and setbacks that the former have in the second, under the auspices of multilevel governance, namely, local, national, and supranational governance.

This book is organized into three parts, after this Introduction. In the first part, the chapters deal with market approaches to climate change mitigation as well as the effects of climate change on market outcomes, namely, economic growth and industrial output. Part 2 addresses deals with climate change policies in an international bilateral and multilateral countries environment, covering the action of global players. Part 3 deals with more micromanagement approaches to climate change mitigation.

With this organization we seek to provide the reader with a comprehensive approach to the challenges of climate change for the international multicountry relationship which broadly bases on market forces, thus providing information and quantification of the market incentives that can help to mitigate the consequences of climate change.

Regarding international relations, climate change can be seen from several standpoints: as a threat to international security and as a potential enhancer of future conflicts, how states negotiate international treaties and the evolution they have known in international law, and also as climate issues present in the states themselves, defending fundamental values so that they can be perceived as benign powers by others.

But to some authors (Jacobsen 1999, p. 206), "IR scholars did not begin writing about the environment until the late 1980s." After this period the debate on climate issues in international relations was dispersed among various epistemological debates, consolidating even new theories that were until then little dominant.

The end of the Cold War ushered in new international actors, "news" interdependencies, and the erosion of the traditional division between the national and international level. The main debates on IR theory seem insufficient to capture all new dynamics between old and new actors, and only weakly explains the new political, economic, social, and environmental demands of the current world. The national dimension seems to be insufficient to solve the new problems that need not only a global understanding but also comprehensive answers, as climate change and the dominant realist theory of international relations during the Cold War period fail to account for the dynamics of alliance formation "outside" the context of *high politics*. To realist theory sovereign states are the main actors in the international system and interact within an anarchic system in the absence of a world-governing authority, although international and regional institutions such as the UN and the EU create norms within the international system for states to abide by. However, these international actors "do not possess the legitimacy and capacity of force that obliges the states to abide by these norms and to fulfil them. In fact, States will ratify international treaties only if they wish, in a totally voluntary way. It is therefore within the framework of this anarchic environment the states will seek to maximize their relative power, how to ensure their survival. For this the states will make alliances with other states to balance the power" (Morgenthau 1948).

In this way, there is often contempt for interstate cooperation and no capacity of international institutions to shape states' behavior toward issues that do not bring them immediate benefits and increased power, which is actually needed for effective action on climate change (Carter 2010, p. 55). But, according to Heffron (2015, p. 17) "If states cannot work together to protect the planet, species, and all of humanity, from the effects of climate change, then a self-interest style approach can provide a rationale for doing so."

If we look to the foreign policy of the EU, Canada, or Japan, we can understand the importance of "low politics" issues, such as the environment and climate change, and how they present themselves as key drives on international system.

Thus, it is not difficult to perceive the deviation in the epistemological field that the climatic changes have been taking over time from the realistic theory. Instead of being analyzed from the point of view of security threats to the international system, what has been appealing to international relations scholars has been the dynamics of the negotiation process for obtaining international agreements. "By and large, neo-liberal approaches have monopolized the research agenda, which has been conceptualized mainly as the management of interdependence in a system of sovereign states lacking the kind of central authorities assumed to be capable of providing order and regulation within domestic societies" (Jacobsen 1999, p. 207).

If we look at climate change from the point of view of international negotiations and how understandings between sovereign states are reached and international agreements are carried out, neoliberal institutionalism is relevant to the discussion because it focuses on the role played by institutions within interstate cooperation. Keohane (2005, p. 82) emphasizes that institutions develop independently of the states that created them, and therefore they take on a permanent character. In the context of climate change, this seems to be an important feature, since global climate negotiations also take place outside of the exclusive realm of interstate relations. Keohane and Nye (1989, p. 55) argue, "institutions within the international system establish a network of interactions that will be difficult either to eradicate or drastically rearrange."

Haas et al. (1993) argues that, in the absence of world government, international institutions have three positive effects on the environment: (1) they can create high levels of governmental "concern," (2) they can function as hospitable "contractual" environments in which agreements can be made and kept, and (3) they can build sufficient political and administrative "ability" in national governments. In fact, once a treaty that governs states' responsibilities vis-à-vis climate change has been set up and institutionalized, actors party to the treaty will be bound to a certain extent by the treaty's content, delimiting the range of legitimate courses of action. In fact, international organizations and nongovernmental organizations can play an especially crucial role as agenda settings, framing the climate change discussion and putting pressure on state actors. Thus, much of the interstate discussion within climate change politics is based on institutions such as the United Nations Framework Convention on Climate Change (UNFCCC) or institutionalized frameworks such as the regular Conferences of the Parties (COP). Since institutions have become the most important actors in the discussion on climate change politics, the analysis of a neoliberal institutionalist framework is obligatory.

The United Nations Framework Convention on Climate Change (UNFCCC) is the first and main international agreement on climate action. It was one of three conventions adopted at the Rio Earth Summit in 1992. Until now, it has been ratified by almost 200 countries. It started as a way for countries to work together to limit global temperature increases and climate change and to cope with their impacts (Blobel and Meyer-Ohlendorf 2006). The Parties to the Convention agreed that the extent to which developing nations can meet their treaty obligations would depend on the extent to which developed countries provide finance and technology and that developed countries "should take the lead in combating climate change and the adverse effects." According to the agreement, "Economic and social development and poverty eradication are the first and overriding priorities of the developing country parties" (Gupta 2010).

The climate change international process revolves around the annual sessions of the COP, which bring together all countries that are Parties to the Convention. Article 7.2 defines the COP as the "supreme body" of the Convention, as it is its highest decision-making authority (United Nations Framework on Climate Change 1998, p. 27).

The first major change came in 1997 at the third Conference of the Parties, when Parties to the UNFCCC adopted the Kyoto Protocol (United Nations Framework on Climate Change 1998). This agreement created the first and only legally binding targets for developed nations to reduce their emissions, as well as important international monitoring, reporting, and verification mechanisms to enforce compliance. To help countries meet their targets, the protocol created "flexibility mechanisms"—

such as carbon trading and the Clean Development Mechanism, which allows industrialized nations to reach their targets by investing in emissions reductions in developing nations.

After Protocol Kyoto, the recent Paris agreement constitutes the most important step in addressing climate change by the international community. It is considered by some authors (Dimitrov 2016) as a political success in climate negotiations international and traditional state-based diplomacy, because the twenty-first Conference of the Parties (COP-21) was a culmination of a 4-year diplomatic process and was a genuine collective effort to reach mutual compromise and contains policy obligations for all countries. The election of Donald Trump as 44th president of the United States and his individual position against international climate accords, and all environmental issues, not surprisingly determined the position of the United States with regard to the Paris Agreements, but does not negate the importance of the agreement.

Indeed, what has been achieved by the United Nations since the United Nations Framework Convention on Climate Change in 1992 to date (as can be seen in Table 1) is truly remarkable and reflects a culture of compromise in international relations between states. It is important not only from the point of view of mere international negotiations but also from the need to respond to a global problem through a "global government" for the climate, approaching a utopian form of Kant's universal cosmopolitanism.

Over the last 25 years, the United Nations has failed to respond to all of the challenges of global development through climate change but have succeeded in creating a "green agenda" in parallel with human rights, from which states today can hardly escape.

To the constructivist approach "We must acknowledge that the debate over climate change, like almost all environmental issues, is a debate over culture, worldviews, and ideology" (Hoffman 2012, p. 32). More specifically, constructivism posits that ideas shape actors' decisions and behaviors: "The central insight of constructivism is that collectively held ideas shape the social, economic, and political world in which we live." The moral imperative to protect and preserve the environment is a norm that developed primarily in the twentieth century. Originating in the endangered species protection movement, the environmentalism movement went through three primary moves, each with influential factors, before it became the norm it is today: firstly, it moved from a national norm to an international norm; secondly, its structure shifted to combine governmental and nongovernmental elements; and, thirdly, it developed into a norm with an international focus, which is different than a national norm spreading to other nations while still having a domestic focus (Epstein 2006, p. 39).

According to Alexander Wendt (1992, p. 398), "The commitment to and the salience of particular identities vary, but each identity is an inherently social definition of the actor [...] which constitute[s] the structure of the social world. Identities are the basis of interests." So one of the key aspects of the development of the environmental movement was the resolution in rhetoric and discourse about

**Table 1** Main international events on climate change

| Date | Event | Achievements/developments |
| --- | --- | --- |
| November 1988 | Intergovernmental Panel on Climate Change (IPCC) established | World Meteorological Organization (WMO) and UN Environment Programme (UNEP) establish the Intergovernmental Panel on Climate Change (IPCC). To this day IPCC assessments are the scientific underpinning of international negotiations while also providing unique insights into managing the risk of extreme events and disasters, for example |
| November 1990 | The IPCC releases the first assessment report | The report says "emissions resulting from human activities are substantially increasing the atmospheric concentrations of greenhouse gases" leading to calls by the IPCC and the second World Climate Conference for a global treaty |
| December 1990 | The UN General Assembly establishes the Intergovernmental Negotiating Committee (INC) for a Framework Convention on Climate Change | The INC held 5 sessions where more than 150 states discussed binding commitments, targets, and timetables for emissions reductions, financial mechanisms, technology transfer, and "common but differentiated" responsibilities of developed and developing countries |
| May 1992 | The text of the United Nations Framework Convention on Climate Change is adopted | The UNFCCC objective is to "stabilize greenhouse gas concentrations in the atmosphere at a level that would prevent dangerous anthropogenic interference with the climate system" |
| June 1992 | The United Nations Framework Convention on Climate Change opens for signature at the Earth Summit in Rio, bringing the world together to curb greenhouse gas emissions and adapt to climate change | The UNFCCC has two sister conventions also agreed in Rio, the UN Convention on Biological Diversity and the Convention to Combat Desertification |
| March 21st, 1994 | The United Nations Framework Convention on Climate Change, spawned 2 years earlier in Rio, enters into force | Countries that sign the treaty are known as "Parties." With 196 Parties, the UNFCCC has near-universal membership. Parties meet annually at the Conference of the Parties (COP) to negotiate multilateral responses to climate change |
| April 1995 | Germany's then environment minister, Angela Merkel, presides over the first Conference of the Parties (COP 1) in Berlin | Parties agreed that commitments in the Convention were "inadequate" for meeting convention objectives. The Berlin Mandate establishes a process to negotiate strengthened commitments for developed countries, thus laying the groundwork for the Kyoto Protocol |

(continued)

**Table 1** (continued)

| Date | Event | Achievements/developments |
|---|---|---|
| December 11th, 1997 | The third Conference of the Parties achieves an historical milestone with adoption of the Kyoto Protocol | The world's first greenhouse gas emissions reduction treaty |
| Bonn, July 2001 | A major breakthrough is achieved at the second part of the sixth Conference of the Parties meeting in Bonn | Governments reaching a broad political agreement on the operational rulebook for the 1997 Kyoto Protocol |
| Marrakesh, November 2001 | The seventh Conference of the Parties results in the Marrakesh Accords, setting the stage for ratification of the Kyoto Protocol | This would formalize agreement on operational rules for International Emissions Trading, the Clean Development Mechanism and Joint Implementation along with a compliance regime and accounting procedures |
| January 2005 | EU Emissions Trading launches | The European Union Emissions Trading Scheme, the first and largest emissions trading scheme in the world, launches as a major pillar of EU climate policy. Installations regulated by the scheme are collectively responsible for close to half of the EU's emissions of $CO_2$ |
| February 16th, 2005 | Kyoto Protocol enters into force | History is made when the Russian Federation submitted its instrument of ratification to the Kyoto Protocol, sealing its entry into force |
| December 2005 | The eleventh Conference of the Parties (COP 11) in Montreal | For the first time held in conjunction with the first Conference of the Parties serving as the Meeting of the Parties (CMP 1) |
| January 2006 | Clean Development Mechanism opens | The Clean Development Mechanism, a key mechanism under the Kyoto Protocol, opens for business |
| November 2006 | Twelfth Conference of the Parties held in Nairobi-Kenya | The Subsidiary Body for Scientific and Technological Advice SBSTA is mandated to undertake a program to address impacts, vulnerability, and adaptation to climate change—the Nairobi Work Programme NWP activities are ongoing |
| December 2007 | The thirteenth Conference of the Parties adopts the Bali Road Map, including the Bali Action Plan, charting the course for a new negotiating process to address climate change | The plan has five main categories: shared vision, mitigation, adaptation, technology, and financing |
| January 2008 | Joint Implementation Mechanism starts | The Kyoto Protocol mechanism "Joint Implementation" starts. This allows a country with an emission reduction or limitation commitment under the protocol to earn emission reduction units |

(continued)

**Table 1** (continued)

| Date | Event | Achievements/developments |
|---|---|---|
| | | (ERUs) from an emission reduction or emission removal project in another country with similar commitments |
| December 2008 | The fourteenth Conference of the Parties in Poznan, Poland | Delivers important steps toward assisting developing countries, including the launch of the Adaptation Fund under the Kyoto Protocol and the Poznan Strategic Programme on Technology Transfer |
| December 2009 | World leaders gather for the fifteenth Conference of the Parties in Copenhagen, Denmark, which produced the Copenhagen Accord | Developed countries pledge up to USD 30 billion in fast-start finance for the period 2010–2012 |
| December 2010 | The sixteenth Conference of the Parties results in the Cancun Agreements, a comprehensive package by governments to assist developing nations in dealing with climate change | The Green Climate Fund, the Technology Mechanism, and the Cancun Adaptation Framework are established |
| December 2011 | At the seventeenth Conference of the Parties, in Durban, South Africa, governments commit to a new universal climate change agreement by 2015 for the period beyond 2020, leading to the launch of the Ad Hoc Working Group on the Durban Platform for Enhanced Action or ADP | Momentum for Change, a special initiative of the UNFCCC, shines a light on innovative and transformative climate action taking place around the world |
| December 2012 | At the eighteenth Conference of the Parties, in Doha, governments agree to speedily work toward a universal climate change agreement by 2015 and to find ways to scale up efforts before 2020 beyond existing pledges to curb emissions | They also adopt the Doha Amendment, launching a second commitment period of the Kyoto Protocol |
| September 2013 | New home for UNFCCC secretariat | The UNFCCC secretariat moves to its new headquarters on the UN Campus in Bonn, next to the former German Parliament building. Following major improvements, the building is now a beacon of environmental performance deploying features like solar power and smart lighting |
| September 27th, 2013 | IPCC releases second part of Fifth Assessment Report | The UN Intergovernmental Panel on Climate Change (IPCC) releases the Working Group 1 contribution to its Fifth Assessment Report (AR5), on the science of climate change |
| November 2013 | The nineteenth Conference of the Parties produces the Warsaw Outcomes, | Parties agreed an "international mechanism for loss and damage," which |

(continued)

**Table 1** (continued)

| Date | Event | Achievements/developments |
|---|---|---|
| | including a rulebook for reducing emissions from deforestation and forest degradation and a mechanism to address loss and damage caused by long-term climate change impacts | recognizes that if mitigation does not take place quickly enough and if countries cannot adapt to the resulting climate change, detrimental impacts will be inevitable. Developing countries want this mechanism to be a channel through which they can seek compensation from countries with high greenhouse gas emissions for this damage<br>The next milestone conference will be COP 21 in Paris in 2015, when parties are meant to agree on a comprehensive, legally binding global agreement |
| March 31st, 2014 | IPCC releases second part of Fifth Assessment Report | The UN Intergovernmental Panel on Climate Change (IPCC) releases the Working Group 2 contribution to its Fifth Assessment Report (AR5), on impacts, adaptation, and vulnerability |
| September 2014 | UN Secretary-General's Climate Summit | UN Secretary-General Ban Ki-moon to host a climate summit in New York, inviting Heads of State and Government, business, finance, civil society, and local leaders to mobilize action and ambition on climate change in advance of COP 21 in Paris in 2015 |
| December 2014 | At the twentieth Conference of the Parties, in Lima | World governments will have the opportunity to make a last collective push toward a new and meaningful universal agreement in 2015 and explore the opportunities that biodiversity and climate market incentives can provide for climate change action and sustainable development |
| December 12th, 2005 | COP 21—Historical Paris Agreement adopted | 195 nations agreed to combat climate change and unleash actions and investment toward a low-carbon, resilient, and sustainable future. The Paris Agreement for the first time brings all nations into a common cause based on their historic, current, and future responsibilities |
| November 2016 | COP 22—Marrakech Partnership for Global Climate Action launched | A crucial outcome of the Marrakech climate conference was to move forward on writing the rule book of the Paris Agreement<br>The conference successfully demonstrated to the world that the implementation of the Paris Agreement is |

(continued)

**Table 1** (continued)

| Date | Event | Achievements/developments |
|---|---|---|
| | | underway and launched the Marrakech Partnership for Climate Action |
| November 2017 | The 2017 UN Climate Change Conference will take place from 6 to 17 November at the World Conference Center in Bonn, Germany, the seat of the Climate Change Secretariat | Bonn will also make history by being the first COP to be presided over by a small island developing state: in this case by the Presidency of Fiji |

*Source*: Elaborated by authors from United Nations Framework Convention on Climate Change, available https://cop23.unfccc.int [accessed 10 March 2018]

"green planet" and measures to sustainable development that at the same time protect the planet and address climate change efficiently.

According to Riviere (2014, p. 92), "The proliferation of the environmentalism norm and its effect on constructing political economies depends on its adaption into states' identity. States' identities influence their behaviours because the identity contains within it norms and ideas that construct the field of possible actions. More specifically, since states interact with each other internationally, a state's identity is partially a social construction."

International agreements and negotiations are, as we have seen, one of the fundamental foundations for international relations whether analyzed from the point of view of state "interests" and "threats" or from the point of view of international cooperation and international organizations or from a point of view of norms, values, and ideas. How can theories of international relations explain some of the dilemmas confronting all actors of international system, and perhaps more importantly, how can international relations theories explain some of the dilemmas facing the international system in the face of climate change, which seem to be the big issue for the next few years?

In fact, if we look back at history and evaluate human mobility, we predict that most human movements had climate factors as the main cause, and most wars have been waged as a result of environmental factors, climate changes, and control of natural resources. After all, it is the assumption of this control that maximizes the power of state actors, among others. It is the assumption of this control that maximizes the power of state actors.

From the economics point of view, there are two main approaches: (1) an assessment of the effects of climate change on economic outcomes and (2) the study of the incentives that can prevent or mitigate climate change. First, empirical and theoretical research has been made to evaluate the effects of climate change (and in particular pollution). In the resources and economic growth literature, the use of limited or nonrenewable resources can challenge permanent economic growth if pollution decreases utility (e.g., Aghion and Howitt 1998, Chap. 5) and if pollution depletes nonrenewable resources that are used in production, because at some time in the future, those resources will limit economic growth. However, technological progress can be the way out of stagnation (Reis 2001; Roseta-Palma et al. 2010). Empirically, climate change has been related to economic growth. However, a clear-cut answer has not been achieved. For example, Dell et al. (2012, 2014) conclude for a negative effect of climate change in economic growth. In particular, Dell et al. (2014) reviews the literature on the effects of climate change on the economy. On the contrary Sequeira et al. (2018) conclude for a nonsignificant effect and a positive effect of rising precipitation in poor countries. This positive effect of rising precipitation is also confirmed for hot and temperate countries.

Second, the market incentives that can prevent or not climate change mitigation are highlighted in Balint et al. (2017). They identify four areas of the literature in which complex models have already produced valuable insights: (1) coalition formation and climate negotiations, (2) macroeconomic impacts of climate-related events, (3) energy markets, and (4) diffusion of climate-friendly technologies. On each of these issues, the authors argued that interactions and disequilibrium dynamics provide a complementary and novel perspective to standard economic models. Furthermore, this research highlights the potential economic benefits of mitigation and adaptation policies and the risk of underestimating systemic climate change-related risks.

Economics and international relations cannot be dissociated given the obvious cross-border nature of climate change. Market incentives may be combined and model national players in order to mitigate climate change's potential effects. This book bridges these two fields of research concerning climate change.

In Part 1, market economies approaches are examined. In fact interactions between agents and governments are deeply determined by the way markets function and incentives are placed. On the economics point of view, it is interesting to evaluate the effects of climate change in both the short and long run. This often involves a trade-off between potential positive effects of losing environmental protection and negative effects of climate change in the long run. Both theoretical and empirical approaches are used to analyze the issue, keeping in mind that on the basis of human decisions, there are (almost) rational agents that interact for their own good but also with some level of altruism regarding future generations. First, a game theoretical approach is followed to build an international society based on global common goods and interests, grounded by providing the greatest good to the greatest number, in which the environment offers not only a challenge but, more importantly, the chance to develop pathways to a sustainable future. An evolutionary game theory framework solution for the global climate change game is provided. Then, another chapter analyzes the macroeconomics of environment from an agent-based perspective, outlines directions for future developments, and identifies several domains that lack more intensive attention from the academic community. The chapter concludes that public policies aimed at preventing climate catastrophes must take into account patterns of sentiment spreading, allowing for the potential economic benefits of mitigation of environmental change-related impacts. Using a specific regression method that allows for nonlinear empirical modelling, another chapter finds that the effect of climate change on growth varies with the level of per capita GDP. In assessing the effect of greenhouse gas emissions to GDP growth, a double-threshold effect and three regimes are discovered. The assessment of the effect of $CO_2$ emissions on GDP growth identifies only a single threshold, indicating that only two regimes exist, confirming that the effect of climate change on growth cannot be regarded as a linear relationship.

Part 2 analyzes international relations and politics. The recognition that *climate change is a global phenomenon* without boarders is essential. Several policy issues are discussed with special focus on the European Union (EU), the Balkans, Africa, and the BRICS (Brazil, Russia, India, China, and South Africa). First, the discussion addresses if the European Union authority in environmental issues is recognized in the diplomatic arena. Related to that, although the several challenges that Balkan states still must face, such as corruption and weak governance, the EU accession process has been the main political driver of change in the region, providing opportunities for improving the environment in different ways. Then the discussion turns to Africa's ability to benefit from sustainable development synergies embedded in the mitigation and adaptation strategies in the INDCs and the fact that this will be greatly limited by institutional and policy environment that hinders funding, capacity building, and technological innovation systems development. Fourth is a discussion of the role of the New Development Bank, known as the BRICS Bank, as the driving force behind clean energy financing and its contribution to sustainable development. Common to these topics are the importance of international institutions and their policies to climate change management and mitigation at the country level. Sixth, in a very interesting chapter about the Amazon Forest, revolutionary technologies of the *fourth Industrial Revolution* are proposed as central to this necessary paradigm shift. Because the Amazon rainforest is a global natural asset for avoiding climate change, Brazil is also advised to follow a diplomatic plan that upholds fundamental ecological principles for creating a safe space for humanity, thereby attracting international investments to manage and preserve the forest.

Part 3 covers a micro approach related to climate change, related to both *technological change* and *marketing* strategies. In fact, technological change may be essential to mitigate climate change if it can be directed to cleaner technologies. Climate change can also be used as a marketing tool for firms in such a way that it can be a reinforcing tool to mitigate climate change. This acts on the demand side. As far as consumers become more concerned with climate change, they also demand more goods produced with cleaner technologies and having reduced environmental effects. On an alternative view, marketing tools may also act to avoid the negative views that customers can acquire from climate change-driven events such as tornadoes, floods, or fires in well-known tourist destinations.

First, from the technological change point of view, a chapter offers a comparison between the environmental performance of two internal combustion engine vehicles (ICEV) with different engines, one with the conventional diesel engine and the other with an electric motor and a diesel engine (hybrid diesel), and the results show that the hybrid ICEV has a generally better environmental performance compared to the diesel ICEV. Second, from the marketing point of view, marketing strategies that are used by several companies to reduce climate change are studied. Therefore, marketing holds potential solutions to some of the challenges of sustainability faced today, such as climate change. Additionally, climate change effects on a destination's image among tourists are analyzed in an application to the island of Madeira.

## Bibliography


Aghion, P., & Howitt, P. (1998). *Endogenous growth theory*. Cambridge, MA: MIT.

Balint, T., Lamperti, F., Mandel, A., Napoletano, M., Roventini, A., & Sapio, A. (2017). Complexity and the economics of climate change: A survey and a look forward. *Ecological Economics, 138*, 252–265.

Blobel, D., and Meyer-Ohlendorf, N. (2006). United Nations framework convention on climate change: Handbook. Bonn: UNFCCC.

Carter, N. (2010). Climate change and the politics of the global environment. In M. Beeson & N. Bisley (Eds.), *Issues in 21st century world politics* (pp. 52–65). London: Palgrave Macmillan.

Dell, M., Jones, B., & Olken, B. (2012). Temperature shocks and economic growth: Evidence from the last half century. *American Journal of Economics: Macroeconomics, 4*(3), 66–95.

Dell, M., Jones, B., & Olken, B. (2014). What do we learn from the weather? The new climate-economy literature. *Journal of Economic Literature, 52*(3), 740–798.

Dimitrov, R. (2016). The Paris Agreement on climate change: Behind closed doors. *Global Environmental Politics, 16*(3), 1–11.

Epstein, C. (2006). The making of global environmental norms: Endangered species protection. *Global Environmental Politics, 6*(2), 32–54.

Gupta, J. (2010). A history of international climate change policy. *Wiley Interdisciplinary Reviews: Climate Change, 1*(5), 636–653.

Haas, P. M., Keohane, R. O., & Levy, M. A. (1993). *Institutions for the earth. Sources of effective international environmental protection*. Cambridge: MIT.

Heffron, D. (2015). *What do realists think about climate change?* London: Centre for Geopolitics & Security in Realism Studies.

Hoffman, A. J. (2012). Climate science as culture war. *Stanford Social Innovation Review, 10*, 30–37.

Jacobsen, S. (1999). International relations and global environmental change review of the burgeoning literature on the environment. *Cooperation and Conflict, 34*(2), 205–236.

Keohane, R. (2005). *After hegemony: Cooperation and discord in the world political economy*. Princeton, NJ: Princeton University Press.

Keohane, R. O., & Nye, J. S. (1989). *Power and interdependence*. New York: Harper Collins.

Morgenthau, H. (1948). *Politics among nations: The struggle for power and peace*. New York: Alfred A. Knopf.

Pereira, J. C. (2015). Environmental issues and international relations, a new global (dis)order – The role of international relations in promoting a concerted international system. *Revista Brasileira de Política Internacional, 58*(1), 191–209.

Reis, A. B. (2001). Endogenous growth and the possibility of eliminating pollution. *Journal of Environmental Economics and Management, 42*, 360–373.

Riviere, L. L. (2014). Towards a constructivist international political economy of climate change. *Issues in Political Economy, 23*, 90–101.

Roseta-Palma, C., Lopes, A., & Sequeira, T. (2010). Externalities in an endogenous growth model with social and natural capital. *Ecological Economics, 69*(2010), 603–612.

Sequeira, T., Santos, M., & Magalhães, M. (2018). Climate change and economic growth: A heterogeneous panel data approach. *Environmental Change and Pollution Research, 25*, 22725–22735.

United Nations Framework on Climate Change. (1998). *Kyoto protocol – Reference manual*. New York: UNFCC.

Wendt, A. (1992). Anarchy is what states make of it: The social construction of power politics. *International Organization, 46*(2), 391–425.

# Part I
# A Market Approach

# Global Development and Climate Change: A Game Theory Approach

**António Bento Caleiro, Miguel Rocha de Sousa, and Ingo Andrade de Oliveira**

## Contents

1 Introduction ........ 18
2 Framing Climate Change Within Our Approach ........ 20
2.1 International Relations (IR) and Climate Change (CC) ........ 20
2.2 Economics and Traditional Game Theory (GT): Setting the Stage for Instrumental Rationality ........ 24
3 From Traditional Game Theory to Agent-Based Simulation: The CC Contribution ........ 26
4 Conclusions, Extensions and Further Research ........ 32
References ........ 33

**Abstract** The increasing concern with climate change is one of the main issues of our time, and thus we aim to theoretically and mathematically analyse its causes. However our approach follows a different stream of thought, presenting the reasoning and decision-making processes between technical and moral solutions. We have resorted to game theory models in order to demonstrate cooperative and non-cooperative scenarios, ranging from the traditional to the evolutionary within

---

The chapter is inspired by a previous paper by Ingo Andrade de Oliveira and Stephánie Rycken. We would also like to thank the extremely constructive comments and suggestions from an anonymous reviewer.

---

A. B. Caleiro (✉)
Department of Economics, University of Évora, Évora, Portugal
e-mail: caleiro@uevora.pt

M. R. de Sousa
Department of Economics, University of Évora, Évora, Portugal

Research Center in Political Science (CICP), University of Évora, Évora, Portugal

Center for Advanced Studies in Management and Economics (CEFAGE), University of Évora, Évora, Portugal
e-mail: mrsousa@uevora.pt

I. A. de Oliveira
University of Évora, Évora, Portugal



T. Sequeira, L. Reis (eds.), *Climate Change and Global Development*, Contributions to Economics, https://doi.org/10.1007/978-3-030-02662-2_2

game theory. In doing so we are able to glimpse the development of modern society and a paradigm shift regarding human control over nature and to what extent it is harmful to the sustainability of our environment and the survival of future generations. Merging different fields of knowledge, we present a theoretical-philosophical approach, combined with empirical-mathematical solutions taking into account the agent-based behaviour guided blindly by instrumental rationality.

**Keywords** Evolutionary game theory · Global development · Game theory · Moral and technical solutions · International relations and politics · Economics · Sustainable development

**JEL Codes** C70 · D70 · F64 · O13 · O19

## 1 Introduction

We live in an age marked by interconnectedness and mutual cooperation, not only in the economical but also in the social field and above all, and inevitably, in the political one. The conceived structures of states are now challenged by multiple agents and interests, motivated not only by self-development goals but by the side effects of each sovereign country's interests. Development is not enough anymore. Sustainable development requires cooperation and multilateral agreements which are based on the logical field of instrumental rationality. Achieving a win-win scenario is essential, or no deal will be reached.

Climate change is a global common goods dilemma (see Hardin 1968)—when one pollutes, the social cost is a burden to all, but the relevant internalized private costs have not yet been factored in. Others, such as Vajpeyi (2013), provide an international relations framework for the assessment of climate change impacts based on the notion of a power-driven discourse, or, for instance, one of the authors of this chapter in Assad et al. (2013) provided an assessment of the impact of climate change on Brazil's agricultural business through physical-atmospherical models.

Climate change is relevant—as the bulk of its consequences are a burden on the entire planet, the term ***ecocide*** is gaining popularity, which we break down as *ecological+suicide*. However, beyond this tragedy-driven approach, we are looking for viable solutions. It has been difficult for states to adhere to and, above all, reach the targets set out in international protocols (e.g. the Kyoto Protocol) on reducing carbon emissions, which is obviously necessary in the face of climate change. The main reason is free-riding, between countries and between generations. Free-riding becomes particularly difficult to combat given the sovereignty that is characteristic of different countries, the equality of all before the law and the fact that it is impossible for countries to manage each other's internal affairs (Nordhaus 2015).

Additionally, participation in international agreements is voluntary. To arrive at a solution, Nordhaus (2015) proposes the creation of a climatic club, with penalties for non-members, where it is in the interest of the states themselves to belong to the club.

In turn, by also recognizing that the international coordination of environmental policies has been difficult to verify, Keohane and Victor (2016) suggest the decentralization of international policy coordination as a first step towards an effective reduction of the problem.

Another optimistic solution is proposed in Buchholz and Sandler (2017). It is sometimes acknowledged and even recommended that there should be countries which take the lead in environmental policy changes aimed at reducing the emissions of greenhouse gases. Indeed, in a traditional model of game theory, this leadership encourages free-riding by followers, giving rise to a crowding-out effect.[1] Buchholz and Sandler (2017) show that it does not necessarily have to be so if a leader's benevolent behaviour signal positively influences the follower's expectations about the leader's actual behaviour.

In his review of the literature, Heal (2017) points out the key role of the appropriate discount rate. When coupled with a sufficiently high-risk aversion to an even small possibility of a disastrous outcome, this may also be the solution to the problem associated with the difficulty in achieving a benevolent change in environmental policies by the various countries.

Our new approach, in this chapter, comes from modelling through game theory and stressing mathematically and theoretically all possible outcomes, as we aim to offer possible scenarios based on moral and technical solutions where each country may be able to fulfil their interests without undermining the common good.

We try to establish a new model, based on two strands of literature, which we briefly sum up in Sect. 2: the first strand from international relations, where we abridge several international relations approaches and set our goal as instrumental rationality led by the states, and the second strand of literature, from mathematics and economics, which we cross with the first one, where we use game theory. We start with the classical Von-Neumann-Morgenstern zero-sum game and Nash equilibria, which, while still relevant 50 years after its creation, seem to lack the robustness required in order to find a truly viable solution to the global climate change game.

Next, we focus on Sect. 3, the newness of the chapter and its contribution, a holistic evolutionary agent-based simulation approach, which we might modestly describe as the climate change evolutionary game theory solution. Indeed, it may be time to build an international society based on global common goods and interests, grounded in providing the greatest good to the greatest number, where the environment presents not only a challenge but, more importantly, a chance to develop pathways to a sustainable future. We provide an evolutionary game theory framework proposal of a solution for the global climate change game.

[1]As recognized in Buchholz and Sandler (2017), this effect can be reinforced if followers are sufficiently risk-averse and/or climate change damage to be uncertain (Auerswald et al. 2011).

Our approach is relevant, not only for its newness but essentially because it provides a viable solution to a global dilemma, establishing new frontiers of research and global policy advice in international fora. We also offer an explanation for the many failed solutions of climate change negotiations, namely, the drawbacks in the Kyoto protocol and the more recent Paris protocol. Our model introduces a retaliation punishment and a positive feedback/cooperation award after many rounds of the game. The innovation of our model is not only in the introduction of game theory to the global climate change game but also the introduction of its time dimension, with a twofold dynamic horizon with a learning environment. A point in time, say $t^*$, if attained yields a point of no return, and the arrival at a disastrous situation, which we parametrize, must be avoided at all costs. A point in time, say $t+$, instead must be determined—a point from which evolutionary cooperation might save us all from the *immersing* trap and thus provide a truly sustainable path out of disaster and avert ecocide, therefore achieving a positive final solution, which we shall all consider as a final best solution.

In Sect. 4, we end with some general conclusions, limitations encountered and further perspectives of research.

# 2 Framing Climate Change Within Our Approach

## *2.1 International Relations (IR) and Climate Change (CC)*

In this section we distinguish two basic approaches from a philosophical and international relations background: rational/technical solutions versus moral solutions, facing the international system and international order.

The writings and concepts that best frame the idea of technical solutions and rationality addressed in this chapter can be identified in the studies of Habermas (1984). Habermas began to analyse the theory of rationalization, left unfinished by Max Weber—or at the least brought it to light again—reconstructing Weber's project as a whole and using it as a guideline to empirically observe the result of a universal-historical process of rationalization (Habermas 1984: 143). This universal-historical process will be analysed through the lens of human behaviour and human interaction with nature, particularly the substantial paradigm shift which began with the Enlightenment and the birth of modern society.

We consider moral solutions as a complementary concept to the concept of technical solutions, where morality, according to Kant, is derived from ontological fields reached only through reasoning. In order to do this, we will walk through other ideas, and authors which demonstrate the gap which exists between theoretical and practical reason on the one hand, and on the other hand, show the limitations of instrumental rationality in solving the problems posed by climate change.

Practical reason can be seen in Weber's contributions in the idea of "Zweckrationalität", or as it is called nowadays, "Purposive/Instrumental Rationality". It initially appeared in the field of sociology, "determined by expectations as to the behaviour of objects in the environment and other human beings; these expectations are used as "conditions" or "means" for the attainment of the actor's own rationally pursued and calculated ends (Weber 1978: 24), and later on is introduced in other fields such as economics. In short, it is a tool used by an actor to reach an end, where rationality is pursued and calculated.

What changed since then? The method of handling problems. In G. Hardin's article "The Tragedy of the Commons", he defines technical solutions as "(...) a solution that requires a change only in the techniques of the natural sciences, demanding little or nothing in the way of change in human values or ideas of morality" (Hardin 1968: 243–253).

Now that we have introduced some key aspects of technical solutions and reasoning, our aim is to introduce the issue of moral solutions and value rationality. Only then we will be able to understand the process of decision-making aided by a game-theoretical approach.

As the twin concept of technical solutions, as it is presented in this chapter, morality and its vast, varied and complex definitions will remain the path to reasoning. It is the way we comprehend climate change and its importance according to the principle of universality.

The acknowledgement of climate change [CC] is the starting point in order to validate morality in attitudes towards the object. We cannot expect certain kinds of behaviour to occur—for religious, cultural, scientific or any other reasons—if there is no common agreement on the subject. The International Society[2] is based on these shared common values, which we address here in relation to the issue of CC. While we do not wish to enter a discussion of that subject here, we do acknowledge the conflicts of interests which arise between economic, political, industrial and other social groups. Assuming CC as a shared common value, our theoretical and game-theoretical approaches will have this starting point while aiming to achieve the best outcome for all players, through reasoning.

Secondly, the groundwork undertaken in order to examine morality in decision-making will be founded on three pillars: (1) "the greatest good to the greatest number",[3] as stated Jeremy Bentham; (2) "act only according to that maxim whereby you can at the same time, will that it should become a universal law",[4] posed by Kant; and (3) "the morality of an act is a function of the state of the system at the time it is performed", according to Joseph Fletcher. These different degrees of morality lead to different approaches, namely, utilitarianism, Kantianism (highlighting the

---

[2]As noted by Hedley Bull (2002: 13), and bringing up his definition according to the issue of CC we perceive, it is worthy to discuss whether our shared beliefs frame the states nowadays as a society of states, a system of states or rather as an international anarchy.

[3]The idea of the greatest good to the greatest number is essential in welfare functions and crucial in all economics decision-making. Our Nash equilibria and evolutionary approach assume this kind of instrumental rationality.

[4]Also, again we are framing global universal concepts, such that in our global economic model, the implications are that we are dealing with universal laws; thus, we are dealing with homogeneous agents in economic terms.

categorical imperative concept) and situational ethics,[5] placing the object—CC as a global common good—in the light of international relations schools and their analysis according to the constructivist, idealist and realist schools, respectively.

Third, as game theory demands the best choice in a given scenario, it is important to understand the categorical imperative [CI] as the basis for moral decisions. A few concepts related to the CI must be clarified: (a) "a priori reasoning" is the way of thinking without the empirical experiences and in its turn is related to (b) "pure practical reason", which is, in short, the action taken based solely on (c) good will. Kant argues that "good will" is related to the superior good, for example, the golden rule: "do unto others as you would have them do unto you". In that way, rationality based on the categorical imperative will be substantially different from rationality in technical solutions, which was defined before as a purposive/instrumental rationality. This introduces the possibility of a different outcome when, for example, the prisoner's dilemma is played repeatedly and each agent seeks their own best results, normally not the best choice for any given scenario and reaching the minor Nash equilibria.

Now that the moral aspects of solutions have been introduced, Weber also frames morality within his definition of rationality. Previously we mentioned the "Zweckrationalität" or purposive/instrumental rationality, although Max Weber also sheds light on a different—and even complementary—sort of rationality, namely, the "wertrational" or the value-oriented reason: "That is, determined by a conscious belief in the value of its own sake of some ethical, aesthetic, religious, or other form of behaviour, independently of the prospects of success; (Weber 1978: 24–25). Decisions based on "wertrational" cannot be fully understood solely through instrumental rationality. Likewise, it shows that not all decisions are purely instrumental.

Decision-making through the lens of morality may show the failings of mathematical models, where technical solutions based on instrumental rationality often fail to reach the desired outcome. It is still part of the reasoning path that the structures of power—the states as rational agents—shall take into account when currently shared beliefs and consequences of CC fail to fully encompass the solution to the problem.

To frame the reasoning process of decision-making in international relations [IR], it is essential to arrive at a consensus on the question of the dominant theory of current foreign affairs and also to acknowledge different degrees of analysis within IR and its fields of knowledge: economics, international law, foreign and domestic policies and so on. As an example, in the field of economics, we can say the dominant theory is neo-liberalism, although it is arguably not the only relevant theory present. Here we will assume that neorealism is the current dominant theory of international politics. In doing so, we aim to analyse and understand the reasoning at an agent-based level, considering states as rational agents.

To better frame the reasoning of states as rational agents, we have divided them in three different groups, according to their level of development (UNDP 2017): developed countries, developing countries and least-developed countries. According

[5]Once it considers "loving ends justify any means", it is framed here as realist tradition.

to this literature (Sen 1999; Stiglitz 2002; Sachs 2005, 2015), we have aimed to predict the behaviour of states based on their level of technical capabilities and their influence in foreign affairs. For example, the developed countries group consists mostly of OCDE countries, whose industry, agriculture and finance sectors rank highest in development. Developing countries, on the other hand, are better represented by the BRICS and their emerging industry, as well as their weight in trade and influence (e.g. demands they are making on the current structures of the UN). Least-developed countries are those which are not significant with regard to industry, economy or technical capabilities but are still included in our approach as moral solutions must take into consideration the lives of their citizens, according to the principles which govern the society of states on the issue of global welfare.

When discussing IP and the process of decision-making among states, one must clarify the common ground, through the bindings of international law (base of behaviour to political agents), the economic goals (where we stress rationality) of a given country and the variables used by these rational agents to reach their objectives.

Currently the legal concerns surrounding treaties[6] over the UNFCCC (United Nations Framework Convention on Climate Change) can be summed, on its binding features and the efficiency of sanctions. For example, it is generally accepted that Kyoto treaty has failed (Nordhaus 2015) due to the US withdrawal (despite being one of the treaty's main architects) and a lack of ratification at state domestic level. Despite the success of the ongoing Paris Agreement[7] which was ratified by 172 of 197 parties, some analysts are concerned over the lack of sanction clauses. Countries are taking action according to their own nationally determined contributions (NDC's), one of the innovative aspects of the Paris Agreement, but there is no sanction in case of failure.

Rational agents here clearly face a deep-complex problem, as presented considering only the fields of IR and game theory. However we consider it is through the welfare and its underlying ethical parameters and empirical facts the variables must converge. Here our game theoretical analysis will focus on a new approach, evolutionary game theory.

We now focus again on the instrumental rationality of the states and look at the bases of game theory decision on the next section, which are at the roots of our economics contribution.

[6]The diplomatic process of making treaties is the primary source of international law, and it goes through different stages before becoming binding/an international law: (1) negotiation (two or more countries design the treaty according to their interests), (2) ratification (the approval of the treaty by their domestic policy institutions), (3) renegotiation (after domestic discussions/parley, treaties may suffer some alterations) and (4) implementation (the approval of a treaty by international institutions).

[7]http://unfccc.int/paris_agreement/items/9485.php–accessed on December 22, 2017.

## 2.2 Economics and Traditional Game Theory (GT): Setting the Stage for Instrumental Rationality

On the economic side, many renowned economists have found common ground in some models despite focusing on different variables. Most of the work on the issue of CC is based on integrated assessment models [IAM's] which represent functions that relate temperature change to economic losses, such as welfare. As the recent study shown in Heal (2017), the default in almost every area of economics comes from intellectual traditions such as utilitarianism (Heal 2017: 1047) and the Ramsey rule (1928) precedent, which in turn converges with our analysis.

The Ramsey rule sets the variables used in economic analysis ranging from pure rate of time preference, the elasticity of marginal utility, rate of growth consumption, social rate of discount and many others. It is though in Heal we can verify the existence of ethical parameters and empirical facts. Key aspects of economic analysis and its underlying philosophical aspects are recalled here as moral and/or technical solutions. Heal and his recent study shows the problem of convergence among economists "an irony that seems to have escaped most of commentators is that while Nordhaus (2007), Stern (2006) and Weitzman (2007) all invoke the Ramsey equation in their choices of parameters, this equations does not in fact apply to the optimal climate-management problem. The reason is, of course, the external effect associated with the emissions of GHGs".

A historical approach to game theory recalls that the first paper on the existence of a solution for a non-cooperative zero-sum game was Von Neumann's (1928) classical paper. The foundations of game theory were introduced by what is now seen as the opus magnum of this field: Von Neumann and Morgenstern's (1944) manual, *Theory of Games and Economic Behaviour*. Nevertheless, throughout its 500 pages, this book only found solutions for non-cooperative games, a traditional conflict framework among agents, but for a very specific case, zero-sum games. A zero-sum game is a kind of game in which the gains of some exactly offset the losses of others, thus establishing that when someone gains is at another's expense, so there are no *win-win* or *lose-lose* equilibriums.

Von Neumann was a Hungarian born polymath genius whose approach was based on cooperation and dialogue; so he always tried to find cooperative solutions to non-cooperative games by forming coalitions. It was an ingenious young American mathematician, John Nash, who in his early 20s, also at Princeton, established through two simple papers (Nash 1950a, b, 1951), the first axiomatic approach to equilibria in economics and a general solution to non-cooperative games, including also non-zero-sum games.[8,9] Khun (1953) was the first to introduce the extensive

[8]Where Von Neumann was a cooperative group leader and discussant, Nash had an individualistic approach (Nasar 1998); thus the different approaches both men dealt with led to quite different conclusions, and both men were playing rivals and not coalitions.

[9]Nash, a bright young genius in his early 20s, would entry into a spiral of mental illness for 30 years to recover quite miraculously and win the 1994 Nobel prize in economics for his solution general

form of a game, and finally, Lloyd Shapley (1953) created the first general coalition solution to cooperative games. Aumann-Shapley value would extend the results of cooperative games to a continuum of players and infinite number of them.

Ken Binmore (1994: Vol. 1, 1998: Vol. 2) would formalize the contractual view approach using game theory in the late 1990s, early 2000s, thus giving a new soundness to the solutions of game theory to social choice, and even political science approach.

Traditional game theory has been used for modelling of climate change and common goods dilemmas. For instance, Nordhaus (2015) defines an optimal solution by avoiding free-riding in international climate policy by creating climate clubs, which would penalize non-members by paying climate fines by polluting. Keohane and Victor (2016) base themselves on the notion whether this global climate game is cooperative or non-cooperative and what kind of non-cooperative approaches can be reached. Buchholz and Sandler (2017) provide a new benchmark trying to stress the role of leadership and global public good provision. Heal (2017) does an extensive survey of the economics of climate change in the *Journal of Economic Literature*, thus stressing the opportune timeliness of this research agenda, focusing on the importance of the appropriate discount rate.

To be more specific, in order to reach a solution, Nordhaus (2015) proposes the creation of a climatic club, with penalties for non-members, where it is in the interest of the states themselves to belong to the club.

In turn, by also recognizing that the international coordination of environmental policies has been difficult to verify, Keohane and Victor (2016) suggest the decentralization of international policy coordination as a first step towards an effective reduction of the problem.

Another optimistic solution is proposed in Buchholz and Sandler (2017). It is sometimes acknowledged and even recommended that there should be countries which take the lead in environmental policy changes aimed at reducing the emissions of greenhouse gases. Indeed, in a traditional model of game theory, this leadership encourages free-riding by followers, giving rise to a crowding-out effect.[10] Buchholz and Sandler (2017) show that it does not necessarily have to be so if a leader's benevolent behaviour signal positively influences the follower's expectations about the leader's actual behaviour.

In his review of the literature, Heal (2017) points out the key role of the appropriate discount rate. When coupled with a sufficiently high-risk aversion to an even small possibility of a disastrous outcome, this may also be the solution to the

bargaining non-cooperative game and, even later, won the prestigious Niels Henrik Abel's Prize in Mathematics in 2015 for his contributions for partial differential equations (related to an embeddedness theorem in topology which joined topology with algebra) (http://www.abelprize.no/nyheter/vis.html?tid=63589). The prize was awarded to Nash in 2015, and he was killed in a taxi car crash on the New Jersey Turnpike, just 2 days after coming back from Norway.

[10]As recognized in Buchholz and Sandler (2017), this effect can be reinforced if followers are sufficiently risk-averse and/or climate change damage to be uncertain (Auerswald et al. 2011).

problem associated with the difficulty in achieving a benevolent change in environmental policies by the various countries.

Crossing game theory, economics and climate diplomacy are not new—for instance, Decanio and Fremstad (2013) do it by defining pure strategy Nash equilibria, min-max equilibria and departing from what is called a New Periodic Table (NPT) in the climate economics diplomacy literature. This study comprises $2 \times 2$ order games first introduced by Robinson and Goforth (2005), where they defined 144 possible combinations of games among two types of states (e.g. Great Powers or coalitions of states). The negotiations can be depicted as a no-conflict game, prisoner's dilemma, coordination game, chicken-type game or cycle, depending on the payoff matrix. One of their most important results is that scientific information regarding the severity of the risks of climate change tends to change the nature of the game from a prisoner's dilemma to a coordination game (Decanio and Fremstad 2013).

Thus, the novelty in our approach is to depart from the classical prisoner's dilemma but take into account the role of learning by doing/playing repeatedly, and therefore the evolutionary trajectories across time change contingent on the penalty reward/fine—which is also in line with the theory of belonging to climate clubs, as Nordhaus (2015) proposes.

We now turn to the finer points and begin our new approach, merging international relations, the instrumental approach and game theory and departing from classical game theory towards an evolutionary modelling approach through agent-based simulation.

## 3 From Traditional Game Theory to Agent-Based Simulation: The CC Contribution

For now, let us consider two agents, each of whom can make two types of decisions: to cooperate or to defect. Tables 1 and 2 represent the payoffs, respectively, of agents 1 and 2, where, as usual, the decisions of agent 1 are represented in rows and the decisions of agent 2 are represented in columns.

Most of the literature believes that the game of climate change can be represented by the particular case of the prisoners' dilemma (see, inter alia, Soroos 1994;

**Table 1** Payoffs matrix of agent 1

| | Cooperate | Defect |
|---|---|---|
| Cooperate | $\beta$ | $\delta$ |
| Defect | $\alpha$ | $\gamma$ |

**Table 2** Payoffs matrix of agent 2

| | Cooperate | Defect |
|---|---|---|
| Cooperate | $\eta$ | $\varepsilon$ |
| Defect | $\lambda$ | $\theta$ |

Gardiner 2006; Pittel and Rübbelke 2008, 2012).[11,12] Thus, we will consider the following hypotheses:

- Hypotheses 1: $\alpha > \beta > \gamma > \delta$ and $\varepsilon > \eta > \theta > \lambda$.

Under the usual assumptions, in a single-shot game, both agents decide to defect, leading to an outcome that is Pareto-inferior for both players (in case of both cooperating). It is also well-known that an infinite repetition of the game may induce the cooperative solution, when both agents possess a sufficiently high rate of time preference/discount (Aumann 1959). This is so because (considering agent 1 as an example), if $\rho_1$ (where $0 \leq \rho_1 < 1$) denotes the discount rate of agent 1, the present value of the cooperative solution, verified from $t = 0$ to $\infty$, will be:

$$\frac{\beta}{1-\rho_1}, \tag{1}$$

which is higher than the present value of the cheating solution,[13] i.e.

$$\alpha + \frac{\rho_1}{1-\rho_1}\gamma, \tag{2}$$

whenever

$$\beta > (1-\rho_1)\alpha + \rho_1\gamma. \tag{3}$$

It is also important to recognize that, in case of a finite repetition of the game, the number of repetitions by both agents being known the defect decision by both agents is to be the solution of the game, as the cooperative solution will, by backward induction, not be considered credible.

From the empirical point of view, it seems that the most polluting countries are those that, in the short term, will suffer the least from climate change (Harrington et al. 2016).[14] With this fact in mind, let us assume, without loss of generality, that

[11]It is interesting to note that, usually, the singular is used to designate this type of game, i.e. the prisoner's dilemma. This may be due to the fact that the preferences of one prisoner are representative of the preferences of all prisoners. In our case, because payment levels may be different, we prefer to consider the plural, i.e. prisoners' dilemma.

[12]In this regard, it is illuminating to quote *The Economist* (2007): "After all, all countries will enjoy the benefits of a stable climate whether they have helped to bring it about or not. So a government that can persuade others to cut their greenhouse-gas emissions without doing so itself gets the best of both worlds: it avoids all the expense and self-denial involved, and yet still escapes catastrophe. [...] The problem, of course, is that if everyone is counting on others to act, no one will, and the consequences could be much worse than if everyone had simply done their bit to begin with. Game theorists call a simplified version of this scenario the 'prisoner's dilemma'".

[13]We will assume infinite retaliation.

[14]From an empirical point of view, there is some support for the idea that, in the short run, the countries that pollute less are those more vulnerable to the malefic consequences of the climate

agent 1 represents the group of countries that are less willing to diminish pollution. This can be translated into different relationships in the payoff matrices, namely:

1. $\gamma > \theta$—The Nash equilibrium is worse for the less polluting countries, in the short run.
2. $\alpha > \varepsilon$—The free-ride solution is better for the more polluting countries, in the short run.
3. $\beta - \gamma < \eta - \theta$—The benefits of cooperation (relative to defection) are higher for the less polluting countries, in the short run. This hypothesis, together with hypothesis 1, leads to.
4. $\beta < \eta$—Cooperative behaviour is less attractive for the most polluting countries, in the short run.

Assuming an infinite repetition of the game, those four inequalities explain the empirical fact of so many failed negotiations to diminish pollution, which may be due to a reduced rate of discount (and/or a short time horizon). This also means that, in order to obtain a full cooperative behaviour, the rate of discount must be sufficiently high for the most polluting countries, which is the truer the more they must be aware that there is a point in time, say $t^*$, where all the signals of the abovementioned four restrictions will change, i.e. when the Nash equilibrium will become worse for the most polluting countries, making cooperative behaviour more attractive to these countries.

In case that point in time is known, $t^*$, an immediate implication is that pollution will increase until that moment. The good news is that it may start decreasing after that, in particular after a certain point in time, say $t+$, where the game of climate change is no longer a prisoners' dilemma but a game of common interests, i.e. when $\beta > \alpha$ and $\eta > \varepsilon$. Plainly, it is crucial that the catastrophic consequences of climate change, associated with the point of no return, do not occur before that point in time, $t+$.

The line of reasoning presented above draws attention to the fact that the characteristics of the game may change over time, as might the characteristics of the players themselves. Therefore, an evolutionary view of the prisoners' dilemma makes it (more) interesting and appropriate to address the situation, particularly when using an agent-based simulation approach (Wilensky 2002).

That more complete view implies a (recent) change of the modelling paradigm of agents as individuals (temporarily) located in space. Thus, a more recent line of approach failed to assume a homogeneous population of individuals with uniformly distributed (perfect) information to increasingly assume a heterogeneous population composed of agents with different levels of information (see e.g. Akerlof 1997) and/or different characteristics. Agents behave in a potentially complex way and interact with each other over time and also with the environment in which they are inserted, in the search of the fulfilment of their goals. As the environment changes,

---

change. See, for instance, the ND-GAIN Country Index (http://index.gain.org/; accessed on August 11, 2017).

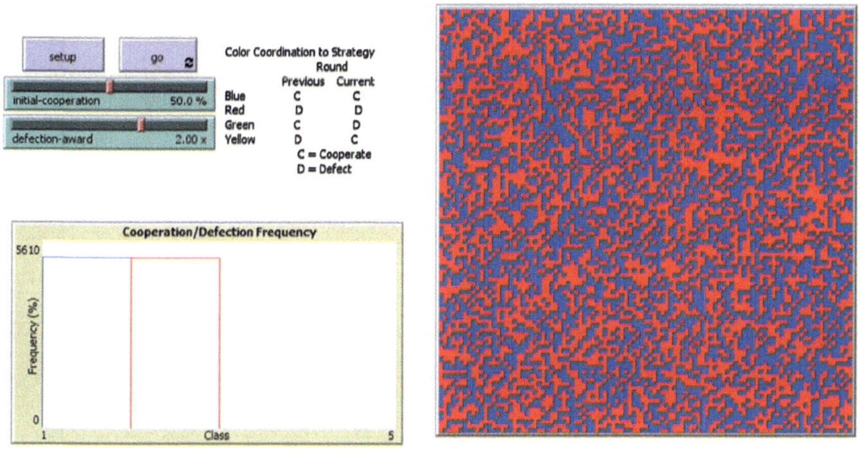


**Fig. 1** The initial situation

the agents' characteristics and consequent decisions also change, often in complex ways. The so-called agent-based simulation intends to tackle these situations of an evolutionary nature.

According to this line of thought, the analysis is increasingly focused on (admittedly) complex systems composed of a large number of agents that interact with each other. As recognized in Goodchild et al. (2000), the aim of this approach is not the determination of allegedly simple principles of general order, since such principles can hardly emerge from the analysis of such complex systems, but rather serve as examples of this approach as standards of comparison with actual behaviour, and this comparison is a starting point for a better understanding of reality (Arthur et al. 1997).

The dynamic process of the strategic interactions of countries playing the prisoners' dilemma in an evolutionary way is to be simulated using a NetLogo model (Wilensky 2002).[15] In the evolutionary prisoners' dilemma simulation, one may consider distinct percentages of cooperators as well as distinct rewards associated to the defect decision. Figure 1 illustrates the case where, at the beginning of the game, half of the players are cooperative in nature and the reward of defection is moderately high. See Wilensky (2002) for the full description of the game.

Plainly, despite the balanced initial-cooperation percentage, a sufficiently high level of defection-award will lead to the disastrous situation where every agent chooses the defect behaviour, as illustrated in Fig. 2. This is to be the case, unless all agents are of the cooperative type, at the beginning of the game. As we know from the reality of climate change negotiations, this decision by every country, not to pollute in excess, does not apply.

[15]NetLogo is a multi-agent programmable modeling environment, freely available at http://ccl.northwestern.edu/netlogo/ with a very wide range of applications, as evidenced by the wide range of models available to users.

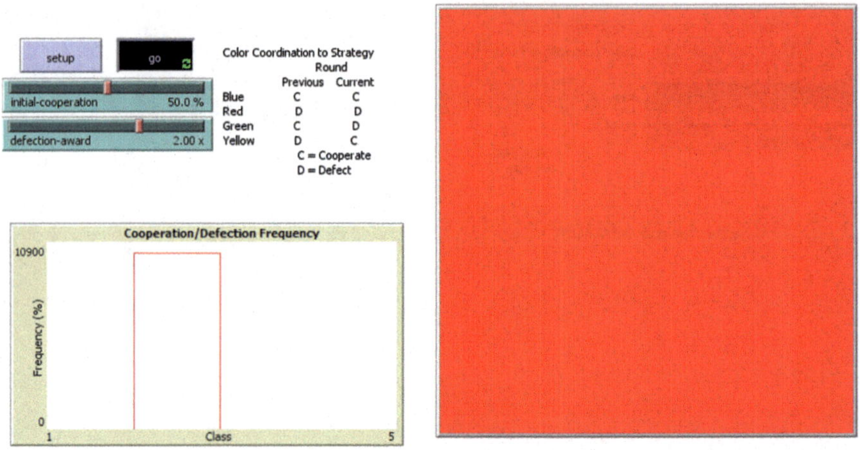


**Fig. 2** The disastrous situation

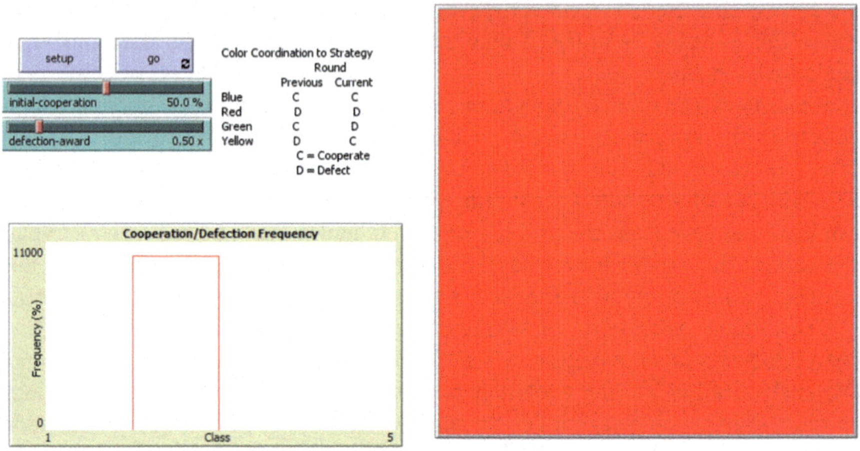


**Fig. 3** The reduced defection-award situation

Taking this disastrous situation as a starting point, let us now consider the point in time, $t^*$, where the defection-award is substantially reduced, given the substantial increase of climate change costs associated to that disastrous situation.[16] Figure 3 illustrates this case.

The reduction of defection-award will eventually lead to the inversion of the disastrous path of increased pollution. Figures 4 and 5 illustrate some intermediate situations after that reduction of free-ride benefits.

[16]We are assuming that this point in time occurs before the point of, say, no return.

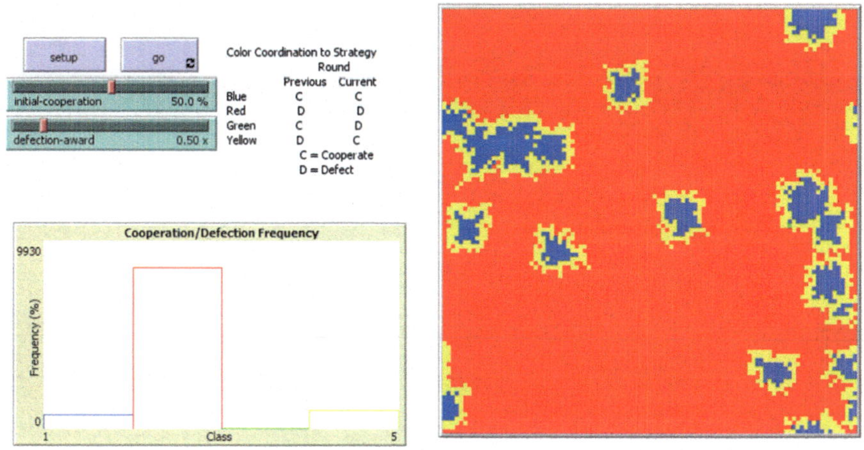


**Fig. 4** A first intermediate situation

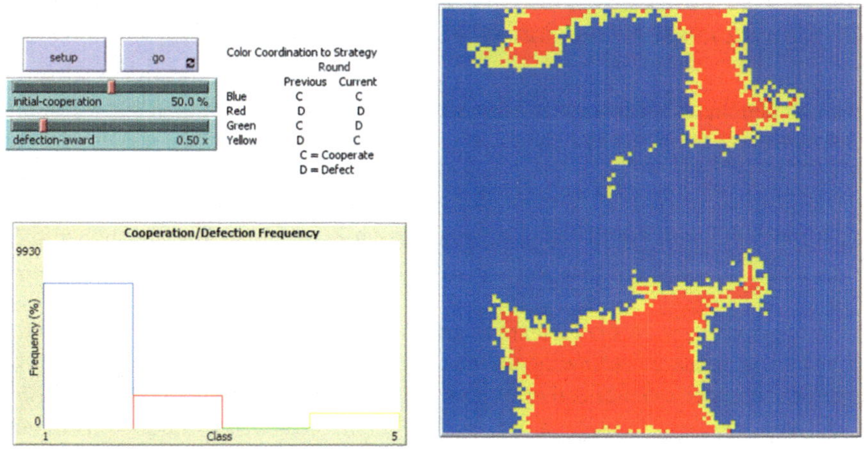


**Fig. 5** A second intermediate situation

The final situation of this benevolent change in what concerns the attitude towards climate change, before identified as moment $t^*$, is illustrated in Fig. 6.

Clearly, the policy implication of this agent-based simulation gives credence to the need for urgent recognition that the sooner the defection-award decreases, the sooner the disastrous path of climate change, and its dire consequences will change towards a (more) benevolent situation. In order to achieve that, either the time horizon of politicians must increase, or, at least, these agents must not ignore what reality has been showing lately.

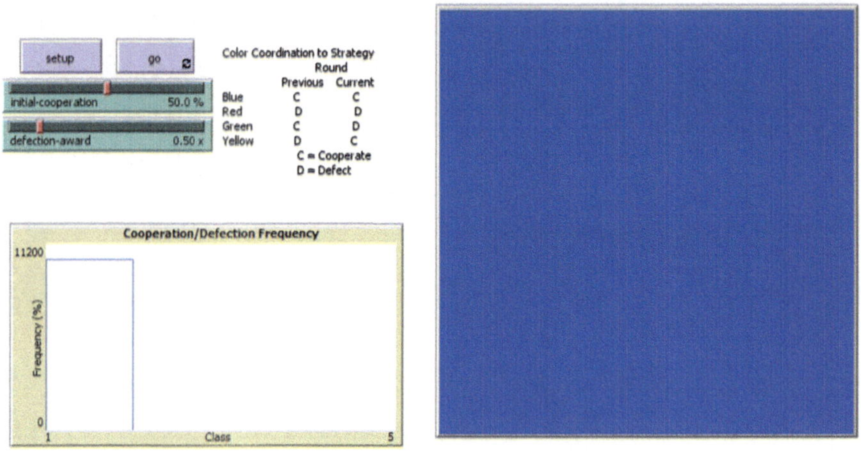


**Fig. 6** The final situation

# 4 Conclusions, Extensions and Further Research

We first conclude that ecocide is likely to occur and we are heading in that direction. Our parametrization of a sufficient low value for the parameter of no punishment for pollution, and thus a high level of defection-award, in our agent-based simulations, shows that, even starting from a cooperative situation, one may easily end in a disastrous situation, the so-called ecocide, which is presented in the pathway from Figs. 1 to 2. This illustrates most of the rationale behind failed climate change negotiations.

There are still signs of hope that this situation might be averted, if the premium for defection is reduced, retaliation is introduced in our model, and the process of learning becomes more effective. With this in mind, we can predict that in international fora, we might change from a non-cooperative solution to a truly global, less free-riding, solution, where the final solution would be a cooperative one—pathway Figs. 3, 4 and 5.

There are some caveats in our approach; pay-offs are exogenous in the sense that they are not reflecting the change completely endogenously. This might be achieved by also using a new kind of approach and extension of game theory, in its inception, known as drama theory (Howard 1994a, b). Drama theory predicts that pay-offs change endogenously with the strategies of players and are correlated. Its name comes from the fact that an agent can dramatize a sufficiently bad or good outcome, influencing each other endogenously in the decision. These new equilibria, which we would call drama Nash equilibria, would be contingent on the correlated effect between agents on the strategies, and endogenous drama pay-offs (Levy et al. 2009a,b) can be applied to climate change. With an interdependent world order and new trends emerging, with more resort to dramatization, namely, by such global leaders as President Trump or President Putin, this deserves further study.

The level of rationality we are using might also be discussed. We could argue for a lesser degree of rationality, namely, in the sense of as described by Herbert Simon, and display that this only worked for a limited time frame, horizon and benchmark of parameters. The so-called implicit bounded rationality. Nevertheless, one could go the other way round and even argue that the solution could be easier and faster, even though all recent evidence tends to point the other way, by postulating what is called hyper-rationality. Players might over-calculate the very long run in a snapshot and increasingly adjust to it. All these discussions are left open, as avenues of research for further study and queries, which are naturally worthwhile.

## References

Akerlof, G. A. (1997). Social distance and social decisions. *Econometrica, 65*, 1005–1027.

Arthur, B. A., Durlauf, S. N., & Lane, D. A. (1997). *The economy as an evolving complex system II*. Reading, MA: Addison-Wesley.

Assad, E. D., Buainain, A. M., Pinto, H. S., Rocha De Sousa, M., & Duarte, V. S. (2013). Climate change: Challenges for Brazil. In D. Vajpeyi (Ed.), *Climate change, sustainable development and international security (Chap. 6)* (pp. 169–199). Lanham, MD: Lexington (An imprint of Rowman and Littlefield). isbn:978-0-7391-8146-1 https://rowman.com/ISBN/9780739181461

Aumann, R. (1959). Acceptable points in general cooperative n-person games. In R. D. Luce & A. W. Tucker (Eds.), *Contributions to the Theory 23 of Games IV, Annals of Mathematics Study* (Vol. 40, pp. 287–324). Princeton, NJ: Princeton University Press.

Auerswald, H., Konrad, K. A., & Thum, M. (2011). *Adaption, mitigation and risk-taking in climate policy* (CESifo Working Paper Series 3320). Munich: CESifo Group.

Binmore, K. (1994). *Playing fair: Game theory and the social contract* (Vol. 1). Cambridge, MA: MIT Press.

Binmore, K. G. (1998). *Game theory and the social contract: Just playing* (Vol. 2). Cambridge, MA: MIT Press.

Buchholz, W., & Sandler, T. (2017). Successful Leadership in Global Public Good Provision: Incorporating Behavioural Approaches. *Environmental and Resource Economics, 67*, 591–607.

Bull, H. (2002). *The anarchical society* (3rd ed.). New York: Palgrave.

Decanio, S. J., & Fremstad, A. (2013). Game theory and climate diplomacy. *Ecological Economics, 85*, 177–187.

Gardiner, S. M. (2006). A perfect moral storm: Climate change, intergenerational ethics and the problem of moral corruption. *Environmental Values, 15*(3), 397–413.

Goodchild, M. F., Anselin, L., Appelbaum, R. P., & Harthorn, B. H. (2000). Toward spatially integrated social science. *International Regional Science Review, 23*(2), 139–159.

Habermas, J. (1984). *The theory of communicative act* (T. McCarthy, Trans.). Boston: Beacon Press.

Hardin, G. (1968). The tragedy of the commons. *Science, New Series, 162*(3859), 243–253.

Harrington, L. J., Frame, D. J., Fischer, E. M., Hawkins, E., Joshi, M., et al. (2016). Poorest countries experience earlier anthropogenic emergence of daily temperature extremes. *Environmental Research Letters, 11*(5), 055007.

Heal, G. (2017). The economics of the climate. *Journal of Economic Literature, 55*, 1046–1063.

Howard, N. (1994a). Drama theory and its relation to game theory. Part 1: Dramatic resolution Vs. rational solution. *Group Decision and Negotiation, 3*, 187–206.

Howard, N. (1994b). Drama theory and its relation to game theory. Part 2: Formal model of the resolution process. *Group Decision and Negotiation, 3*, 207–235.

Keohane, R. O., & Victor, D. G. (2016). Cooperation and discord in global climate policy. *Nature Climate Change, 6*, 570–575.

Khun, H. (1953). Extensive games and the problem of information. In H. W. Kuhn, & A. W. Tucker (Eds.), *Contributions to the theory of games* (Vol. II, pp. 193–216). In H. W. Kuhn (Ed.). (1996). *Classics in game theory* (pp. 46–68). Princeton: Princeton University Press.

Levy, J. K., Hipel, K. W., & Howard, N. (2009a). Advances in drama theory for managing hazards and disasters. Part I: Theoretical foundation. *Group Decision Negotiation, 18*, 303–316.

Levy, J. K., Hipel, K. W., & Howard, N. (2009b). Advances in drama theory for managing hazards and disasters. Part II: Coping with global climate change and environmental catastrophe. *Group Decision Negotiation, 18*, 317–334.

Nasar, S. (1998). *A beautiful mind*. London: Faber and Faber.

Nash, J. (1950a). Equilibrium points in n-person games. *PNAS: Proceedings of the National Academy of Sciences, 36*, 48–49. In H. W. Khun (Ed.). (1996) *Classics in game theory* (pp. 3–4). Princeton: Princeton University Press.

Nash, J. (1950b). The bargaining problem. *Econometrica, 18*, 155–162. In H. W. Khun (Ed.). (1996) *Classics in game theory* (pp. 5–13). Princeton: Princeton University Press.

Nash, J (1951) Non-cooperative games. *Annals of Mathematics, 54*, 286–295. In H. W. Khun (Ed.). (1996) *Classics in game theory* (pp. 14–26). Princeton: Princeton University Press.

Nordhaus, W. D. (2007). A review of the Stern review on the economics of climate change. *Journal of Economic Literature, 45*(3), 686–702.

Nordhaus, W. (2015). Climate Clubs: Overcoming Free-Riding in International Climate Policy. *American Economic Review, 105*, 1339–1370.

Pittel, K., & Rübbelke, D. T. (2008). Climate policy and ancillary benefits: A survey and integration into the modelling of international negotiations on climate change. *Ecological Economics, 68* (1), 210–220.

Pittel, K., & Rübbelke, D. T. (2012). Transitions in the negotiations on climate change: from prisoner's dilemma to chicken and beyond. *International Environmental Agreements: Politics, Law and Economics, 12*(1), 23–39.

Robinson, D., & Goforth, D. (2005). *The topology of 2x2 games: A new periodic table*. New York: Routledge.

Sachs, J. D. (2015). *The age of sustainable development*. New York, Columbia University Press.

Sachs, J. (2005). *The end of poverty: How we can make it happen in our lifetime*. London: Penguin.

Sen, A. K. (1999). *Development as freedom*. Alfred Knopf: New York.

Shapley, L. (1953). A value for n-person games. In H. W. Kuhn, A. W. Tucker (Eds.), *Contributions to the theory of games* (Vol. II, pp. 307–317). In H. W. Kuhn (Ed.). (1996). *Classics in game theory* (pp. 69–79). Princeton: Princeton University Press.

Soroos, M. S. (1994). Global change, environmental security, and the prisoner's dilemma. *Journal of Peace Research, 31*(3), 317–332.

Stern, N. (2006). *Stern review: The economics of climate change* (Vol. 30). London: HM treasury.

Stiglitz, J. E. (2002). *Globalization and its Discontents* (Vol. 500). New York: Norton.

The Economist. (2007, September 27). *Playing games with the planet*. Accessed August 11, 2017, from http://www.economist.com/node/9867020

UNDP. (2017). *United Nations Human Development Report*, NY.

Vajpeyi, D. K. (Ed.). (2013). *Climate change, sustainable development and international security*. Lexington (An imprint of Rowman & Littlefield). isbn:978-0-7391-8146-1 https://rowman.com/ISBN/9780739181461

Von Neumann, J. (1928). Zur Theorie der Gesellschaftsspiele. *Mathematische Annalen, 100*(1), 295–320. https://doi.org/10.1007/BF01448847

Von Neumann, J., & Morgenstern, O. (1944)[1994]. *The theory of games and economic behavior. 50th Anniversary Edition*. Princeton, NJ: Princeton University Press.

Weber, M. (1978). In G. Roth & C. Wittich (Eds.), *Economy and society*. Berkeley: University of California Press.

Weitzman, M. L. (2007). The Stern review on the economics of climate change. *Journal of Economic Literature, 45*(3), 703–724.

Wilensky, U. (2002). *NetLogo PD Basic Evolutionary model.* Evanston, IL: Center for Connected Learning and Computer-Based Modeling, Northwestern University http://ccl.northwestern.edu/netlogo/models/PDBasicEvolutionary

# On the Edge of Climate Change: In a Search of an Adequate Agent-Based Methodology to Model Environmental Dynamics

**Mariya Gubareva and Orlando Gomes**

## Contents


1 Introduction ........ 38
2 Survey on Agent-Based Modeling of One-Way Climate Influence on the Economy ...... 39
  2.1 Agent-Based Modeling of Flood Risk and Insurance ........ 39
  2.2 Agent-Based Modeling Approach to Climate Shocks and Migration ........ 42
  2.3 Agent-Based Modeling of Climate, Population Growth, and Water Shortages ....... 45
3 Surprise: Agent-Based Alternative Futures Modeling for Studying Wild Fires ........ 47
4 Discussing Agent-Based Modeling of Integrated Economy-Climate Systems ........ 49
  4.1 Agent-Based Modeling of Personal Beliefs and Climate Change Interrelation ....... 49
  4.2 Agent-Based Climate-Economic Model to Assess Green Investment Effects ........ 53
5 Concluding Remarks ........ 55
References ........ 56



**Abstract** Climate change is one of the most challenging and threatening uncertainties of our days, impacting mankind with a continuously growing intensity. In our chapter, we rise an alert flag and provide evidence of the nonexistence of a dominant or consensual approach to correctly account for the interconnectedness between the decision-making process by economic agents and the environmental damages that affect them. Our survey analyzes a series of distinct contemporary contributions on the subject, trying to create awareness on the state of art in the



M. Gubareva (✉)
ISCAL—Lisbon Accounting and Business School, Instituto Politécnico de Lisboa, Lisbon, Portugal

ISEG—Lisbon School of Economics and Management, Universidade de Lisboa, Lisbon, Portugal

SOCIUS—Research Centre in Economic and Organizational Sociology, CSG—Research in Social Sciences and Management, Lisbon, Portugal
e-mail: mgubareva@iscal.ipl.pt

O. Gomes
ISCAL—Lisbon Accounting and Business School, Instituto Politécnico de Lisboa, Lisbon, Portugal

Business Research Unit (BRU/IUL), Lisbon, Portugal





T. Sequeira, L. Reis (eds.), *Climate Change and Global Development*, Contributions to Economics, https://doi.org/10.1007/978-3-030-02662-2_3

agent-based modeling applied to climate change. We analyze the macroeconomics of the environment from an agent-based perspective, outline directions for future developments, and identify several domains that lack more intensive attention from the academic community. We conclude that public policies aimed at preventing climate catastrophes must take into account patterns of sentiment spreading, allowing for the potential economic benefits of mitigation of environmental change-related impacts.

**Keywords** Agent-based model · Climate change · Economic development · Economic growth · Integrated economy-climate modeling

## 1 Introduction

In the course of the last few years, agent-based economy-climate models began to play an important role in the contemporaneous discussions on environmental and socioeconomic dynamics. Though this research line is in its early stage of development, up to the time being, it has produced lots of valuable intakes on how economies could face climate shocks. Still there is a long way to do both to talk the talk and to walk the walk, in order to arrive at the stage when agent-based models (ABMs) applied to the climate change will serve the functions of well-designed economic model.

The most challenging aspect of the agent-based modeling, which still is on the agenda of academic community, is to provide insights into the complex interrelation between global economic development and climate change in a broad sense. It would be much better if we had precise and thoroughly tested ABMs, which would serve as a solid base for elaborating adequate polices to withstand diverse climate-related threads as well as for preparing swift and strait-to-the-point responses to unexpected environmental shocks on economic systems.

From a social perspective, moderate weather changes make little difference on human society, but climate catastrophes like floods, droughts, hurricanes, large forest fires, etc. result in profound consequences for individual households, in particular, and for the regional and/or global economy, in general. In this chapter, we rise an alert flag and provide evidence of the nonexistence of a dominant or consensual approach to correctly account for the interconnectedness of the decision-making process by the economic agents and the environmental damages affecting levels of wealth of individual households through shocks on the overall economic output.

Differently from a rather broad scope of the recently published survey on complexity and the economics of climate change by Balint et al. (2017), we solely focus on the agent-based economy-climate modeling aimed at interrelations between the economic growth and development, on one hand, and the surrounding environmental ecosystem, on the other. Our chapter surveys a set of selected contemporary contributions on the subject, finalized in the form of already published papers. Our target is to create awareness of the most recent 2016–2017 advances of the state of art in the ABM applied to climate change.

We analyze the macroeconomics of environment from an agent-based perspective, outline directions of future developments, and identify several domains that lack more intensive attention from the academic community, namely, demographic dynamics, modeling of environmental shocks on agents decision-making process, individual energy consumption choices, and social contagion impact on sentiments propagation.

These issues require adequate methodologies to correctly account for heterogeneity of agents, diversity of their interactions, and out-of-equilibrium dynamics in order to provide a complementary and novel perspective to conventional economic analysis, where equilibrium conditions are pre-specified and the respective determinacy is commonly assured by the full rationality assumption. In this context we demonstrate that public policies aimed at preventing climate catastrophes must take into account patterns of sentiment spreading allowing for the potential economic benefits of mitigation of environmental change-related impacts.

Our chapter is structured as follows. First, we discuss original works on how it could be modeled a one-way climate influence over the economy. This discussion covers three papers: first, studying flood impacts on economic growth; second, investigating diverse climate shocks on migration; and, third, focusing on how climate dynamics affects population growth and water shortages.

In parallel, we present a very interesting, insightful, challenging, and rather unusual research, based on ABM of alternative futures, which assesses how economic development on a regional scale through growing urbanization affects climate in the form of the occurrences of surprise forest fires.

And last, but not least, we discuss two related studies on the integrated treatment of economy-climate ecosystem. These works advance a commonly used ABM methodology incorporating heterogeneous adaptive agents, whose capacity to learn in battles of beliefs and perspectives affects the outcome of the model. We believe that such integrated frameworks should gain a wider spread in the field.

## 2 Survey on Agent-Based Modeling of One-Way Climate Influence on the Economy

### *2.1 Agent-Based Modeling of Flood Risk and Insurance*

The problem of flooding, which has occupied human minds since old times, is a subject of this innovative contemporaneous work. The research selected to illustrate the main issues of these topics by Jenkins et al. (2017) focuses on how implementation of the flood insurance mechanisms could eventually mitigate related financial and economic damages affecting the society as a whole as well as create incentives to improve efficiency of the flood risk management. This study refines the preceding work by Dubbelboer et al. (2017), counting with the participation of Jenkins and Hole, who continued the research in this field.

The authors of the chosen most advanced work by Jenkins et al. (2017) related to agent-based approach applied to the flood risk management, namely, Jenkins, Surminski, Hall, and Crick (JSHC), develop an ABM of flood risk and insurance and apply it to study the Greater London, namely, the London Borough of Camden. This particular UK region is characterized by the augmented risk of inundation. Along with the characteristic summer storms, historically occurred here, the flood risk is attributed to the growth in population and exuberated urban developments, which bring under great pressure the drainage systems and the surface.

The JSHC ABM consists of the five following types of agents, each with its inherent features, and interacting with each other: people, insurer, bank, developer, and a local government.

The first type of agents is homeowner. Homeowner at each moment in time could be either a buyer or a seller. Homeowner is obliged to subscribe for the flood insurance. In other words, each property is insured against flood. The house could be sold (i.e., homeowner turns to be the seller) due to the financial or dislocation reasons. If the owner does not sell his property, he decides whether to invest in property-level protection measures, which envisage preventing flood damages. Buyers make decisions to acquire the house based on their reserve price (i.e., the maximum amount that they would be willing to pay to acquire it) and 3-year-long flood history (i.e., more recent flood events). The stronger the flood occurrence, the more influential it is on the decision by individuals not to buy a home. The quantity of buyers is assessed based on projected population trends to 2041 taking into account migration processes.

The second category of agents is insurer. The main function of insurer is to provide each agent who owns a home (i.e., people, developer, and bank) with flood insurance. Insurer charges insurance premium calculated personally for each agent-house owner. The value of premium is a positive function of the flood risk estimates, which could be mitigated by the investments in sustainable drainage systems and property-level protection measures. In its turn, insurer pays reinsurance premium for the government in order to pass the responsibility to cover the costs of flood damages, which is higher than those insurers can support. Thus, authors simulate recently implemented by UK authorities flood insurance scheme, Flood Re.

Developers represent the third class of agents. The main objective of the developer is to make profit though construction and selling the houses. The construction ability is limited by intensity of already existing houses and government's targets. Half of the new construction possesses flood defense features. For simplicity reasons, the model assumes that the developer cannot be involved in more than one construction. Another simplification is related to the consideration that newly constructed buildings and construction activities itself do not influence flood risk. This simplification in fact represents a weak point of the model.

The fourth agent is the government. Its main role is to protect people from floods through the investments in flood protection mechanisms. Such funding occurs in two ways. First, the government distributes among homeowners the target grants with the maximum limit of £5000 per house. Second, the government invests in sustainable drainage system projects. Additionally, the government evaluates the development projects presented by developer for the authorities' approval. The rate of success is equal to 70%.

The fifth, and the last agent, is the bank. The purpose of the bank is to put on sale the house by its market price in the case in which the owner cannot support house fee payments during 3 consecutive years.

Researchers, in their agent-based simulations of floods and the respective damages, employ UK geographic information system data, UK residential building data, and surface water flood depth maps. In order to introduce the effect of climate change into the model, authors estimate the change in probability of surface water flood events of a given magnitude. The annual flood risk is calculated separately for each house using the formula developed in Beven and Hall (2014).

The ABM of flood risk and insurance developed in the paper allows performing several studies. The first type of analysis is related to the question of how agents could contribute to surface water flood risk reduction. The second class of problems focuses on the specific role of insurance mechanisms in flood damage mitigation.

Applying their ABM, authors demonstrate that investment in protection mechanisms is beneficial for the diminishing of flood risk, but nevertheless the trend on flood risk increase prevails with and without flood protection measures. Researchers attribute this evidence to the continued development of houses in areas of flood risk.

In respect of the flood insurance analyses, the paper advocates the implementation of insurance mechanism, which has a positive feedback on investments in flood protection arrangements. JSHC ABM, applied to Greater London case, shows that reinsurance procedure, performed by the government for the insurer, allows to diminish significantly the insurance premium payments for the most flood risky areas.

Concluding, the developed JSHC ABM gives an indication on constantly growing flood risk due to the socioeconomic developments and detects positive economic, social, and climate features due to the public–private partnership in the flood insurance schemes.

Still, assuming that newly constructed buildings and construction activities themselves do not influence flood risk, JSHC, in fact, break the feedback loop between human activity and the climate, focusing solely on how the climate changes, herein alterations in flood levels, impact human society.

To be on the edge of climate change, the ABM should be expanded to include the whole feedback loop from the climate to the agents of the society and then back to climate, etc. We believe that such kind of research is highly desirable and that the described above work could/should be continued in this direction. Next subsection discusses ABM of climate-economy relationship, assessing influence of floods and other climate disasters on migration activity.

### 2.2 Agent-Based Modeling Approach to Climate Shocks and Migration

Herein we focus on the research by Entwisle et al. (2016), which studies climate change influence on migration processes in rural agricultural areas. This work develops empirically grounded ABM, which appeals for reassessing the neoclassical economic paradigm advocating the large migration response to climate shocks.

Authors perform their weather-migration analysis employing the extensive ethnographic, qualitative, spatial, and social survey data from the Nang Rong district of northeast Thailand. Nang Rong is a rural agricultural area, specialized on rice and upland crops cultivation, with monsoonal climate.

Migration is treated as a strategy that Nang Rong farmers allegedly have followed for a long time in order to respond, from one side, to the climate-posted challenge of rain-fed paddy rice cultivating with minimal irrigation and, from another side, to the national and international economic shocks affecting the demand for agricultural products.

The recent studies on rainfall patterns of the Nang Rong region evidence the annual rainfall downward trend with increased variance year-to-year. The possibility to uncover the mechanisms of relationship between climate change and migratory decisions incentivizes the authors to study migration responses to the proposed four baseline weather scenarios. They are (1) a reference scenario, reflecting normal timing and normal amount of rainfall, (2) a 7-year period of extremely dry weather, (3) a 7-year period of extremely wet weather, and (4) a 7-year period of an excessively volatile weather, frequently switching between the extrema of (2) and (3) scenarios.

Entwisle et al. (2016) develop ABM composed of (1) individual agents forming (2) households, which in their turn are embedded in (3) social networks, and (4) village-type settlements. Households own or/and rent (5) land parcels. Each of the five mentioned above modeled agents is endowed with the following specific characteristics and assumptions.

Individuals are ascribed with the age, gender, and marital status. They participate in demographic, social, and economic processes, such as birth, death, and migration. The eligible age for migration belongs to interval between the 10 and 29 years old, which is in line with range of migration in Nang Rong region. Authors calculate migration probability for two age groups separately, namely, for 10–19 years old and 20–29 years old.

Households are characterized by their land ownership, by the position in the village, and the connection with the wealthy households. Under environmental and climate constraints, household agents make decisions on what kind of crop to cultivate and what inputs (e.g., fertilizer), and its quantity should be utilized. Such decisions, combined with the land quality and weather scenarios, affect the crop productivity level, and therefore household revenues quantified as a cumulative difference between income and expenses.

Authors model social network (i.e., connectivity) of village-aggregate individuals and households. They are represented in the model by such variables as population size, migration prevalence, and ties to current migrants among others. Land parcels are described by their size, distance from the village, probability of flood, soil quality, etc.

Authors apply the proposed ABM to analyze migration as a function of four climate scenarios: normal, droughts, floods, and the unstable flood-drought alternations. Surprisingly, Entwisle et al. (2016) find small difference in emigration between the normal, flood, drought, and severely volatile weather scenarios despite of the occurring large drops in the crop production under the climate shock scenarios. Authors estimate that migration in the flood, drought, and excessive variability scenarios belong to the 101 and 103% interval of the normal climate scenario.

In fact, it is not as surprising as authors claim at least due to the following reason. The historic genetic memory, or so to say collective memory, tells agents that on average the weather tends to be somewhat fair. That is why they still live there at their usual habitats. Thus, impacts on the agents' decisions from poor weather provoked drops in the crop are counterbalanced by the agents' implicit expectation of the improved, better than the average weather during the times to follow.

Not surprisingly, authors highlight the necessity of further investigation of such result, which contradicts to the neoclassical economics theory based on the standard regression-based model. Namely, in accordance with the regression-based estimations, there are larger increases in migration for the drought scenario (associated with the largest drops in income and assets) and smaller increases in migration for the flood scenario (associated with the negligible decreases in assets).

The difference between the draft scenario and the flood scenario is also not surprising and quite comprehensible, as the soil quality is improving after the flood due to the deposited sludge and sediments. Thus, floods are potentially benefic to the future crops. This circumstance is likely to attenuate immediate attempts to migrate due to the flood occurrence. On the contrary, there is no mitigation in the draft-provoked negative impacts on crops and household incomes.

In order to explore the possible causes of unexpected outcome of the proposed ABM of migration, researchers perform four additional experiments. First, Entwisle et al. (2016) eliminate from their migration baseline model all individual characteristics such as age, gender, marital status, and ties to current migrants. Authors justify this procedure by the importance of these features, which could overwhelm the climate change in determining migration outcomes. Nevertheless, the first version of the ABM approach arrives to the conclusion that individual characteristics do not have differential effects for the three extreme climate cases and thus do not help to explain the smaller than expected migration response to climate change in a baseline model.

Second, authors consider the hypothesis of the migration as the inverse function of the remittances during climate shocks. Thus remittances could eventually provide a clue of the weak migration response. Hoverer, testing this second hypothesis by employing the appropriate alteration in the baseline ABM, researchers evidence that there is no meaningful influence of remittances, which could explain the absence of a larger migration response to the climate scenarios.

In the third modification of the ABM, authors investigate the significance of social network in a lack of migration response to the climate scenarios. Researchers, excluding from the analysis variables responsible for social networks, come to another unexpected surprising result. Under flood, drought, and flood-drought scenarios, migration rates are lower than in the case of the baseline model. Entwisle et al. (2016) explain this finding by the complexity of social network interactions, which simultaneously could amplify and diminish migration. For example, from one side, ties to migrants increase migration probability, but from other side, migration violates connections between different households in a village which and thus decreases the migration.

The fourth, and final, test envisages studying the migration dependence upon household wealth. Individuals adapt their behavior if their household loses income. The more prolonged the period of loses is, the higher will be the probability to emigrate. Authors evidence much higher migration rates in this simulation if compared with the original model.

Summarizing, authors argue that pre-existing condition of high migration rates is the most important factor. Thus, climate events could have little impact on contemporaneous migration rates. We posit that to influence significantly migration rates, the generation and even intergeneration collective memory should be affected in such a manner that, on average, agents would not anymore judge the weather as reasonably fair but rather as generally poor. In this case, the mechanisms of counterbalancing drops in the crops by expectations of weather improvement will stop working. But it takes time, perhaps generations, to produce changes in the collective memory of individual agents and households.

There is also another possible explanation of why pre-existing condition of high migration rates is the most important factor. It is due to contagion mechanisms. The higher the previous migration rates, the stronger might be the social contagion effect among the individual agents and households. Perhaps, the possibility to include appropriate contagion mechanism in the considered ABM would improve its adherence to observed reality. Still, this study is very challenging as it motivates academic community to reassess and rethink the climate change effects on migration.

The authors conclude their paper with intriguing questions. Why do people continue living in their areas after the devastating climate events occurred there (e.g., in New Orleans after Hurricane Katrina and in New York City after Hurricane Sandy)? Developing further this question, we may guess that it may be due to the popular wisdom saying that "bombs never hit twice in the same spot," do they? Maybe superstition, religiosity, everyday mind, and common sense consideration need to be explored in more detail in ABM?

Why does the Netherlands continue to be the most densely populated nation in the world, while the most part of its territory is under sea level? Perhaps income level, righteous society, tolerance culture, social justice, fair legal system, and insurance policies along with cultural customs and traditions are eventually overweighting existing risk of future negative impacts due to the breach of dams and following inundation of land? Should these factors be included in ABM description of the regions potentially subjected to natural disasters and climate changes? For more insights, see Ermolieva et al. (2017).

Why does Maldives also remain one of the most densely populated archipelago while suffering severe land loss due to sea level rise? Perhaps, individual agents and households underestimate tale risks of sudden rise of sea level? Maybe there are several barriers to emigration: material, cultural, etc.? Maybe individual agent and households on islands should be modeled somewhat differently from the continental population? Are the alternatives fair enough for them? And finally, according to Plato's philosophy, "no one can escape his destiny," so why should Maldivians emigrate?

At the time of this chapter writing, similar questions keep arriving. Whether or not the Hurricane Harvey impacted Texas, the Typhoon Hato affected the southern provinces of China or the Hurricane Irma passed through and heavily damaged the Florida State and the Caribbean region augment the migration processes from these areas and how the authorities should react on such kind of uncertainty. Certainly, the research in this field is to be carried further on. In the following subsection, we describe ABM accessing the role of population among other factors in the climate-economy relationship and the effectiveness of water-related policies.

## 2.3 Agent-Based Modeling of Climate, Population Growth, and Water Shortages

In this subsection we present a study by Mashhadi et al. (2017), which contributes to the literature investigating the problem of sustainable urban water supply under the pressure of continuous expansion of the city areas along with the adverse climate change causing negative impacts. The authors develop ABM, which integrates traditionally used supply and demand management approaches with the dynamic interactions between environment, technology, and society.

The proposed model is applied to simulate the water supply system of the city of Raleigh, North Carolina, USA. This particular North American region has been affected by severe hydrological droughts in 2002, 2005, and 2007, which led to record low water levels and large economic losses due to water restrictions.

The model framework describes interactions between household agents and water utility manager agents connected to a water supply reservoir. Households are considered as agents, to which the individual characteristics such as a number of members in a household, house built year, and irrigated area are ascribed. Each household has indoor and outdoor water demands.

The water use behaviors of households are represented as rules, while water planner is endowed with the right to introduce restrictions, which, in their turn, depend upon changes in the level of water in the storage reservoir. Water utility managers make decisions on restrictions in accordance with the reservoir model simulations, which indicate surface water dynamics. Restrictions are adopted in response to decreasing volume of water in the reservoir and removed when the water level returns to its normal range of values.

The possibility of restrictions imposed by utility manager agents allows preserving water supply in case of water shortages or droughts. It is assumed that household agents must comply with the restrictions immediately for the case of outdoor water use, while indoor water consumption remains unaffected.

In order to estimate the daily indoor water necessities, authors employ two data sources. The first source is the Vickers' (2001) report on water use at the household appliance level. Second is the De Oreo et al.'s (2011) probabilistic distributions of bathtubs water consumption and leakages in California. Outdoor water demand is estimated using a theoretical irrigation model. This model, conceived quantifying irrigation requirements, includes such variables as the size of land dedicated to each type of farmed cultures, efficiency of irrigation technologies, and crop coefficient for each type of cultivated plants.

Authors perform two types of water supply simulations with the proposed water supply ABM. The first simulation is conducted for the historic data related to the period from 1983 to 2014. The objective of this first simulation is to explore the accuracy of the modeling prediction power through the back-testing analyses. The model generates monthly values of the water demand, reservoir storage, and reservoir release. In order to obtain one of these three values, the historic data for the remaining two is used as input. The proposed model passes the test and reveals an acceptable level of performance.

The second simulation is performed with the purpose to test water management strategies for the projected 19-year period under alternative climate circumstances. Authors consider four management scenarios: (1) "no management," when households do not modernize water equipment; (2) "retrofitting" scenario, when no drought restriction is imposed while household agents upgrade water appliances; (3) "relaxed drought response," when drought restrictions are not applied immediately; and (4) "strict drought response" scenario, which corresponds to immediate reaction to small losses in reservoir storage.

Although the water deficit, i.e., when demand is superior to supply, is observed under "no management" strategy for all-weather scenarios, the three other scenarios incorporating active management strategies contribute to maintain relatively high levels of water reservoir.

For example, under active management strategies, a considerable diminishing of water storage becomes observable only after climate conditions turn to be circa 60% drier than the normal ones, as water demand is adequately attenuated through retrofitting and drought restrictions, applied when needed.

Authors also estimate the sustainability index for each management option. For the "no management" case, sustainability abruptly decreases while the climate becomes arid, while for the (2), (3), and (4) management strategies, it equals 1. Based on their simulations, authors report that there is little risk of water deficit for the analyzed region until the year 2032 if no considerable climate change occurs.

Summarizing, the proposed model illustrates the importance of researching population growth and their behavior, water shortages, climate change, and management strategies in the effectiveness of the reservoir storage. Authors argue that the developed water supply ABM has potential to be improved in several ways.

The model in its current stage of development does not have a predictive power. Still, the results of the authors' work demonstrate that the use of the ABM approach could be an important instrument and suitable for projecting the effectiveness of management actions and recommend drought policies for improving the sustainability of water resources in urban areas for the holistic and adequately adherent to reality "all-weather" scenario.

Additionally, in order to better estimate potential effects of climate change on water storage, a hydrologic model could be embedded in the base model to simulate expected precipitation and temperature time series. Apart from this, complementary water management strategies could be developed and included in the proposed model.

## 3 Surprise: Agent-Based Alternative Futures Modeling for Studying Wild Fires

The work by Hulse et al. (2016) is chosen herein to shed light on probabilistic aspects of ABM, which frequently remain without a due attention of academic community involved in climate change ABM. The discussed study offers a valuable methodological contribution to the literature on probabilistic treatment of surprising phenomena, in general, as well as on wildfire surprise forecasting, in particular. It provides a condensed survey of diverse surprise prediction approaches emphasizing their importance in designing strategies capable to diminish negative impacts of climate- and economy-related surprise events on human society.

Additionally, authors provide a profound revision of definitions and qualitative descriptions of surprise and discuss relevant prediction methodologies proposed in literature from 1984 to 2013. This aspect of the paper also represents a considerable input into the systematization of knowledge on the role of surprise in the complex social, economic, and environmental systems.

Authors develop an innovative approach, which integrates the ABM of land use change by Guzy et al. (2008), climate-sensitive successional model of vegetation, and the mechanistic wildfire model driven by climate inputs. For a contemporaneous comprehensive survey of non-agent-based components of quantitative and qualitative simulation scenario modeling, we suggest further reading of Symstad et al. (2017).

The resulting integrated ABM, presented by Hulse et al. (2016), relies on surprise event definition elaborated in a well-known seminal work by Holling (1986). In accordance with this definition, surprise is a local phenomenon when perceived reality departs qualitatively from expectation.

Researchers apply their integrated agent-based approach to study wildfire-surprising phenomenon in Willamette Valley eco-region, which represents one of the most urbanized and fastest-growing eco-regions in Oregon State of the USA. The size of the studied area is 81,000 ha. This area contains a considerable variety of residential developments, such as big urban center, smaller urban centers, and rural housing.

This territory, chosen by the authors for testing the proposed integrated ABM, is characterized by a relatively high exposure to fire risk. Almost half of the considered eco-zone is qualified as a wildland-urban interface, where areas of urban constructions are intertwined with undeveloped wildland, thus making fire risk increase. This feature justifies the appropriateness of choice of the Willamette Valley eco-region for studying fire surprises in the wide context of climate change and socioeconomic impacts of land resources usage.

Hulse et al. (2016) simulate the territory inhabited by the agents making decisions based on both their individual preferences and the level of production generated by the land owned by each agent. The agents' decisions are related to the new building initiatives, timber production, commercial developments, environmental recovery, and fuel reduction procedures. In their turn, agents' decisions, along with the biophysical events, provoke the landscape changes, which take place at annual frequency within 2007–2056 time interval.

In general, landscape transformations depend on (1) climate change, (2) development patterns, and (3) fire risk management. In the integrated model, climate change projections influence simulated fire weather and determines vegetation succession in the analyzed ecosystem.

The climate-related part of the proposed model uses the climate data from the Hadley (Johns et al. 2003) and MIROC 3.2 model for interdisciplinary research on climate (Hasumi and Emori 2004). On one hand, the MIROC is but coupled general circulation model, which consists of five component models: atmosphere, land, river, sea ice, and ocean. On the other hand, the Hadley permits to model global climate change, warming, under diverse emissions scenarios.

Due to the global scale of climate projection outcomes of these future scenario-modeling tools and due to the local regional scale of the present study by Hulse et al. (2016), authors perform downscaling of Hadley and MIROC data in order to use them as inputs in their agent-based integrated approach.

Authors employ two types of downscaled climate data, which is compared with the observed regional variations in temperature and precipitation during the twentieth century and project in this manner future climate impacts on vegetation and wildfire in the Pacific Northwest climate.

Modeling the urbanization development activities of agents, authors consider the two following alternatives. First, a compact urban development scenario replicates the contemporaneous practices of land usage preserving the farm and forest areas. Second, the dispersed urban development scenario adopts opposite policies to the current practices, thus promoting more dispersed rural development.

In respect to fire hazard management, authors also propose two strategies. The first is conventional fuel management supporting rapid suppression of fire with a little attention to overall resiliency of land to wildfire. Second mixed fuel-biodiversity strategy emphases the overall landscape resiliency to fire allowing fire to move through the areas with low risk to people, structures, and ecosystems, such as oak savanna and woodland.

In order to be able of mapping the surprising fire occurrences, Hulse et al. (2016) quantify the deviation of the modeled future fires from the reference benchmarks determined using Oregon wildfire historical records from 1960 to 2011, which obey a power law property. The benchmark fire for the high climate change scenario is equal to 6000 ha fire, which corresponds to the largest fire in the previous 50+ years from the historical fire data. Additionally, authors employ the 600 ha fire threshold, which corresponds to the low climate change scenario.

Mapped differences between the future fire projections and fire thresholds allow obtaining the probability of surprising fire under each modeled scenario. On the other side, the surprising fire likelihood depends on relationship between weather, the landscape agent's activity, and their response to perceived fire risk and topography, among other factors.

The proposed integrated ABM application provides the reader with several insights. For example, model's simulations show that the higher probability of fire is associated with the mixed fuels management strategy, accounting for more than 65% of large fire occurrences for different climate change scenarios. It is observed that 4 of the large fires affect more than 12,000 ha each, and 18 of the large fires involve more than 9000 ha of territory each. Authors also highlight that large fires usually occur after the relatively small fires, which affect less than 100 ha.

Additionally, model's application demonstrates that the likelihood of surprise fire augments while the population density, i.e., housing constructions, increases. Nevertheless, dispersed rural developments are associated with more than 70% of early surprising fires.

Summarizing, this paper provides the advanced methodology, based on ABM pillar, which allows predicting and localize fire occurrences. This work is especially interesting as many researchers performing the ABM somehow disrespect the probability nature of modeled future paths, while this work tries to raise a kind of alert flag focusing and correctly interpreting the inherently probabilistic nature of integrated climate-economy-climate interactions.

# 4 Discussing Agent-Based Modeling of Integrated Economy-Climate Systems

## *4.1 Agent-Based Modeling of Personal Beliefs and Climate Change Interrelation*

The research selected herein for detailed discussion, Geisendorf (2016), addresses the problem of climate sensitivity to the economic activity and individual beliefs on a global scale and in a general form. Typically agent-based climate-economic models are developed in order to analyze very specific topics related to the particular regions. Differently from such studies, Geisendorf (2016) reassesses and advances the original "battle of perspectives" model proposed by Janssen (1996) and Janssen and de Vries (1998), which until nowadays has been the only model approaching the global climate change issues from the generic perspective introducing the human biases interactions.

Geisendorf (2016) connects the original core "battle of perspectives" model to the ABM. This linkage allows simulating the individual adjustment processes in accordance with their understanding of the observed model-generated world performance. It is important to note that the "real-world" performance is hypothetical. The author combines the traditional macroeconomic growth model with both economy-on-climate (via emissions) and climate-to-the-economy (via economic losses due to climate change and the costs of preventing it) feedbacks.

In order to analyze how different perspectives influence the economy and climate change, the researcher introduces the classification composed of the four types of economic agents, be them firms or single persons. All agents have double impact on the modeled economy: (1) they decide what percentage of output to be invested into traditional capital in order to support economic growth; (2) agents determine the speed of shift from fossil to alternative climate-friendly fuels in accordance with the deviation of temperature change from their expectations, reflecting their climate policy, but within the bounds of their rooted beliefs.

The first type of agents believes in a self-correction ability of the climate-economic system, and, thus, economic activity should not be constrained. The author defines them as "free market advocates," which are characterized by the high level of $CO_2$ emissions and relatively low level of willingness to limit economic growth. It is assumed that "free market advocates" seek at least 3.2% annual economic growth, which corresponds to the average world GDP growth for the 2005–2014 period with the exclusion of the 2008 and 2009 financial crisis years. If the realized economic growth rate is less than the minimum benchmark rate, the agents increase their investments to stimulate economy. The transition from fossil to regenerative fuels is considered very slow, with a half-life time of 1000 years.

The second group of "scientifically informed" agents believes in a possibility of restrained abuse of nature. They base their perspectives on the Intergovernmental Panel on Climate Change (IPCC) reports and advocate stable economic growth. Thus, "scientifically informed" agents generate considerable level of $CO_2$ emissions and possess reasonably strong willingness to squeeze economic growth if necessary according to their environment-related beliefs. Specifically, for this group, investment adjustments represent a function of both previous investments and economic growth changes. In respect of climate change awareness, "scientifically informed" agents aim at preventing +2° (and higher) temperature increase compared to 1900. If temperature increases by +0.5°, they accelerate the transition from fossil fuels aspiring to reach a half time of at least 100 years or even down to 20 years in the event of extreme overheating on a planetary scale.

The third type of agents, so-called environmentalists, is risk-averse. They prefer absence of economic growth and, thus, low levels of wealth, associated with the safe natural environment, to the growing economy, which leads to the high living standards and exposes natural habitat to a very high risk of a climate disaster. The author associates "environmentalists" with the low level of $CO_2$ emissions and extremely high degree of readiness to contract the economic growth. Thus, for all the possible scenarios, they are attributed always with the highest transition speed to green fuels (with a half-life time of 20 years) and capital investments to solely compensate for its depreciation.

The fourth, and the last, category of agents includes those who do not possess any position in respect of climate change and economic growth. The author calls them "fatalists" as they believe that it is impossible to manage the economy and the natural environment, and, thus, desired outcomes are unachievable. The researcher attributes "fatalists" with the low level of $CO_2$ emissions simultaneously with the low willingness to limit economic growth. Author explains this by the following logic: on one hand, there is no intention to prevent climate change by "fatalists"; on the other hand, they have no intention to contribute to economic growth, and, thus, there is low degree of negative impact on climate.

The four described above agent groups govern the world conjointly. The world climate-economic policy is determined as the average of the proposed actions, weighted by the number of agents implementing them. The author performs two types of simulations with the starting point in 2005. The first scenario of experiments assumes that agents are nonadaptive and do not have learning ability. The second range of simulations is performed for the case of the adaptive agents, who learn from each other and can change their beliefs in the case when the beliefs of other agents fit better the world reality.

The non-learning agents' model simulation aims to study the three following situations: (1) world fitting the agent's beliefs, (2) no fitting world to the agent's beliefs, and (3) free-riding effects within mixed populations. The test (1) reveals impressive economic growth for the "free market advocates" until 2070, and since then the growth rate remains stable at a target level of 3.2%. Temperature increases over time, but the economy is not affected, thus the necessity to decrease $CO_2$ emissions arises only in 2100. The "scientifically informed" and "environmentalists" agents find themselves much more sensitive to the climate change: they manage to implement the fuel transition policies until 2095 and 2035, respectively. Also both of them experience a moderate economic growth.

The test (2) shows that the "free market advocates" type of agents, placed in nonconforming worlds, generate very low economic growth, which begins to decline from 2040 for the environmentalist conditions case. Since 2091 the GDP level drops below its 2005 initial value. Climate change affects the economy much earlier, leading to the fossil fuel substitution until 2065. In their turn, the agents, "environmentalists," placed in the free market world, experience a moderate economic growth, more than doubling their environmentalists' GDP over the next century. Still, the growth of the aggregate GDP of environmentalists is six times lower than the growth they could have in the free market world should they not share environmentalists' beliefs.

The experiment (3) evidences the beneficial free-riding effect for the "free market advocates" living in "environmentalist" or/and "scientifically informed" world(s) due to the climate polices implemented by the agents-"environmentalist."

The learning agents model simulations study the case when "free market advocates" are placed in a "scientifically informed" world and vice versa, adapting their perspectives. The bigger the difference between the agents' perception and the surrounding reality, the higher is the probability of changes in agents' beliefs and, hence, in agents' behavior. The author calculates that the economic growth rate for "scientifically informed" learning agents under the free market conditions is approximately slightly below the growth rate of the innate incumbent "free market advocates." Over the century the former are not much worse than the latter as the GDP of the initially "scientifically informed" but then adapted agents is only 12.5% below the GDP of the innate "free market advocates."

Differently from the initially "scientifically informed" but then adapted agents placed into the free market world, who perform worse than the incumbent "free market advocates," the performance of the latter continues to be superior to the performance of other types of agents, also when "free market advocates" agents are adapting being placed in a scientifically informed world. They perform better for a long time in comparison with the innate worldview of incumbent "scientifically informed" agents. Thus, the performed simulations do not allow for finding drawbacks and disadvantages of the free market strategy showing it superiority relative to strategies based on other worldviews and beliefs.

Additionally, the author analyzes the influence of "fatalist" agents on economic growth and climate change. Geisendorf (2016) advocates the importance of "fatalists" to be included in the model as in real world many people behave as "fatalists," and, in fact, they adapt beliefs over time and affect the future growth paths. The author estimates that there is at least 10% of the population, which has features of "fatalists." In accordance with the model output, when the percentage of "fatalists" in a society is high (50%), the learning process is slowed down, economic growth is weak, and, consequently, the climate change is negligibly small. On the other hand, when the "fatalists" are in minority in a modeled society (10%), the "free market advocates" agents overinvest, and economic growth becomes much higher as it would be without "fatalists." The increases in $CO_2$ emission and temperature are also much more pronounced if compared with the no "fatalists" world.

Overall, Geisendorf (2016) proposes the general agent-based climate-economic model on a world scale. The developed model allows analyzing the global climate change through the prism of agents' perceptions and their contradictions under uncertainty. The paper alerts on the strong effects of human beliefs and perceptions on climate policy. This model allows deriving general conclusions on climate change issues, simultaneously considering two-side relationship: economy influence on climate and vice versa. The complete economy-climate-economy feedback loops proposed in this seminal paper make it an important benchmark for future research. The next subsection is dedicated to the modified Geisendorf (2016) model, which aims to analyze how different types of investment impact economic development and climate change.

## 4.2 *Agent-Based Climate-Economic Model to Assess Green Investment Effects*

Further developments of the dynamic agent-based climate-economic model by Geisendorf (2016) are undertaken in the most recent selected work by Geisendorf and Klippert (2017). In this paper, the authors modify the original model in order to study the complex relationship between the perspectives of heterogeneous groups of adaptive agents, economic growth, and climate change. The novelty of this research resides in the assumption that expenditure for climate protection is classified as an investment in technology and not as a cost with a negative effect on production like many macroeconomic integrated assessment models consider.

The paper extensively criticizes macroeconomic growth models for their negligence in differentiating the investments mitigating climate change from the investments in its conventional sense (e.g., infrastructure, machinery, etc.). In general, such models consider traditional investments as a catalyst of capital stock accumulation, which supports economic growth. Nevertheless, these models fail by not recognizing and not accounting for the capacity of green investments to diminish $CO_2$ emissions. They also fail in a sense of not augmenting simultaneously the stock of capital, which then should result in a stronger economic growth.

Additionally, the authors address another common drawback of macroeconomic integrated assessment models by accounting for the heterogeneity and adaptive behavior of economic agents. Traditional models and their extensions assume homogeneous agents with rational behavior, which hardly correspond to how the real world is functioning. The authors fill this gap by introducing in their agent-based climate-economic model the limited rationality assumption and taking into consideration that individuals and groups of individuals may be affected by the variety of perceptions and constantly changing economic and climate perspectives.

In the proposed model, agents are faced with the choice to allocate their wealth into conventional investment, into green investment, or equally into both types. The agents' investment preferences depend on their individual perceptions. Thus, the agents with high economic growth perspectives, so-called "free market economist" agents, invest in traditional capital stock, as they believe in low productivity of green capital in comparison with the traditional one. On the contrary, the agents concerned most about climate change negative impacts, i.e., "environmentalists," invest in green capital, for example, low-carbon energy, renewable energy, energy-saving measures, water infrastructure, and other technological know-how related to climate protecting mechanisms. "Scientifically informed" agents invest equally in conventional and green capital stocks though they believe in a higher productive capacity of the latter.

In order to allow economic agents to be influenced by economic and climate factors, the authors create the possibility of the "free market economist" to invest in green capital if damage originated by the climate change is higher than 1% of GDP. "Environmentalists" increase green investments at the expense of traditional capital stock under conditions when observed $CO_2$ emissions exceed its desired level by more than 1%. "Scientifically informed" agents are enforced to augment successively their investments in climate change mitigation actions from the existing 2% of GDP targeted by the threshold level if a temperature increase is higher than 0.5°C.

Furthermore, authors assume that $CO_2$ emissions are negatively related to the green capital stock accumulation. In parallel, Geisendorf and Klippert (2017) introduce the inverse GDP dynamic function of temperature, which accounts for decreasing economic output under temperature increases and vice versa.

Following the procedures proposed in the approach developed by Geisendorf (2016) and described earlier in this chapter, the authors, at the first stage, perform two simulations, with starting point at year 2005. At the beginning, they place the agents in an environment functioning in accordance with their perceptions and then place the agents in a world ascribed different from their beliefs' features and dynamics.

The second step is concerned with the assessment of how agents' beliefs influence economic growth and preventive measures aimed at climate protection. Authors investigate two opposed modes of agents' learning capacity: the absence of learning ability of non-adapting agents and the existence of learning potential of the adapting ones.

Under the non-adapting agent hypothesis, authors consider three situations: (1) "free market economist" world, (2) "scientifically informed" agents universe, (3) and "environmentalists" reality. "Free market economist" environment implies the climate robustness to the $CO_2$ emission increase and, thus, low productivity of green capital. The "environmentalist" agents in such reality experience decline in economic output, as they heavily invest in green stock by cutting investments in conventional stock, which becomes exhausted due to depreciation.

In a world working in accordance with "scientifically informed" agents' views, "environmentalists" achieve almost the same GDP growth rate as the "free market economists" do, and higher than that of "scientifically informed" agents. Also the economic performance of "environmentalists" is better in the Geisendorf and Klippert (2017) model if compared to the reference Geisendorf (2016) model, which does not consider the effect of green investments. In "environmentalists" world, "scientifically informed" agents perform better by all, climate and economic, parameters if they have a choice to invest in green capital comparing to the absence of such possibility. "Free market economists" in "environmentalists" world generate GDP, which is 60% lower than "scientifically informed" and "environmentalists" produce.

In order to study economic and climate dynamics of the world populated by heterogeneous adaptive agents, authors perform a comparative analysis of nonadaptive versus adaptive agents placed in three different realities. First, while "environmentalist" and "scientifically informed" adaptive agents live in a "free market economists" reality, the economic performance and climate conditions of this world improve in comparison with the nonadaptive agents' condition.

Second, when the "environmentalists" world is mainly populated by the adaptive "free market economist" agents, they also outperform nonadaptive agents in output growth and climate parameters.

Third, "scientifically informed" world with heterogeneous adaptive agents could achieve a win-win situation, as the influence of two other extreme perspectives is weakened by "scientifically informed" believes. Nevertheless, the overall performance in respect of output, $CO_2$ emissions, and temperature is moderate.

Summarizing, this paper provides the important tool for the analysis of a complex relationship between human perceptions, economic development, and climate change. The authors, based on their modified agent-based climate-economic model, advocate that wrong perceptions about the real state of the world could originate strong effects on the climate and on the economy. The authors show that even after such wrong assumptions are corrected, it takes a long time to mitigate negative economic and climate impacts. Hence the attempts to cultivate adequate beliefs adherent to the surrounding reality are very important. Perhaps the effects of positive contagion of such correct perceptions could be further investigated in future developments of this field of knowledge.

## 5 Concluding Remarks

As demonstrated across the chapter, the ABMs are capable of capturing all the complexity of interactions between the heterogeneous agents and hence represent appropriate tools for focusing on specific areas of interrelations potentially insightful in a search for answers to the problems of economy-environment ecosystems. Still, all models in general are based on assumptions and employ simplifications. The most important, hence, is to assure that assumptions are appropriate and that they do not precondition and/or distort the outcomes of climate change agent-based modeling.

For instance, the discussed research employs simultaneously both stochastic and dynamic approaches. Still outcomes of stochastic analyses depend on type of statistical distributions used to generate series of random variables. If employed distributions do not have considerable fat tails, as the normal distribution, such models will be hardly suitable for modeling and forecasting the most important extreme economic crisis-like events and/or natural climate-related catastrophes, be they provoked or not by human activity and vice versa. So, a due caution is needed in order to use the ABM outcomes as an urgent and unquestionable guidance to elaborate appropriate climate-friendly policies.

Another important drawback on direct applicability of the ABM outcomes to hands-on management of climate-economy relationship is the limited depth and accuracy of historical records on climate. Not every aspect of climate change has sufficiently long data series. Among rather rare exceptions is the study of Nile floods as water level in the river was recorded along several millennia (see Hurst (1954)). The short data histories on climate parameters, along with complex system features of agent-based modeling, make a calibration of ABMs a quite complicate issue with a questionable reliability (see Fagiolo and Roventini 2017).

This alert is shared in the recent working paper by Lamperti et al. (2017), which acknowledges that the magnitude and the uncertainty associated to climate change impacts increase over time and thus turn shaping future scenarios a rather complicated task. So, certainly, the academic community will adequately address with time many of these issues. It is especially so, as the use of ABMs for studies of economy-climate relationship is still a quite recent phenomenon.

We believe that the role of ABMs in studies of environmental dynamics will keep increasing, as ABMs allow for assessing the full range from micro- to macroeconomic influences on climate change as well as impacts on micro-, meso-, and macro-scale climate shocks on economic development. What is also important is that ABMs allow considering endogenous technological and structural changes continuously occurring in the human society.

It is worth noting that ABMs on macro- and microscales are not excluding each other; rather, they are complementary as the former allow for checking the outcomes of the later and vice versa. It is also important to mention that ABMs already started to influence policymakers even though their explanatory power still deserves to be improved.

The main challenge that should be raised in front of ABMs is how to develop more accurate and precise approaches to modeling and reflecting economy-climate interrelations. Perhaps one of the solutions can be found in incorporating into ABMs the interrelations between climate change and social inequality. It seems like ABMs also lack a due accounting for relationship between financial markets and the productive economy, not fully uncovering the potential of green investment to withstand negative aspects of climate change. We hope that in the near future ABMs will help us to reduce the magnitude and frequency of human disgrace caused by environmental catastrophes through an asymptotic diminishing of knowledge gaps relative to interactions of economic development and climate change.

**Acknowledgments** Authors gratefully acknowledge the financial support by the IPL, Instituto Politécnico de Lisboa (Lisbon Polytechnic Institute), under the Projects IPL/2017/MacroTools/ISCAL and IPL/2018/MacroViews/ISCAL, and by the Portuguese national funding agency for science, research and technology (FCT) under the Project UID/SOC/04521/2013.

## References

Balint, T., Lamperti, F., Mandel, A., Napoletano, M., Roventini, A., & Sapio, A. (2017). Complexity and the economics of climate change: A survey and a look forward. *Ecological Economics, 138*, 252–265.

Beven, K., & Hall, J. (2014). *Applied uncertainty analysis for flood risk management*. London: Imperial College Press.

De Oreo, W. B., Mayer, P. W., Martien, L., Hayden, M., Funk, A., Kramer-Duffield, M., et al. (2011). *California single-family water use efficiency study*. Report prepared for the California Department of Water Resources.

Dubbelboer, J., Nikolic, I., Jenkins, K., & Hall, J. (2017). An agent-based model of flood risk and insurance. *Journal of Artificial Societies and Social Simulation, 20*(1), 6. https://doi.org/10.18564/jasss.3135

Entwisle, B., Williams, N., Verdery, A., Rindfuss, R., Walsh, S., et al. (2016). Climate shocks and migration: An agent-based modeling approach. *Population and Environment, 38*(1), 47–71.

Ermolieva, T., Filatova, T., Ermoliev, Y., Obersteiner, M., Bruijn, K., et al. (2017). Flood Catastrophe Model for designing optimal flood insurance program: Estimating location-specific premiums in the Netherlands. *Risk Analysis, 37*(1), 82–98.

Fagiolo, G., & Roventini, A. (2017). Macroeconomic policy in DSGE and agent-based models redux: New developments and challenges ahead. *Journal of Artificial Societies and Social Simulation, 20*(1), 1.

Geisendorf, S. (2016). The impact of personal beliefs on climate change: The "battle of perspectives" revisited. *Journal of Evolutionary Economics, 26*, 551–580.

Geisendorf, S., & Klippert, C. (2017). The effect of green investments in an agent-based climate-economic model. *Environmental Modeling and Assessment, 22*(4), 323–343.

Guzy, M., Smith, C., Bolte, J., Hulse, D., & Gregory, S. (2008). Policy research employing agent-based modeling to assess future impacts of urban expansion onto farm and forest lands. *Ecology and Society, 13*(1), 37.

Hasumi, H., & Emori, S. (2004). *K-1 coupled GCM (MIROC) description* (K-1 Technical Report 1). Tokyo, Japan.

Holling, C. (1986). The resilience of terrestrial ecosystems: Local surprise and global change. In W. C. Clark & R. E. Munn (Eds.), *Sustainable development of the biosphere*. Cambridge: Cambridge University Press. isbn:0-521-32369-X.

Hulse, D., Branscombb, A., Enrighta, C., Johnsonc, B., Eversa, C., Bolted, J., et al. (2016). Anticipating surprise: Using agent-based alternative futures simulation modeling to identify and map surprising fires in the Willamette Valley, Oregon USA. *Landscape and Urban Planning, 156*, 26–43.

Hurst, H. (1954). Measurement and utilization of the water resources of the Nile basin. *Proceedings of the Institution of Civil Engineers, 3*(Part 3), 1–26.

Janssen M. A. (1996). *Meeting targets: Tools to support integrated assessment modelling of global change*. PhD thesis, University of Maastricht, The Netherlands. ISBN 90-9009908-5

Janssen, M. A., & de Vries, B. (1998). The battle of perspectives: A multi-agent model with adaptive responses to climate change. *Ecological Economics, 26*(1), 43–65.

Jenkins, K., Surminski, S., Hall, J., & Crick, F. (2017). Assessing surface water flood risk and management strategies under future climate change: Insights from an agent-based model. *Science of the Total Environment, 595*, 159–168.

Johns, T., Gregory, J., Ingram, W., Johnson, C., Jones, A., et al. (2003). Anthropogenic climate change for 1860 to 2100 simulated with the HadCM3 model under updated emissions scenarios. *Climate. Dynamics, 20*, 583–612.

Lamperti, F., Dosi, G., Napoletano, M., Roventini, A., & Sapio, A. (2017). *Faraway, so close: Coupled climate and economic dynamics in an agent-based integrated assessment model* (LEM Working Paper 2017/12). Available at SSRN: https://ssrn.com/abstract=2944328 or https://doi.org/10.2139/ssrn.2944328

Mashhadi, A., Shafiee, M., & Berglund, E. (2017). Agent-based modeling to simulate the dynamics of urban water supply: Climate, population growth, and water shortages. *Sustainable Cities and Society, 28*, 420–434.

Symstad, A., Fisichelli, N., Miller, B., Rowland, E., & Schuurman, G. (2017). Multiple methods for multiple futures: Integrating qualitative scenario planning and quantitative simulation modeling for natural resource decision making. *Climate Risk Management* https://doi.org/10.1016/j.crm.2017.07.002

Vickers, A. (2001). *Handbook of water use and conservation*. Amherst: WaterPlow Press.

# Greenhouse Gas, Carbon Dioxide Emissions and Economic Growth: Empirical Evidence from Threshold Effect

**Qaiser Munir and Sook Ching Kok**

## Contents


1 Background ..... 60
2 Literature Review ..... 62
2.1 Environmental Kuznets Curve ..... 62
2.2 Climate Change Effect to Growth ..... 64
3 Data and Methodology ..... 66
3.1 Data Source and Variable Descriptions ..... 66
3.2 Methodology ..... 67
4 Results ..... 69
4.1 Descriptive Statistics ..... 69
4.2 Estimation Results for Greenhouse Gas Emissions ..... 70
4.2.1 Test Results for Threshold Effects ..... 70
4.2.2 Regression Results ..... 72
5 Estimation Results for Carbon Dioxide Emissions ..... 73
5.1 Test Results for Threshold Effects ..... 74
5.2 Regression Results ..... 75
6 Summary and Conclusions ..... 76
Appendix: GHG Emissions and CO2 Emissions in Countries ..... 78
References ..... 80


**Abstract** Climate change poses the serious challenge of greenhouse gas emissions reduction with much concern over carbon dioxide (CO2) which is the main component of overall greenhouse gas emissions. Emissions control by developing countries is becoming a key for effective mitigation of climate change, as these countries now account for more than a half of global emissions and are still expanding their energy infrastructures. One solution to this is by reducing greenhouse gas emissions to achieve the target of mitigating global warming. However, reducing CO2 emissions

Q. Munir (✉)
Department of Economics, Institute of Business Administration, Karachi, Pakistan
e-mail: qmunir@iba.edu.pk

S. C. Kok
Faculty of Business, Economics and Accountancy, Universiti Malaysia Sabah, Kota Kinabalu, Sabah, Malaysia
e-mail: emily@ums.edu.my

 
T. Sequeira, L. Reis (eds.), *Climate Change and Global Development*,
Contributions to Economics, https://doi.org/10.1007/978-3-030-02662-2_4

means decreasing a country's industrial production, and it is likely to decrease gross domestic product (GDP) growth, employment, etc. In other words, achieving a higher growth will induce more emissions of CO2 thus deepening the climate crisis with disturbing long-run consequences, both economically and socially. Therefore, how to effectively control greenhouse gas emissions especially the emissions of CO2 and to achieve a sustainable economic growth at the same time is a serious problem in policy-makings. Although a large amount of academic researches have been devoted to climate change, the overall climate change effect to growth is not conclusive for developed and developing countries. Previous studies suggest that there is a linear effect of climate change to growth, neglecting the non-linearity issue between climate change and growth. Through this present study, we address the issue of the existence of threshold effects in the relationship between climate change and growth rate of GDP in the context of global data, using new endogenous panel threshold autoregressive (TAR) model. Our main interest is to provide answer to the question whether climate change can affect long-run economic growth. With threshold regression models, we find that the effect of climate change on growth varies with the level of GDP per capita. In assessing the effect of greenhouse gas emissions to GDP growth, double threshold effect and three regimes are discovered. The assessment of the effect of CO2 emissions to GDP growth has detected single threshold indicating only two regimes exist.

**Keywords** Climate change · Sustainable economic growth · Carbon dioxide emissions · Endogenous panel threshold autoregressive model

## 1 Background

It is well-accepted that economic development can incur pollution and lead to environmental degradation. One enormous aspect of pollution and its impact on environmental quality is with respect to air pollution and the resulted climate change. In more details, greenhouse gas emissions lead to the greenhouse effect and finally global warming. Specifically, fossil fuels like coal, natural gas, and oil have high carbon content thus the burning of fossil fuels is responsible for a large part of greenhouse gas emissions. In many countries, the primary source of energy is relying on fossil fuels; therefore, CO2 emissions are expected. In addition, the transportation sector has an intensive usage of fossil fuels. It has been indicated by some past studies that the transportation sector contributes significantly to the increase in CO2 emissions, for example, in South Africa (Prozzi et al. 2002), in the USA (Davies et al. 2007) and in China (Tian et al. 2014). Apart from fossil fuels, greenhouse gas emissions are attributed to land use change or the so-called carbon debt. For instance, Searchinger et al. (2008) address the problem of greenhouse gas increase in the USA due to land use change that converts croplands for biofuels in providing substitute for gasoline. Furthermore, cement production is another important source of CO2 emissions. In fact, economic development ultimately contributes to economic growth, but it involves negative externality in terms of environmental degradation. The largest emitting countries can be both developed and developing countries. In year 2015, the six largest emitting countries were China (29%), the USA (14%), the European Union (EU) (10%), India (7%), Russian Federation (5%) and Japan (3.5%) (Olivier et al. 2016). Hence, a new challenge faced by global nations is to cope with the effect of climate change. In this, the effect to growth deserves more attention because this is the primary goal for an economy.

Literature shows that climate change is closely linked with growth. There are two main directions in this area of research. One direction of study is to postulate an inverted U-shaped relationship or to determine the turning point between pollutants and per capita income, based on the environmental Kuznets curve (EKC) hypothesis. The hypothesis states that environmental quality tends to get worse with growth until reaching certain level of higher income, and then environmental quality tends to improve with growth. Numerous past studies in this direction are found. Some examples are Stern et al. (1996), Lean and Smyth (2009), Borhan et al. (2012), Linh and Lin (2014), Alam et al. (2016), Özokcu and Özdemir (2017), and so forth. Another direction of study is to investigate the effect of climate change to growth, much differing from the first direction which emphasizes on the effect of growth to environmental quality. In view of the hazardous impact of climate change on growth, the second direction of study has recently attracted a fast-growing interest among researchers. The existing studies are including Fankhauser and Tol (2005), Estrada et al. (2015), and Taylor et al. (2016) which focus on global economies; Du et al. (2017) concentrate on developed countries, the USA and European Union (EU); Lu et al. (2015) look into the case of China, then Alagidede et al. (2016) focus on Sub-Saharan Africa, and the study by Elshennawy et al. (2016) is on Egypt. Thus far, the investigations of climate change effect to growth are apparently still limited. More research works are required to provide empirical evidence and to offer recommendations for policy-makings.

The objective of our study is to examine the existence of threshold effects in the relationship between climate change and growth rate of real gross domestic product (GDP) in the context of global data from 1970 till 2015, using new endogenous panel threshold autoregressive (TAR) model. How to effectively control greenhouse gas emissions especially the emissions of CO2 and to achieve a sustainable economic growth at the same time is a serious problem in policy-making. Even though a large amount of studies are devoted to investigate the effect of climate change to growth, the overall climate change effect to growth is not conclusive for developed and developing countries. Previous studies suggest that there is a linear effect of climate change to growth, neglecting the non-linearity issue between climate change and growth. In this present study, the main question to be answered is whether climate change affects the long-run economic growth. In the presence of threshold effect, the effect of climate change to growth is expected to change with the level of per capita income. Our study will extend the existing literature and come up with some recommendations for better policy-makings.

The remainder of this chapter is organized as follows: Sect. 2 is literature review; Sect. 3 discusses data and methodology; Sect. 4 reports the empirical results; the last section summarizes and concludes.

## 2 Literature Review

This review of literature is concentrating on the link between climate change and economic growth aligned with the objective of our present study. This area of research is of particularly great interest due to the concern of global warming and the challenge to achieve a sustainable economic growth. The existing studies can be divided into two categories: environmental Kuznets curve and the effect of climate change to growth.

### 2.1 *Environmental Kuznets Curve*

The first category is for those studies aiming to postulate an "inverted U-shaped" relationship or to determine the turning point between pollutants and per capita income, based on the environmental Kuznets curve (EKC) hypothesis. Clearly, the studies in this direction emphasize on the effect of growth to environmental quality. Differently, studies in the second category are intended to investigate the effect of climate change to growth. The latter category gains a fast-growing interest due to increasing awareness of the tremendous climate change impact on growth.

Specifically, the non-linear relationship between pollutants (CO2 emission is typically used as a proxy for climate change impact measure) and per capita income (a proxy of economic growth) is postulated by the EKC hypothesis. EKC is named for Kuznets (1955). The hypothesis postulates an "inverted U-shaped" relationship between pollutants and per capita income. That is, environmental quality degrades with higher level of income at the early stage of economic development until reaching the threshold turning point then environmental quality improves with higher level of income, in which this occurs at the later stage.

Stern et al. (1996) point out that as the EKC hypothesis presumes growth will reduce the environmental impact of economic activity, the concept is dependent on a model of economy where there is no feedback from environmental quality to production possibilities as well as trade. Thus, violating these assumptions may lead to fundamental problems in estimating the parameters of the model. It is recommended to use econometric and qualitative historical analysis for individual countries. The existence of EKC is empirically examined by a number of past studies through different econometric applications.

Linh and Lin (2014) apply the Johansen (1991) cointegration approach and Granger causality in the VECM context to examine the relationships between CO2 emissions, energy consumption, foreign direct investment, and economic growth for the case of Vietnam. This country is one of the fastest-growing emerging markets in Asia. The annual data used are GDP per capita, foreign direct investment inflows and stocks per capita, energy consumption measured in kilotonnes (kt) oil equivalence, and CO2 emissions measured in kt CO2 emissions. All data cover the period from 1980 to 2010. It is found that the variables are cointegrated. The regression using OLS shows there are two income levels which correspond to a positive coefficient and a negative coefficient. However, the estimated coefficients are statistically insignificant. The results are unable to conclude on the EKC hypothesis. Granger causality test in the context of VECM is used, and it reveals the evidence of short-run bidirectional causal relationships between income and foreign direct investment, as well as between energy consumption and foreign direct investment, long-run bidirectional causations between CO2 emissions and income, energy consumption and income, energy consumption and foreign direct investment, and income and foreign direct investment.

The EKC hypothesis in the context of ASEAN has been re-examined by recent studies using some econometric techniques. Lean and Smyth (2009) analyse the interactions between CO2 emissions, energy consumption, and economic growth on a panel of ASEAN-5 countries, namely, Indonesia, Malaysia, the Philippines, Singapore, and Thailand over the period of 1980–2006. The study employs annual data on CO2 emissions measured in metric tons per capita, electricity consumption measured in million kWh per capita, and income level measured by real GDP per capita. By using the Johansen Fisher panel cointegration test proposed by Maddala and Wu (1999), we obtain results that indicate a single cointegration vector implying a long-run cointegration between the variables. Panel long-run estimation using the panel version of dynamic OLS proposed by Pedroni (2001) gives the results that indicate a positive relationship in the long-run relationship between CO2 emissions and electricity consumption and a non-linear long-run relationship between CO2 emissions and economic growth. The results from Granger causality test indicate unidirectional causality in the short-run from CO2 emissions and electricity consumption to economic growth, consistent with the EKC hypothesis. Borhan et al. (2012) contribute empirical evidence of the relationship between CO2 emissions and GDP per capita in ASEAN-8 over the period of 1965–2010 from a three-equation simultaneous model. In the first equation, pollutant is treated as a function of income and population density. In the second equation, income is a function of pollutant, labour, government expenditure, foreign direct investment, fixed capital investment, and net export. In the third equation, population density is a function of pollutant. We determine to use the fixed-effect one-way model for panel data based on the Hausman specification test result employing annual data. The regression results of simultaneous equation method reveal that an increase in income of 1 percent can lead to 2,31,41,576 percent increase in pollutant, then a decrease in pollutant of 76.86 percent, and after that a decrease in pollutant of 7.12 percent. Besides that, pollutant has a negative effect on population density. Hence, the evidence is in favour of the EKC hypothesis.

Alam et al. (2016) employ time series data of four highly populated countries, namely, Brazil, China, India, and Indonesia and use the ARDL bounds test approach of long-run and short-run relationships to investigate the impacts of income, energy consumption, and population growth on CO2 emissions for the period of 1970–2012. The results support the EKC hypothesis for Brazil and Indonesia in the long run and short run; China in the long run, however, does not support the hypothesis for India. The relationship between population growth and CO2 emissions is found empirically valid only in Brazil and India. Özokcu and Özdemir (2017) employ the data of 26 OECD and 52 emerging countries in panel data estimations to investigate the relationship between CO2 emissions and income over the period of 1980–2010. The results of the two separate models do not support the EKC hypothesis.

## 2.2 Climate Change Effect to Growth

Next, another strand of studies is in the attempt on investigating the climate change effect to growth. Taylor et al. (2016) apply a demand-driven model to analyse the relationships among greenhouse gas accumulation, global warming, and economic growth in the long run and their relationships with employment, distribution, and labour productivity in the short run. Their findings reflect that climate change mitigation policy is able to stabilize economy at higher income levels and lower carbon levels at low cost. Specifically, a decrease in investments of around 1 percent of the world's GDP is expected to reduce almost all of the net carbon emissions over time.

Several studies have applied the standard Ramsay-Cass-Koopmans growth model where a social planner maximizes the utility of identical consumers through intertemporal optimization. Using the intertemporal approach, it allows for endogenous anticipative adaptation responses to climate change impacts. It is different from the standard recursive-dynamic approach where climate shocks hit agents in the model by surprise (Elshennawy et al. 2016). More specifically, endogenous growth may be affected by the impact from climate change, capital accumulation effect, savings effect, and the possible influences from the changes in labour productivity and different rates of technical progress which can exacerbate the capital accumulation and savings effects. By assuming one with a closed economy, i.e. excluding the possible effects from international trade and capital flows, Fankhauser and Tol (2005) based on the simulations for developed and developing countries conclude that the capital accumulation effect is more pronounced and the savings effect is relatively weaker in an endogenous growth model. If savings rate is constant, a decrease in output which is caused by climate change will reduce investments proportionately and thus depress future production via the capital accumulation effect. On the other hand, if savings rate is endogenous, the model suggests that forward-looking agents will adapt their savings behaviour to the impact of future climate change. Hence, growth is suppressed in absolute and per capita terms via the savings effect. Elshennawy et al. (2016) explore the long-run growth prospect for Egypt using an intertemporal general equilibrium model. The model used incorporates the impact estimates based on agricultural productivity, labour productivity, and potential losses due to sea-level rise, to simulate the effects of climate change on aggregate consumption, investment, and income up to the year 2050. Forward-looking agents, population growth, and technical progress are taken into account. The simulation results indicate that in the absence of policy-led adaptation investments, real GDP is about 6.5 percent which is lower than in a hypothetical baseline without climate change impact. With adaptation investments, the GDP loss in 2050 can be reduced to around 2.6 percent.

As mentioned by Estrada et al. (2015), the integrated assessment models (IAMs) usually assume climate change affects economic growth only when it occurs; in other words, the impact has no persistence. The study looks into the persistence of climate change shocks in the GDP series for individual regions including Europe, Latin America, South Asia, North Asia, North America, Africa, China, and the world. By adding the memory parameter in the impact function of IAM, it is possible to capture the indirect impact of climate change to growth which is the aftershocks from past shocks or the persistence of shocks to growth. Meantime, the direct impact of climate change is new shocks in that year. Two scenarios, RCP4.5 and RCP8.5, are examined by using the modified IAM model with different values of the memory parameter. Apparently, the estimates of the costs of climate change are different from the estimates produced when the memory parameter is not included, which suggests the regional disparities may be larger than previously estimated.

A few studies have attempted to determine the optimal level between climate change and economic growth. Lu et al. (2015) apply a two-stage least squares (TSLS) method and employ the annual data of China from 1979 to 2012 to analyse the relationships among climate threshold, financial hoarding, and economic growth. Findings suggest the existence of positive cumulative effects between financial hoarding and economic growth, and the effects are restraint by climate threshold. One of the policy implications derived is to stress on climate change mitigation policy by giving emphasis on technology innovation and adaptive adjustment in climate-sensitive industries, through the improvement in financial derivatives as means of risk transfer mechanism. For the case of Sub-Saharan Africa (SSA), Alagidede et al. (2016) use a sample of 18 SSA countries and a panel dataset from 1970 to 2009 that consists of real GDP per capita and two climate change data: temperature and precipitation. The empirical model used is developed from the Cobb-Douglas production function. In terms of methodology, panel cointegration technique is utilized for analysing the long-run and short-run climate change effects to economic growth. Both the pooled mean group (PMG) and mean group (MG) estimators are utilized. PMG estimator uses the panel extension of autoregressive distributed lag (ARDL) model which can provide information about the contemporaneous impacts and the speed of adjustment towards equilibrium after a shock. We manage to establish the long-run relationships between temperature and economic growth, as well as between precipitation and economic growth. The findings also indicate temperature has a more destructive effect to growth in the short run. Furthermore, in the case of SSA, the increases in temperature beyond 24.9 °C can reduce growth significantly. Du et al. (2017) analyse the impact of temperature increases on economic growth for the cases of the USA and EU. The empirical methodology used is based on a linear spline model derived from the Cobb-Douglas production function. It is found that the optimal temperature is at 6 °C in each case. There exists a non-linear relationship between temperature and economic growth, in which growth increases as temperature rises until reaching the optimal level then growth decreases with higher temperature. Hence, the use of more proactive climate mitigation policy is suggested to counteract the devastating effect of climate change to growth.

The existing studies that have investigated the effect of climate change to growth are still limited. Among these studies, only a few of them have examined a non-linear relationship between climate change and growth. We note that determining the optimal level between climate change and economic growth may provide important inference regarding growth sustainability. However, this aspect is paid little attention so far.

# 3 Data and Methodology

## *3.1 Data Source and Variable Descriptions*

We employ the panel data starts from year 1970 to 2015. All the data are extracted from the World Development Indicators (2017) of the World Bank. The dependent variable is real GDP growth (GDP_GROWTH). The data of GDP are in constant 2010 US dollars. GDP is defined as the total of gross value added by all resident producers in an economy, plus any product taxes and deducting any subsidies not included in the value of products. Real GDP per capita (LN_GDP) is the threshold variable. It is obtained from dividing real GDP by midyear population. Real GDP equals GDP in constant 2010 USD. Thus, the threshold variable is measured by real GDP per capita. We choose to use real GDP per capita as the threshold variable in order to take into account the influence of population on income. This is to allow for a comparison between countries to show the relative performance of the countries, specifically, to capture the variations in performance between low- and middle-income and high-income countries.

The proxies for climate change are (1) total greenhouse gas emissions (GHG) in kt of CO2 equivalent composing of CO2 totals, which includes other biomass burning (e.g. forest fires, post-burn decay, peat fires, and decay of drained peatlands), all anthropogenic CH4 sources, N2O sources, and F-gases (HFCs, PFCs, and SF6) but excludes short-cycle biomass burning (e.g. agricultural waste burning and Savannah burning), and (2) carbon dioxide emissions (CO2 KT) from the burning of fossil fuels; the manufacture of cement; the consumption of solid, liquid, as well as gas fuels; and flaring. The graphs of GHG and CO2 emissions based on high-income, middle-income, and low-income countries are provided in the Appendix.

In terms of control variables, we follow the standard variables in growth regression as suggested by Sala-i-Martin (1997). First, we include gross fixed capital formation (GFCF % GDP). The GFCF % GDP includes land improvements, the purchases of plant, equipment and machinery purchases, as well as the construction of roads, railways, commercial and industrial buildings, hospitals, schools, and private residential dwellings. Second, general government final consumption expenditure (GGFCE % GDP) is the sum of all government current expenditures for the purchases of goods and services, which cover the compensation of employees and most of the expenditures on national defence and security. It excludes government military expenditures as these are considered as part of government capital formation. Third, we add trade as a control variable. Trade (TRADE % GDP) is calculated by adding the values of exports and imports of goods and services, and it is measured as a share of GDP. Further, we use broad money (M2) as a proxy for financial deployment. The variable broad money (BM M2 % of GDP) is calculated by adding currency outside banks; demand deposits other than those of the central government; savings, time, and foreign currency deposits of resident sectors other than the central government; bank and traveller's checks; and other securities (e.g. commercial paper, certificate of deposit). Lastly, we add annual population growth (POP growth). The annual population growth rate (POP growth) for year $t$ is the exponential rate of growth of midyear population from year $t - 1$ to $t$, expressed as a percentage. Population is based on the de facto definition of population, which includes all residents regardless of legal status or citizenship. The measures HFCF %GDP, GGFCE%GDP, TRADE%GDP, M2%GDP, and POP growth are the data directly extracted from the World Development Indicators (2017) of the World Bank.

## 3.2 *Methodology*

We propose to use the linear equations below to examine the relationship between real GDP growth (GDP_GROWTH) with each proxy for climate change: (1) total greenhouse gas emissions (GHG KT) and (2) carbon dioxide emissions (CO2 KT).

$$\text{GDP\_GROWTH}_{it} = \beta \text{GHG}_{it} + \gamma X_{it} + \varepsilon_{it} \tag{1a}$$

$$\text{GDP\_GROWTH}_{it} = \beta \text{CO2}_{it} + \gamma X_{it} + \varepsilon_{it} \tag{1b}$$

where $\text{GDP\_GROWTH}_{it}$ is the growth rate of real GDP, $\text{GHG}_{it}$ is total greenhouse gas emissions, $\text{CO2}_{it}$ is carbon dioxide emissions, $X_{it}$ is a vector of control variables, $\varepsilon_{it}$ is an error term, $i = 1,\ldots,N$ represents country, and $t = 1970,\ldots,2015$ represents year.

A non-linear inverted U-shaped relationship between pollutants and per capita income is postulated by Kuznets (1955). It has been examined in a number of past studies. Recently, a few empirical studies have found the optimal level in the relationship between climate change and growth. In the case of SSA, the increases in temperature beyond 24.9 °C can decrease growth significantly (Alagidede et al.

2016). In the cases of the USA and EU, growth increases as temperature rises until reaching the optimal level and then growth decreases with higher temperature (Du et al. 2017). Therefore, we have strong reasons to believe that the regression functions for the relationship between real GDP growth and the proxies for climate change may not be identical across all observations in a sample. The observations in a sample may fall into discrete classes. This motivates us to examine the existence of threshold effects between real GDP growth and the proxies for climate change. For this, we follow the procedure of Hansen (1999) and start by setting up a single threshold model. Our structural equation is expressed as follows:

$$\mathrm{GDP\_GROWTH}_{it} = \mu_i + \beta_1 x_{it} I(q_{it} \leq \gamma) + \beta_2 x_{it} I(q_{it} > \gamma) + e_{it} \tag{2}$$

where GDP _ $\mathrm{GROWTH}_{it}$ is the growth rate of real GDP for country $i$ at year $t$, $x_{it}$ is a set of explanatory variables including a proxy for climate change and all the control variables used, $I(\cdot)$ represents the indicator function, $q_{it}$ is the threshold variable real GDP per capita, $\gamma$ is the threshold parameter which separates observations in a sample into subsamples, and $e_{it}$ denotes the error term which is a single realization of white noise that is independent and identically distributed and with zero mean and finite variance $\sigma^2$. The structural equation in Eq. (2) can also be written as follows:

$$y_{it} = \begin{cases} \mu_i + \beta_1 x_{it} + e_{it}, & q_{it} \leq \gamma \\ \mu_i + \beta_2 x_{it} + e_{it}, & q_{it} > \gamma \end{cases} \tag{3}$$

where Eq. (3) indicates the separation of observations into regime I and regime II depending on whether the real GDP per capita threshold $q_{it}$ is smaller or larger than the threshold parameter $\gamma$. These regimes are determined by the regression slopes, $\beta_1$ and $\beta_2$, and for this, it requires that all the explanatory variables $x_{it}$ and threshold variable $q_{it}$ are not time invariant. There is no threshold if the two regression slopes $\beta_1$ and $\beta_2$ appear to be equal. In that case, Eq. (1) becomes a special case of the panel threshold model [Eq. (3)].

The threshold parameter $\gamma$ is estimated using ordinary least squares (OLS) through the two-step procedure following Hansen (2000). The first step involves computing the sum of squares of errors separately for any given $\gamma$. The second step allows minimizing concentrated sum of squared errors for any given $\gamma$, in order to derive the function:

$$\hat{\gamma} = \arg\min_\gamma S_1(\gamma) \tag{4}$$

where $S_1(\gamma) = \hat{e}^*(\gamma)'\hat{e}^*(\gamma) = Y^{*'}\left(I - X^*(\gamma)'\left(X^*(\gamma)'X^*(\gamma)\right)^{-1}X^*(\gamma)'\right)Y^*$ and $I$ is the identify matrix. Once $\hat{\gamma}$ is obtained, we then calculate the residual vector $\hat{e}^* = \hat{e}*(\hat{\gamma})$ and the square of residuals $\hat{\sigma^2} = \hat{\sigma^2}(\hat{\gamma}) = 1/n(T-1)\hat{e}^*(\gamma)'\hat{e}^*(\gamma) = 1/n(T-1)S_1(\gamma)$, $n$ is the number of countries and $t$ is the number of years.

To detect the threshold effect, we test the null hypothesis of no threshold effect $H_0 : \beta_1 = \beta_2$, which is set in Eq. (3). The test statistic is based on the likelihood ratio of $F_1 = \left(S_0 - S_1(\hat{\gamma})\right)/\hat{\sigma}^2$, where $S_0$ is sum of squared residuals under null hypothesis $H_0$. Under the null hypothesis, the value of threshold $\gamma$ is unrecognized. Therefore, the distribution of the likelihood test is nonstandard. To address this issue, Hansen (1999) proposes the heteroscedasticity-consistent Lagrange multiplier (*LM*) bootstrap procedure to compute *p*-value for the null hypothesis and indicates that the *p*-value based on this structure is asymptotically valid. Upon the existence of threshold effect, we should test for the asymptotic distribution of threshold estimate, $H_0 : \gamma = \gamma_0$, and apply the likelihood test calculating as $LR_1(\gamma) = \left(S_1(\gamma) - S_1(\hat{\gamma})\right)/\hat{\sigma}^2$ with the asymptotic confidence intervals $c(\alpha) = -2\ \log\left(1 - \sqrt{1-\alpha}\right)$.

The same procedure is replicated in testing for double or multiple threshold effects. For the case of double threshold effects or three regimes, the structural equations can be written as follows:

$$y_{it} = \begin{cases} \mu_i + \beta_1 x_{it} + e_{it}, & q_{it} \le \gamma_1 \\ u_i + \beta_2 x_{it} + e_{it}, & \gamma_1 < q_{it} \le \gamma_2 \\ \mu_i + \beta_3 x_{it} + e_{it}, & \gamma_2 < q_{it} \end{cases} \tag{5}$$

where the threshold value is $\gamma_1 < \gamma_2$. For the case of multiple thresholds, the above structural equations [Eq. (5)] can be extended through similar process $(\gamma_1, \gamma_2, \gamma_{3,\ldots,}\gamma_n)$.

# 4 Results

## 4.1 Descriptive Statistics

In this section, we report the summary statistics for real GDP growth (GDP_GROWTH), real GDP per capita (LN_GDP), greenhouse gas emissions (GHG KT), carbon dioxide emissions (CO2 KT), gross fixed capital formation (GFCF % GDP), general government final consumption expenditure (GGFCE % GDP), trade (TRADE % GDP), broad money (BM M2 % of GDP), and annual population growth rate (POP growth). We notice that some of the continuous variables have extreme values. To make sure these extreme values do not bias our results, we use the winsorization method to deal with this issue.

As reported in Table 1, the mean for real GDP growth is around 2.551 and for real GDP per capita around 8.478 over the whole sample period. Greenhouse gas (GHG) and carbon dioxide (CO2) emissions are on average 16.202 and 5.477, respectively. The control variables, gross fixed capital formation (GFCF), general government final consumption expenditure (GGFCE), trade (TRADE), and broad money M2 (BM M2) are measured as shares of GDP. The estimated mean values for these variables are 24.164, 14.254, 42.833, and 67.309, respectively. Obviously, broad

**Table 1** Summary statistics

| Variable | Mean | Std. dev. | Min | Max |
|---|---|---|---|---|
| GDP_GROWTH | 2.551 | 2.038 | −4.122 | 8.272 |
| LN_GDP | 8.478 | 1.414 | 6.530 | 10.635 |
| GHG (KT) | 16.202 | 0.461 | 15.069 | 16.951 |
| CO2 (KT) | 5.477 | 4.598 | 0.553 | 12.373 |
| GFCF (% GDP) | 24.164 | 3.186 | 14.802 | 31.741 |
| GGFCE (% GDP) | 14.254 | 2.649 | 10.037 | 19.200 |
| TRADE (% GDP) | 42.833 | 12.451 | 16.738 | 65.696 |
| POP_GROWTH | 1.367 | 0.641 | 0.477 | 2.468 |
| BM (M2 % GDP) | 67.309 | 31.073 | 22.662 | 142.298 |

money M2 has the largest share of GDP on average over the sample period, as compared to other variables. GGFCE has the smallest contribution to GDP on average over the period. The average annual growth rate of population growth rate is around 1.367 percent. The standard deviations (Std. dev.), minimum values (Min), and maximum values (Max) of the variables are provided in Table 1.

## 4.2 *Estimation Results for Greenhouse Gas Emissions*

There are two proxies for climate change used in this study, namely, total greenhouse gas emissions (GHG KT) and carbon dioxide emissions (CO2 KT). We first examine the relationship between GHG emissions and real GDP growth by applying the Hansen (1999) panel threshold model.

### 4.2.1 Test Results for Threshold Effects

The number of threshold is determined using the bootstrapping method following Hansen (1999) allowing to obtain an approximation of the $F$-statistics and then calculate the $p$-values. The $F$-statistics contain $F_1$, $F_2$, and $F_3$ to access the null hypothesis of none, one, and two thresholds, respectively. We use 500 bootstrap replications for each of the bootstrap tests performed.

Table 2 presents the test results for threshold effects based on GHG emissions using real GDP per capita (LN_GDP) as threshold variable. The test results are reported for single, double, and triple threshold effects, $F_1$, $F_2$, and $F_3$, along with their bootstrap $p$-values. Both the test statistics $F_1$ and $F_2$ are highly significant at the 1% level with a bootstrap $p$-value of 0. For triple threshold effect, the test statistic $F_3$ with a $p$-value of 0.616 is not statistically significant, which suggests the non-existence of a third threshold. The sequential tests for single, double, and triple threshold effects indicate the existence of two thresholds and thus three regimes in our model.

**Table 2** Tests for threshold effects based on GHG emissions

| *F-test* | *p*-value | Critical values 1% | 5% | 10% | Threshold value ($\hat{\gamma}$) | 95% Confidence interval |
|---|---|---|---|---|---|---|
| Test for single threshold ($H_0$: no threshold) | | | | | | |
| 31.96 | 0.000*** | 10.248 | 9.005 | 8.511 | 7.7968 (2432.929) | [7.6483, 7.7721] |
| Test for double threshold ($H_0$: at most one threshold) | | | | | | |
| 9.54 | 0.000*** | 5.052 | 5.002 | 4.463 | 8.1063 (3315.441) | [8.0754, 8.1388] |
| Test for triple threshold ($H_0$: at most two thresholds) | | | | | | |
| 6.66 | 0.616 | 20.389 | 17.081 | 15.089 | 6.8588 (952.194) | [6.8331, 6.8591] |

Note: This table reports the threshold estimates $(\hat{\gamma})$. *F*-statistics and *p*-values are from repeating bootstrap procedures 500 times for each of the three bootstrap tests. ***, **, and * indicate significance at the 1%, 5%, and 10% levels, respectively. The threshold variable is natural logarithm of GDP per capita

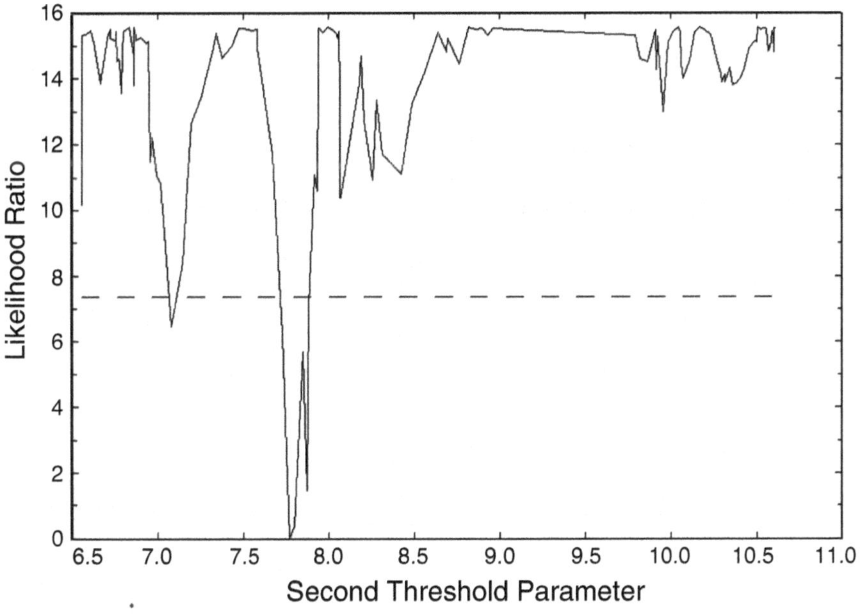


**Fig. 1** Confidence interval construction in single threshold model GHG emissions

The estimated values for the two real GDP per capita thresholds $(\hat{\gamma}_1, \hat{\gamma}_2)$ are reported in Table 2. The point estimates of the thresholds are 7.7968 and 8.1063, respectively, and these values are the log of real GDP per capita. The estimated thresholds in real GDP per capita are around 2432.929 and 3315.441, respectively. The asymptotic 95% confidence intervals ([7.6483, 7.7721]) and ([8.0754, 8.1388]) are narrower and more precise in both the first and second threshold estimates. It indicates little uncertainty about the nature of this division. Figures 1 and 2 are then produced to show the computed likelihood ratio statistics for the single and double

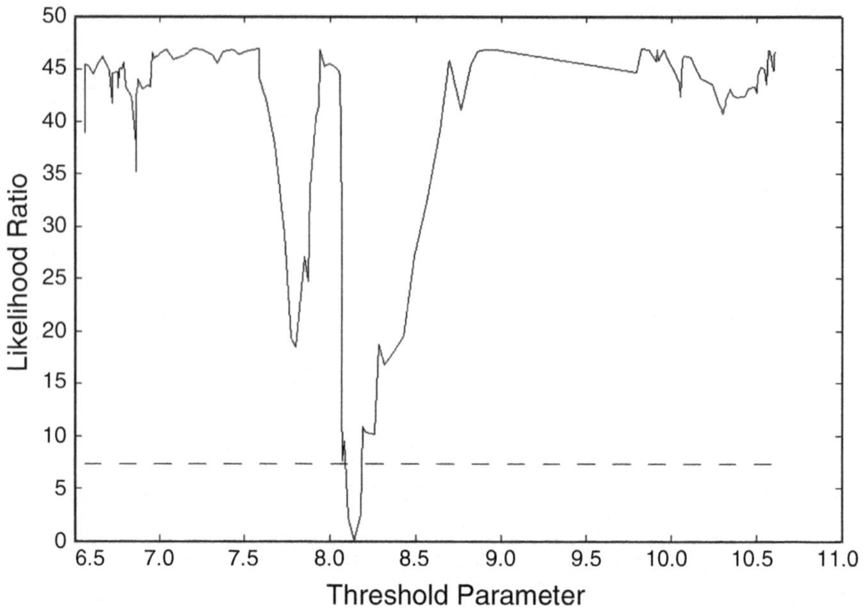


**Fig. 2** Confidence interval construction in double threshold model GHG emissions

threshold models. These figures can confirm the estimates for the first and second thresholds.

### 4.2.2 Regression Results

Having obtained the points of the two thresholds 7.7968 equivalents to real GDP per capita 2432.929 and 8.1063 equivalents to real GDP per capita 3315.441, therefore, the observations in our sample are divided into three regimes. The division of observations into differing regimes is depending on whether the threshold variable is lower or higher than the threshold value ($\gamma_1,\ \gamma_2$). We then define the three regimes as follows: regime I for low per capita income ($q \leq 7.7968$ ) which is equivalent to or less than per capita income 2432.929; regime II for moderate per capita income ($7.7968 < q \leq 8.1063$) which is between per capita income 2432.929 and 3315.441; and regime III for high per capita income ($8.1063 < q$) which is more than per capita income 3315.441.

Splitting the sample into separate regimes allows us to estimate our model for each observation class, thus allowing for a comparison between different groups. Table 3 presents the estimated coefficients, conventional OLS standard errors, and White-corrected standard errors. As indicated by the regression results, the estimated coefficients for GHG emissions in regimes I (2.3159), II (2.5888), and III (−2.3902) are all highly significant at the 1% level. For regime I ((ln GDP) $\leq$ 2432.9) and

**Table 3** Regression estimates for the effect of GHG emissions to GDP growth

| Regressors | Coefficient | Robust standard errors | $p$-values |
|---|---|---|---|
| GFCF (% GDP) | 3.3962 | 0.3567 | 0.011 |
| GGFCE (% GDP) | −22.2199 | 4.5885 | 0.040 |
| TRADE (% GDP) | 0.6786 | 0.0405 | 0.000 |
| POP_GROWTH | −5.0041 | 0.0740 | 0.021 |
| BM (M2 % GDP) | 0.6496 | 0.0451 | 0.000 |
| GHG (KT) | 2.3159 | 0.1912 | 0.007 |
| Regime I (ln GDP) ≤ 2432.929 | | | |
| GHG (KT) | 2.5888 | 0.1951 | 0.006 |
| Regime II 3315.4 ≥ (ln GDP) > 2432.929 | | | |
| GHG (KT) | −2.3902 | 0.2071 | 0.007 |
| Regime III (ln GDP) > 3315.441 | | | |

Notes: White SE refers to White standard errors (considering heteroscedasticity)

regime II (3315.4 ≥ (ln GDP)>2432.9), the estimated coefficients for GHG emissions have positive signs suggesting that real GDP growth increases with GHG emissions in the low per capita income regime and in the middle per capita income regime. Specifically, such positive effect to growth occurs in regime I or the low per capita income regime when per capita income is below 2432.929 (equivalents to the log of real GDP per capita around 7.7968) and occurs in regime II or the middle per capita income regime when per capita income is between 2432.929 and 3315.441 (equivalents to the log of real GDP per capita between 7.7968 and 8.1063). The third regime ((ln GDP) > 3315.4) is when the log of real GDP per capita turns to be greater than 8.1063 or when a value is over real GDP per capita 3315.441. The estimated coefficient for GHG emissions in regime III or the high per capita income regime is having a negative sign (−2.3902). This result suggests the emissions negatively and significantly affect growth when per capita income rises over 3315.441.

The estimated coefficients for all the control variables considered in this study are found to be statistically significant at least at the 5% level. Both general government final consumption expenditure (GGFCE % GDP) and annual growth rate of population (POP_GROWTH) have significant negative effect to growth. Gross fixed capital formation (GFCF % GDP), trade (TRADE % GDP), and broad money (BM M2 % GDP) have significant positive effect to growth.

## 5 Estimation Results for Carbon Dioxide Emissions

In previous Sect. 4.2, the estimation results for total greenhouse gas emissions are discussed in detail. Here, we proceed with the estimation for carbon dioxide emissions (CO2 KT).

**Table 4** Tests for threshold effects based on CO2 emissions

| *F-test* | *p*-value | Critical values 1% | 5% | 10% | Threshold value $(\hat{\gamma})$ | 95% Confidence interval |
|---|---|---|---|---|---|---|
| Test for single threshold (*H*0: no threshold) | | | | | | |
| 31.04 | 0.000*** | 8.5219 | 8.1503 | 7.7556 | 8.1063 (3315.441) | [8.0735, 8.1388] |
| Test for double threshold (*H*0: at most one threshold) | | | | | | |
| 5.88 | 0.3360 | 9.9255 | 8.4686 | 8.0047 | 7.7318 (2279.777) | [7.6748, 7.7721] |

This table reports the threshold estimates $(\hat{\gamma})$. *F*-statistics and *p*-values are from repeating bootstrap procedures 500 times for each of the three bootstrap tests. ***, **, and * indicate significance at the 1%, 5%, and 10% levels, respectively. The threshold variable is natural logarithm of GDP per capita

## 5.1 Test Results for Threshold Effects

To examine the relationship between GDP growth and CO2 emissions, we apply the Hansen (1999) panel threshold model approach. The number of threshold is determined using the bootstrapping method that allows an approximation of the *F*-statistics and then calculating the *p*-values. The *F*-statistics contain $F_1$ and $F_2$ to access the null hypothesis of none and one threshold. Each of the bootstrap tests performed is by using 500 bootstrap replications.

Table 4 reports the test results for threshold effects between CO2 emissions and real GDP growth using real GDP per capita as the threshold variable. It can be seen that the test statistic $F_1$ is highly significant at the 1% level with a bootstrap *p*-value of 0 for single threshold effect. For double threshold effect, the test statistic $F_2$ is statistically not significant with a bootstrap *p*-value 0.3360. These results allow us to conclude that there is only one threshold and two regimes in the regression relationship for CO2 emissions and real GDP growth.

The point estimate of single threshold is 8.1063 which is the log value of real GDP per capita 3315.441. In comparison, this current estimated single threshold for the case CO2 emissions is higher than the previously estimated single threshold in the case of GHG emissions (log GDP per capita 7.7968 or real GDP per capita 2432.929), but it is the same as the estimated double threshold in the case of GHG emissions. In the previous case of GHG emissions, there are two thresholds and three regimes detected: low per capita income regime, middle per capita income regime, and high per capita income regime. For this current case of CO2 emissions, there is only one threshold. We define the two regimes as low and middle per capita income regime and high per capita income regime. Accordingly, we split our observations into low and middle per capita income regime ($q \leq 8.1063$) which is equivalent to or less than per capita income 3315.441, and high per capita income regime ($8.1063 < q$) exceeding per capita income 3315.441.

Figure 3 displays the computed likelihood ratio statistics as a function of the threshold variable for our model which confirm the estimate for single threshold. However, the asymptotic 95% confidence intervals with [8.0735, 8.1388] are narrower indicating little uncertainty about the nature of this division.

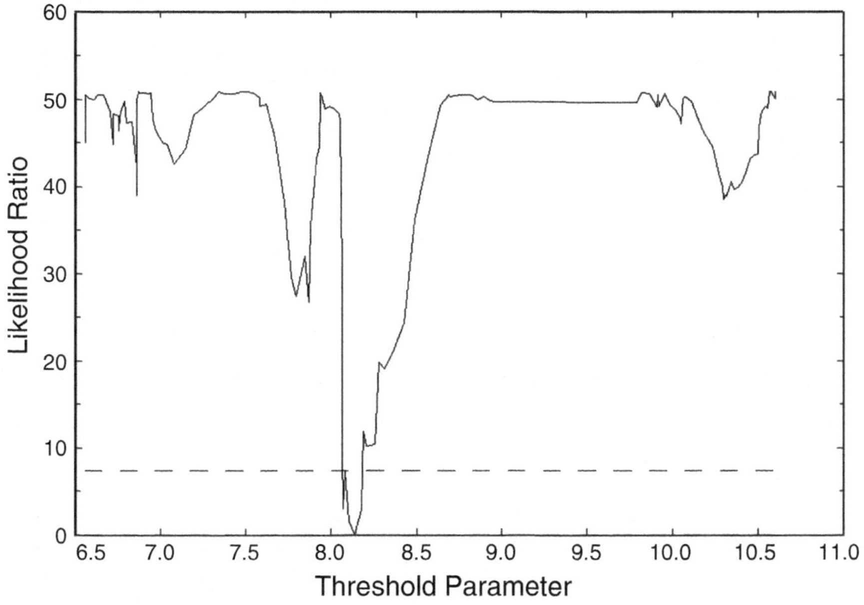


**Fig. 3** Confidence interval construction in single threshold model of CO2

**Table 5** Regression estimates for the effect of CO2 emissions to GDP growth

| Regressors | Coefficient | Robust standard errors | $p$-values |
|---|---|---|---|
| GFCF (% GDP) | 0.2016 | 0.0861 | 0.021 |
| GGFCE (% GDP) | −19.568 | 2.2651 | 0.013 |
| TRADE (% GDP) | 1.1744 | 0.0902 | 0.000 |
| POP_GROWTH | −3.7501 | 0.9488 | 0.058 |
| BM (M2 % GDP) | −1.0251 | 0.7419 | 0.301 |
| (CO2 KT) | 0.3629 | 0.0398 | 0.000 |
| Regime I (ln GDP) $\leq$ 3315.441 | | | |
| (CO2 KT) | −0.8361 | 0.1846 | 0.045 |
| Regime II (ln GDP)>3315.441 | | | |

Notes: White SE refers to White standard errors (considering heteroscedasticity)

## 5.2 *Regression Results*

We estimate our model for each observation class. Table 5 presents the estimated coefficients, conventional OLS standard errors, and White-corrected standard errors of each of the variables for the single threshold model. The estimated coefficient for CO2 emissions (CO2 KT) in regime I ((ln GDP) $\leq$ 3315.441) is 0.3629, which is positive and significant at the 1% level. The result indicates that in the low and middle per capita income regime, real GDP growth increases with CO2 emissions. Meanwhile, we find that the estimated coefficient for CO2 emissions in regime II ((ln GDP) > 3315.441) is −0.8361 and significant at the 5% level. The negative sign of the coefficient indicates that real GDP growth is adversely affected by CO2 emissions in the high per capita income regime. Majority of the control variables considered in this study are found to be statistically significant at least at the 10% level, except for broad money (BM M2 % GDP). General government final consumption expenditure (GGFCE % GDP) and annual growth rate of population (POP_GROWTH) adversely affect real GDP growth. Gross fixed capital formation (GFCF % GDP) and trade (TRADE % GDP) have a positive effect on real GDP growth.

## 6 Summary and Conclusions

Our main objective in this study is to examine the existence of threshold effects in the relationship between climate change and growth rate of real GDP in the context of global data for the period of 1970–2015, using new endogenous panel threshold autoregressive (TAR) model as proposed by Hansen (1999). The critical advantage of this endogenous threshold regression technique over the traditional approach is that it does not require any specified functional form of non-linearity. The way of determining the existence of threshold effects between two variables is different from the traditional approach, in which the threshold level is determined exogenously. If the threshold level is chosen arbitrarily or is not determined within an empirical model, it is not possible to derive confidence intervals for the chosen threshold. Therefore, the number and location of thresholds are endogenously determined by the data. Moreover, asymptotic theory applies, which can be used to construct appropriate confidence intervals. A bootstrap method to assess the statistical significance of the threshold effect, in order to test the null hypothesis of a linear formulation against a threshold alternative, is also available.

We first analyse the effect of total greenhouse gas emissions (GHG KT) to real GDP growth. The number of threshold is determined using the bootstrapping method following Hansen (1999). The test results for threshold effect based on GHG emissions suggest there is double threshold effect and thus three regimes are discovered. We define the differing regimes as follows: regime I for low per capita income ($q \leq 7.7968$) which is equivalent to or less than per capita income 2432.929, regime II for moderate per capita income ($7.7968 < q \leq 8.1063$) which is between per capita income 2432.929 and 3315.441, and regime III for high per capita income ($8.1063 < q$) which is more than per capita income 3315.441. The panel threshold regression results suggest the effect of GHG emissions to real GDP growth varies with the level of real GDP per capita. For regime I ((ln GDP) ≤ 2432.9) and regime II (3315.4 ≥ (ln GDP) > 2432.9), the estimated coefficients for GHG emissions have positive signs suggesting that real GDP growth increases with GHG emissions in the low per capita income regime and in the middle per capita income regime. The third regime ((ln GDP) > 3315.4) is when the log of real GDP per capita turns to be greater than 8.1063 or exceeding real GDP per capita 3315.441. For this regime, the estimated coefficient for GHG emissions has a negative sign (−2.3902) suggesting that GHG emissions negatively and significantly affect growth when per capita income exceeds 3315.441.

Then, we analyse the effect of carbon dioxide emissions (CO2 KT) to real GDP growth. Using the bootstrapping method, we are able to detect the existence of one threshold and discover two regimes. As the value of single threshold is equivalent to the value of the second threshold in the case of GHG emissions, we can define the current regime I and regime II as follows: regime I for low and middle per capita income ($q \leq 8.1063$) which is equivalent to or less than per capita income 3315.441 and regime II for high per capita income ($8.1063 < q$) which is more than per capita income 3315.441. Based on the panel threshold regression results, we find that real GDP growth increases with CO2 emissions in the low and middle per capita income regime. In the high per capita income regime, real GDP growth is adversely affected by the emissions.

We also added several control variables to incorporate in our panel threshold regression models. These variables are found to be useful, including gross fixed capital formation (GFCF % GDP), general government final consumption expenditure (GGFCE % GDP), annual growth rate of population (POP_GROWTH), trade (TRADE % GDP), and broad money (BM, M2 % GDP).

The overall findings indicate that the effect of climate change on real GDP growth varies with the level of real per capita income, suggesting that the relationship between climate change and growth is non-linear with the presence of threshold effects. This is consistent with the past studies that have attempted to determine the optimal level in the relationship between climate change and growth, such as Alagidede et al. (2016) and Du et al. (2017). However, our results from panel threshold regression indicate that both GHG emissions and CO2 emissions affect real GDP growth negatively in the high per capita income regime and contribute to growth in the lower per capita income regime. This finding is inconsistent with the EKC hypothesis which explains that environmental quality tends to get worse with growth until reaching certain level of higher income and then environmental quality tends to improve with growth. It is reflected by an inverted U-shaped relationship between pollutants and per capita income.

Our findings point towards the urgent need for climate change adaptation policies to be implemented in high-income countries. As noted by Du et al. (2017), mitigation over long run does not make developed economies not negatively affected by climate change; therefore, more proactive climate policy is suggested. Even though countries in the groups of low income and middle income may have advantage in terms of growth from the economic activities that cause GHG emissions, these countries can promote clean technology and start to adopt climate change adaptation policies instead of relying on emissions control policies. China is an example. According to Ng and Ren (2017), the country thus far has focused on climate change mitigation; however, the worries over more unpredictable extreme weather have prompted the need for climate change adaptation measures. Moreover, CO2 emissions in China are expected to be on a rising trend as urbanization and industrialization are still ongoing. A set of revised policies for addressing climate change in

China was published in 2013. The new policies are towards adaptation emphasizing on disaster prevention and mitigation, monitoring and early warning, agriculture and water resources, coastal areas and ecosystems, and public health. Obviously, climate change adaptation has become a new direction in policy-makings as well as in research.

Lastly, we remark that our findings suggest climate change adaptation is necessary for high-income countries. For low- and middle-income countries, adaptation can be a better choice of policy for long term because these countries are transforming to a high-income country. Meantime, climate change mitigation should be continued, but each country should reach a balance between mitigation and growth.

**Acknowledgement** This research was supported by a project code SBK0336-2017 from the SGPUMS, Universiti Malaysia Sabah.

## Appendix: GHG Emissions and CO2 Emissions in Countries

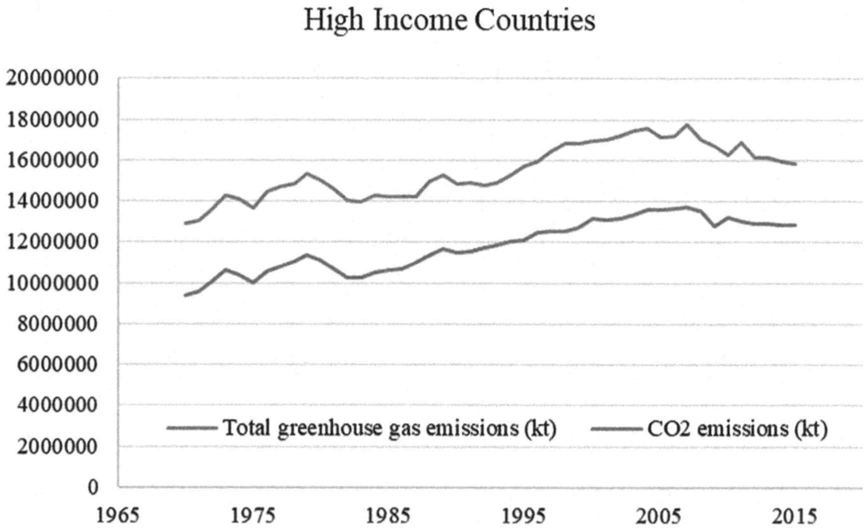

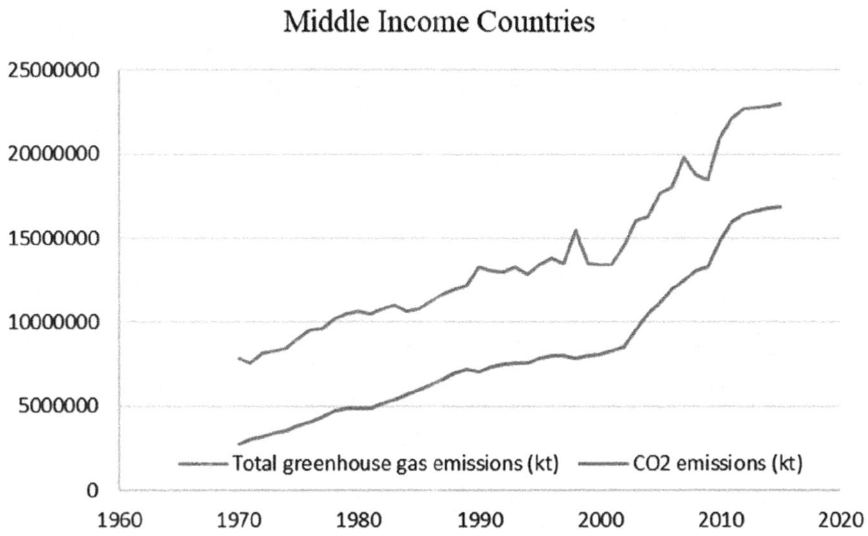
Middle Income Countries
25000000
20000000
15000000
10000000
5000000
0
1960
1970
1980
1990
2000
2010
2020
Total greenhouse gas emissions (kt)
CO2 emissions (kt)

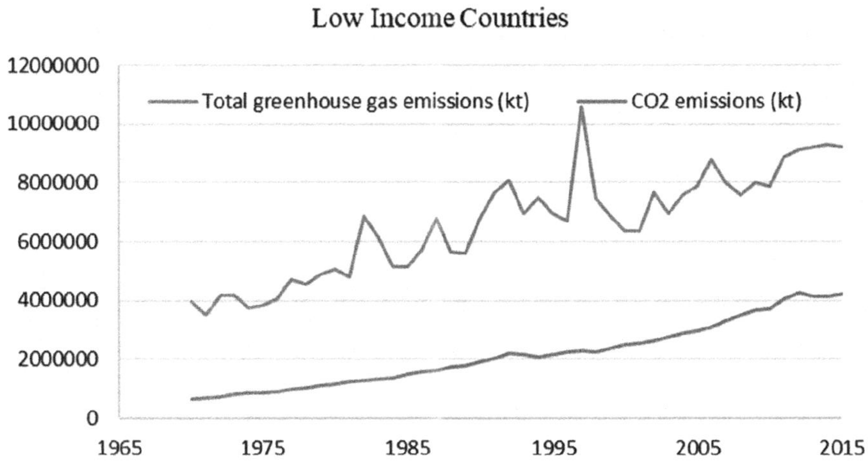
Low Income Countries
12000000
10000000
8000000
6000000
4000000
2000000
0
1965
1975
1985
1995
2005
2015
Total greenhouse gas emissions (kt)
CO2 emissions (kt)

## References


Alagidede, P., Adu, G., & Frimpong, P. B. (2016). The effect of climate change on economic growth: Evidence from Sub-Saharan Africa. *Environmental Economics and Policy Studies, 18*, 417–436.

Alam, M. M., Murad, M. W., Noman, A. H. M., & Ozturk, I. (2016). Relationships among carbon emissions, economic growth, energy consumption and population growth: Testing environmental Kuznets curve hypothesis for Brazil, China, India and Indonesia. *Ecological Indicators, 70*, 466–479.

Borhan, H., Ahmed, E. M., & Hitam, M. (2012). The impact of $CO_2$ on economic growth in ASEAN 8. *Procedia-Social and Behavioural Sciences*, 35, 389–397. In Asia Pacific International Conference on Environment-Behaviour Studies, Salamis Bay Conti Resort Hotel, Famagusta, North Cyprus, December 7–9, 2011.

Davies, J., Grant, M., Venezia, J., & Aamidor, J. (2007). Greenhouse gas emissions of the U.S. transportation sector: Trends, uncertainties, and methodological improvements. *Transportation Research Record: Journal of the Transportation Research Board, 2017*, 41–46.

Du, D., Zhao, X., & Huang, R. (2017). The impact of climate change on developed economies. *Economics Letters, 153*, 43–46.

Elshennawy, A., Robinson, S., & Willenbockel, D. (2016). Climate change and economic growth: An intertemporal general equilibrium analysis for Egypt. *Economic Modelling, 52*, 681–689.

Estrada, F., Tol, R. S. J., & Gay-García, C. (2015). The persistence of shocks in GDP and the estimation of the potential economic costs of climate change. *Environmental Modelling and Software, 69*, 155–165.

Fankhauser, S., & Tol, R. S. J. (2005). On climate change and economic growth. *Resource and Energy Economics, 27*, 1–17.

Hansen, B. E. (1999). Threshold effects in non-dynamic panels: Estimation, testing, and inference. *Journal of Econometrics, 93*, 345–368.

Hansen, B. E. (2000). Sample splitting and threshold estimation. *Econometrica, 68*(3), 575–603.

Johansen, S. (1991). Estimation and hypothesis testing of cointegration vectors in Gaussian vector autoregressive models. *Econometrica, 59*(6), 1551–1580.

Kuznets, S. (1955). Economic growth and income inequality. *American Economic Review, 49*, 1–28.

Lean, H. H., & Smyth, R. (2009). *CO2 emissions, electricity consumption and output in ASEAN* (Development Research Unit Discussion Paper DEVDP 09-13) (pp. 1–31). Monash University.

Linh, D. H., & Lin, S.-M. (2014). $CO_2$ emissions, energy consumption, economic growth and FDI in Vietnam. *Managing Global Transitions, 12*(3), 219–232.

Lu, W., Huang, J., & Musshoff, O. (2015). Climate threshold, financial hoarding and economic growth. *Applied Economics, 47*(42), 4535–4548.

Maddala, G. S., & Wu, S. (1999). A comparative study of unit roots with panel data and a new simple test. *Oxford Bulletin of Economics and Statistics, 61*, 631–651.

Ng, E., & Ren, C. (2017). *China's adaptation to climate and urban climate changes: A critical review* (Urban Climate, UCLIM-00325) (p. 21).

Olivier, J. G. J., Janssens-Maenhout, G., Muntean, M., & Peters, J. A. H. W. (2016). *Trends in global $CO_2$ emissions 2016 report*. PBL Netherlands Environmental Assessment Agency, European Commission Joint Research Centre.

Özokcu, S., & Özdemir, O. (2017). Economic growth, energy, and environmental Kuznets curve. *Renewable and Sustainable Energy Reviews, 72*, 639–647.

Pedroni, P. (2001). Purchasing power parity tests in cointegrated panels. *Review of Economics and Statistics, 83*, 727–731.

Prozzi, J. P., Naude, C., & Sperling, D. (2002). *Transportation in developing countries greenhouse gas Scenarios for South Africa*. Prepared for the Pew Center on Global Climate Change.

Sala-i-Martin, X. (1997, May). I just ran two million regressions. *American Economic Review, 87*, 178–183.

Searchinger, T., Heimlich, R., Houghton, A. A., Dong, F., Elobeid, A., Fabiosa, J., et al. (2008). Use of U.S. croplands for biofuels increases greenhouse gases through emissions from land-use change. *Science, 319*, 1238–1240.

Stern, D. I., Common, M. S., & Barbier, E. B. (1996). Economic growth and environmental degradation: The environmental Kuznets curve and sustainable development. *World Development, 24*(7), 1151–1160.

Taylor, L., Rezai, A., & Foley, D. K. (2016). An integrated approach to climate change, income distribution, employment, and economic growth. *Ecological Economics, 121*, 196–205.

Tian, Y., Zhu, Q., Lai, K.-H., & Lun, Y. H. V. (2014). Analysis of greenhouse gas emissions of freight transport sector in China. *Journal of Transport Geography, 40*, 43–52.

World Bank. (2017). *World development indicators 2017*. Washington, DC.

# Effects of Climate Change to Industrial Outputs and Employment in Asian Emerging Economies

**Sook Ching Kok and Qaiser Munir**

## Contents

1 Background .................... 84
2 Literature Review .................... 86
3 Data .................... 90
4 Methodology .................... 91
4.1 Autoregressive Distributed Lag (ARDL) Cointegration Method .................... 91
5 Results and Discussion .................... 93
5.1 Descriptive Statistics .................... 93
5.2 Augmented Dickey-Fuller Unit Root Test Results .................... 94
5.3 Bounds Tests for Cointegration .................... 95
5.4 ARDL Long-Run Results of the Selected Models .................... 97
5.5 ARDL Short-Run Results of the Selected Models .................... 102
6 Summary and Conclusions .................... 102
Appendix .................... 104
References .................... 107

**Abstract** It is well-established that climate change can be the result of human activities that create greenhouse gas emissions, which causes the greenhouse effect and further lead to the net effect of global warming. As far as the effects of climate change to human health and outputs of economic sectors are concerned, we can expect there will be negative impacts on output and employment. The objective of this study is to investigate the climate change effects to industrial output and employment in the context of ASEAN, by focusing on six emerging economies, namely, Cambodia, Indonesia, Malaysia, the Philippines, Thailand, and Vietnam for the period of 1989–2016. We use temperature and precipitation as the proxies for climate change. We apply the bounds testing procedure proposed by Pesaran et al.

S. C. Kok (✉)
Faculty of Business, Economics and Accountancy, Universiti Malaysia Sabah, Kota Kinabalu, Sabah, Malaysia
e-mail: emily@ums.edu.my

Q. Munir
Department of Economics, Institute of Business Administration, Karachi, Pakistan
e-mail: qmunir@iba.edu.pk

 
T. Sequeira, L. Reis (eds.), *Climate Change and Global Development*, Contributions to Economics, https://doi.org/10.1007/978-3-030-02662-2_5

(Journal of Applied Econometrics 16:289–326, 2001) to analyse the cointegration relationship and the autoregressive distributed lag (ARDL) modelling approach of Pesaran and Shin (An autoregressive distributed lag modeling approach to cointegration analysis. In: Strom S (ed) Econometrics and economic theory in the 20th century: the Ragnar Frisch centennial symposium. Cambridge University Press, Cambridge, 1999) and Pesaran et al. (Journal of Applied Econometrics 16:289–326, 2001) for the long-run and short-run relationships between industrial output and employment with the climate change variables. We found long-run relationship between climate change, industrial output and employment in all the countries analysed, except for the industrial output in Vietnam. Further, the long-run and short-run results show some similarities and variations. Our findings allow us to suggest different policy implications for long run and short run based on our results.

**Keywords** Greenhouse gases · Greenhouse effect · Global warming · Temperature · Precipitation · Industrial output · Employment · ASEAN emerging economies

## 1 Background

Greenhouse gases are referred to as the gases in atmosphere that let the sun's heat in but then block it from escaping, same as glass does in a greenhouse. Among the major greenhouse gases are carbon dioxide, methane, nitrous oxide, hydrofluorocarbons, perfluorocarbons, sulphur hexafluoride, and nitrogen trifluoride; see more details in the Greenhouse Gas Progress Report (2016) prepared by the Environmental Commissioner of Ontario. The greenhouse gas emissions induced by human activities are arguably the main source of climate change. Specifically, the emissions can stem from the burning of fossil fuels, energy use, transportation, deforestation, land use change, cement production, rice cultivation, fertilizer use, and industrial processes. These human activities in turn lead to the greenhouse effect and the net effect of global warming. The common signs of climate change are rising temperature and the shifting trends of rainfall and snow. Climate change poses threats to human health as well as livestock, agriculture, and fisheries products; thus, it potentially affects food security. According to Dey et al. (2016), fisheries resources are vital to the food security of Solomon Islands. As climate change may impact on the fisheries production, food security is likely to be fragile. When concerning human health and outputs of various economic sectors, there can be negative climate change impacts such as industrial output decreases and fall in employment. Policy-makers have thus been suggested for a great urgency to implement effective environmental policies to curb the harmful implications of climate change.

With regard to the effects of climate change to output and employment, ASEAN is an interesting case study. Based on the actual data from 2014 till 2015 and the latest projections for the period of 2016–2018, most countries in this region seem to be able to realize a stunning growth in annual real GDP, for instance, Cambodia

(6.8–7.1%), Indonesia (4.9–5.3%), Lao PDR (6.7–8.0%), Malaysia (4.2–6.0%), Myanmar (6.3–8.0%), the Philippines (5.9–6.9%), and Vietnam (6.0–6.7%). These appealing achievements have highlighted the increasing importance of ASEAN to the world economy. However, there is a serious concern in terms of the region's contribution to the global working-age population. This region faces the problems of slowing population growth and rapid increase in the old-age population. Based on available information about the demographic classification using 2015 demographic characteristics, Malaysia, Singapore, and Vietnam are the cases of late-dividend economies in which the working-age population is decreasing as a share of total population, but it is still growing in absolute numbers in the next 15 years; Thailand is a case of post-dividend economy, where its working-age population is declining in terms of its share in the total population and in absolute numbers in the subsequent 15 years; only Indonesia and the Philippines are the cases of early-dividend economies, in which the share of the working-age population will increase as a share of the total population as well as in absolute terms in the following 15 years. Furthermore, when using the 2030 demographic characteristics, all the aforementioned ASEAN economies become either a post-dividend or a late-dividend economy, except for the Philippines which has remained as an early-dividend economy. It means that this region is moving towards slowing population growth and rapidly increasing in the old-age population: see more details in the Regional Economic Outlook: Asia and Pacific prepared by International Monetary Fund (IMF) (2017). In relation to the above issue, we note that mortality has a direct influence on population growth. Extreme rises in temperature as a result of climate change can lead to higher heat-related mortality rate. According to Ebi et al. (2017), human health is threatened by excessive heat, both when one is exposed to severe heat wave and when heat is a long-term condition. Heat stroke and death are the most serious health consequences of climate change. The heat-related mortality has been empirically proven by Lee and Kim (2016) in South Korea, Weinberger et al. (2017) in ten US metropolitans, and Chen et al. (2017) in Jiangsu Province, China. Most of the ASEAN economies exhibit great potentials in growth, but these economies face the problem of shrinking working-age population. In this regard, it is indeed interesting to find out how climate change may impact industrial output and employment in ASEAN emerging economies.

From the macro perspective, literature shows that climate change tends to have significant impact on economic growth and there are a number of studies in this direction (e.g. Lu et al. 2015; Elshennawy et al. 2016; Du et al. 2017, etc.). Very few past studies which have emphasized on the climate change effect to employment as suggested by Fankhauser et al. (2008). Some past studies have focused on the effect of climate change to industrial output (e.g. Chalise et al. 2017; Hsiang 2010; Somanathan et al. 2014, etc.), but the context of ASEAN is less studied.

The objective of this study is to investigate the climate change effects to industrial output and employment across six ASEAN emerging economies, including Cambodia, Indonesia, Malaysia, the Philippines, Thailand, and Vietnam for the period of 1989–2016. Following literature, temperature, and precipitation are selected as the proxies for climate change. We apply the bounds testing procedure proposed by

Pesaran et al. (2001) to analyse the cointegration relationship and the autoregressive distributed lag (ARDL) approach of Pesaran and Shin (1999) and Pesaran et al. (2001) for the long-run and short-run relationships between industrial output and employment with the climate change variables. This present study is different from the past studies in a significant way: both industrial output and employment are likely to be affected by climate change, but the case of ASEAN emerging economies is less studied in the past. We expect to provide a better understanding of how climate change (i.e. temperature, precipitation) may affect industrial output and employment in these economies and to offer useful policy recommendations.

The remainder of this chapter is organized as follows: Sect. 2 is literature review; Sect. 3 is data description; in Sect. 4, we explain the methodology used; empirical results are discussed in Sect. 5; the last section provides summary and conclusion.

## 2 Literature Review

Our review of literature reveals several major aspects that are often considered for the effects of climate change, but we are more concentrating on the effects to industrial output and employment. In general, it can be noticed that the issue of climate change has been a main focus of economists since the past few decades. In terms of the climate influence on economic development, a number of studies indicate that climate change can significantly affect economic growth. See, for example, Lu et al. (2015), Elshennawy et al. (2016), and Du et al. (2017). Lu et al. (2015) analyse empirically the relationship between climate threshold, financial hoarding, and economic growth in China. The study found a positive cumulative effect between financial hoarding and economic growth that is restraint by the climate threshold. Elshennawy et al. (2016) take into account the impact of climate change in evaluating the long-run growth prospect of Egypt. It is found that adaptation investment reduces the country's GDP loss by approximately 2.6% in 2050. Through the study of Du et al. (2017), a nonlinear relationship between temperature and economic growth is established for each of the cases in European Union (EU) and the USA, implying that growth increases with higher temperature up to the optimal level and then any further rise in temperature is associated with a reduction in growth.

Apparently, the climate change effect to employment has drawn little attention in the past. A remarkable contribution to literature is made by Fankhauser et al. (2008) who suggest three distinguished employment effects due to climate change. A short-term effect is also known as direct employment effect. It occurs when jobs shift from high-carbon activities to low-carbon activities. The latter is usually more labour-intensive. In other words, jobs will move from the affected sectors to the replacing sectors. A medium-term effect is the higher-order, economy-wide effect. An economy is affected by climate policy, as there are job lost and creation along the value chains of the impacted sectors. Further, a long-term effect is also called dynamic effect. Innovation leads to creative destruction which involves structural adjustment. As a result, new investment opportunities, growth, and jobs are created.

There are past studies regarding the industrial output effect of climate change. Agriculture production is directly affected by rising temperature, the changing trends of precipitation, and increasing exposure to solar radiation. Chalise et al. (2017) perform a general equilibrium impacts assessment to capture the economy-wide effects of climate change in the context of Nepal agriculture productivity. The climate change induced productivity lost in the agriculture sector has significant negative impact on the economy of Nepal. The agriculture sector plays an important role in generating economic growth such that the sector contributes about 36% of GDP. The main finding based on simulation results suggests that by year 2080, the real GDP of Nepal will decrease by about 2.49% in the lowest-impact scenario, 6.56% in the medium-impact scenario, and 10.03% in the highest-impact scenario, respectively. Therefore, there is a need for adaptation in spite of mitigation (i.e. by reducing carbon emission). Luo et al. (2017) propose an eco-efficient and climate-smart agriculture framework featuring the adaptation-oriented approach. The framework encompasses two important dimensions of sustainable development, namely, ecological or environmental sustainability and economic sustainability which is related to productivity and profitability. Furthermore, it is geared towards a green economy. Wesseh and Lin (2017) conduct simulations up to year 2060 using a general equilibrium model for the impacts of climate change ($CO_2$, temperature, precipitation, adaptation) on prices [factor prices: labour wages, capital rent, land rent; consumer price index (CPI)], sectoral productions (services, manufactures, processed agricultural commodities, livestock, wheat, rice, other crops, other grains), exports, and also income and welfare (household income). The main findings derived from the simulations include the following: land rents in lower-middle- and upper-middle-income countries and in China decline with the climate change effect, while a reverse situation is observed in low-income and high-income countries and also in the USA which may due to the increase in demand for farmlands. Further, declines in capital rent and labour wage are observed in low-income and lower middle-income countries and in the USA. CPIs do not reflect much the influence of climate change. The reason could be the prices of services and manufactures are not much affected. In terms of sectoral production, high-income countries in general and China show increases in both agricultural and nonagricultural productions but not manufactures. For the USA, the finding indicates the increases in wheat, rice, and manufactures and the declines in other productions. For low-income, lower-middle-income, and upper-middle-income countries, productions decrease as a result of climate change. In high-income countries and in China, agricultural exports increase, while nonagricultural exports decrease. A complete reverse situation is observed in lower-middle-income and upper-middle-income countries. In high-income countries and in China, household incomes increase. In low-income, lower-middle-income, upper-middle-income countries, and in the USA, household incomes decrease.

An interest of research is to find out how forests are vulnerable to climate change. Forests become important as having an essential role in climate change mitigation.

Nitschke et al. (2017) address the decisions in selecting tree species for the Melbourne's urban forest in order to adapt to the changing climate. Employing 163 tree samples which represent five European tree species, London planes, English elm, green leaf elm, wych elm, and field elm, the study gives a finding that the European tree species are vulnerable to the mid-range predicted climate change. Therefore, it is suggested to find alternative tree species. Susaeta et al. (2017) conduct simulations to capture the impacts of climate change on forest growth, optimal harvest age, and land expectation value for the US slash pine forests. For each of the following three scenarios: baseline scenario assuming no changes in climate and $CO_2$ concentration, RCP4.5 scenario that represents low-to-medium greenhouse gas emissions, and RCP8.5 scenario that represents high greenhouse gas emissions, the historical data on temperature and precipitation from 1950 till 2005 are used to generate future predictions for a period of 60 years, 2050–2100. Then, the data are employed as inputs for simulating slash pine growth from planting up to age 60 in 11 sites in the USA including Covington AL, Polk FL, Alachua FL, Jefferson FL, Santa Rosa FL, and Brantley GA.

Next, we shift our attention to the fishing sector. Dey et al. (2016) assess the impacts of three climate change adaptation strategies: aquaculture, fish aggregating devices, and natural resource management, on price, production, and consumption of the Solomon Islands' fishing sector. The study utilizes country-specific partial equilibrium model that incorporates six subsectors including tuna, other oceanic finfish, coastal finfish, coastal invertebrates, freshwater finfish, and freshwater invertebrates. It is found that through the use of fish aggregating devices, the supply of oceanic fish can be increased by around 9–10% in year 2035 and 14–15% in year 2050. In addition, the strategy is expected to reduce the real price of tuna in year 2035 and year 2050. Meantime, aquaculture development may reduce the real price of coastal invertebrates, while the projected real price rises of coastal finfish and invertebrates require the remedy of natural resource management. As postulated by Rodríguez-Rodríguez and Ramudo (2017), production decrease as a result of climate change may or may not affect producer income depending on the available substitutes. If there are substitutes, a decrease in quantity is unlikely to incur price increase; therefore, production decrease may cause lower producer income. The study is an attempt to identify the existence of complements or substitutes for the fresh mussel productions from several European markets including Spain, France, Italy, the Netherlands, Sweden, the United Kingdom, and Belgium. Cointegration approach is utilized to verify the long-run relationships between two or more variables. In terms of data, weekly retail prices for mussel across the markets are collected for analysis. Retail price is considered as it combines wholesale price, tariffs, transportation costs, and other transaction costs. The analysis covers a period of 92 weeks starting from 2016:1 but there are missing data of Italy. As indicated by the cointegration test results, the French and Spanish markets are partially integrated. This may imply the limited substitutes for the fresh mussel productions from European markets. A fall in the production in one of these markets may induce price increase but may not affect producer income.

Climate variation can lead to output lost via a mechanism in which thermal stress in human adversely affects labour productivity. It is believed that high temperatures cause heat stress which decreases productivity and output (Hsiang 2010; Somanathan et al. 2014). Hsiang (2010) applies a multivariate panel regression using ordinary least squares (OLS) to estimate the dependence of production on temperature change and cyclone exposure in 28 Caribbean basin countries across different industries: wholesale, retail, restaurants, and hotels; other services; transport and communication; construction; manufacturing; agriculture, hunting, and fishing; mining; and utilities. Over the period of 1970–2006, the region's output decreased by approximately 2.5% in response to a temporary increase of 1 degree Celsius. From the 2.5%, only 0.1% was attributed to the reduction in agricultural production, and the large part 2.4% was the output losses in nonagricultural industries. This finding indicates the global economic cost of climate change can be larger than previously estimated, as the impact of labour response to heat stress in nonagricultural industries is usually ignored. Agriculture and tourism are the two industries most affected by cyclones, because of the geographical locations of the sectors' productions. As noted above, high temperatures can lead to heat stress which adversely affects productivity and output. The relationship between temperature and output is possibly nonlinear since labours may react differently in response to low, moderate, and high levels of temperature. According to Somanathan et al. (2014), the level of labour comfort should increase with low levels of temperature, receive little impact at moderate levels of temperature, and severely affected by heat stress at high levels of temperature. In the study of Somanathan et al. (2014), it is hypothesized that heat stress should largely depend on the exposure to high temperatures. The study attempts on approximating the nonlinear response of manufacturing workers in India to temperature by using a stepwise linear function of production. The empirical analysis is by using annual panel dataset of the large manufacturing plants employing over 100 workers and daily microdata of case study firms. It is concluded that high temperatures can decrease the manufacturing productivity and output in India. In more detail, the output from labour-intensive manufacturing plants in India decrease by around 3% per degree Celsius. Wang (2017) assesses the impacts of five climate change mitigation technologies on the unified performance of US firms, namely, the eco-efficiency, low-carbon energy, green design, pollution control, and management system technologies. The unified performance integrates operational and environmental performances. The analysis is based on annual periods from 2011 to 2013 and includes a sample that consists of 170 S&P 500 manufacturing firms. Three measures of unified performance are developed using data envelopment analysis (DEA), including unified efficiency (UE), unified efficiency under natural disposability (UEN), and unified efficiency under managerial disposability (UEM). The impacts from technology adoption as well as technology share on UE, UEN, and UEM are analysed using regression. The regression models use UE, UEN, and UEM as dependent variables, and the independent variable is either technology adoption or technology share which represent the technology-related variables. Each regression model uses a similar set of control variables: age of assets, leverage, advertising intensity, and year dummies. The study shows a positive impact from the low-carbon energy technology and a negative impact from the pollution control technology on the unified performance of US firms.

Current studies indicate the tourism industry is significantly affected by climate change. In fact, the interaction between climate change and tourism is not new. As mentioned by Weir (2017), tourism has had received the impact from climate change throughout human history. Meantime, Hewer and Gough (2017) provide a review of 30 research works showing the assessments on climate change impact on tourism and outdoor recreation in the case of Canada. It is concluded that warm-weather activities in Canada benefit from climate change, while cold-weather activities tend to be negatively affected by climate change. To give an example of quantitative study using survey, Rutty and Scott (2010) assess the perceptions of "too hot" for comfortable tourism activities at beach and urban destinations within the Mediterranean region. Based on a survey conducted with 850 university students from Austria, Germany, the Netherlands, Sweden, and Switzerland, the threshold of "unacceptable hot" is identified. The threshold is compared to different thermal conditions with regard to humidity and temperature including a baseline climate change scenario between 1961 and 1990, an early century climate change scenario between 2011 and 2035, a mid-century climate change scenario between 2046 and 2065, and a late century climate change scenario between 2080 and 2099, providing ten destinations within the Mediterranean region. By the mid-century during the peak summer months, two additional beaches and one additional urban destinations are found to be associated with "unacceptable hot". By the late century, the threshold of "unacceptable hot" is exceeded in four additional beaches and five additional urban destinations during summer months. To provide another example, Rosselló et al. (2011) empirically analyse the sensitivity of outbound flows in the United Kingdom to the short-term weather anomalies based on local weather conditions. It is found that the variables of mean temperature, air frost, heat wave, and sunshine days can significantly affect the British outbound flows.

Apart from studying the direct impact of climate change, current research in this area has been extended to explore the impact from climate change adaptation strategy as well as the impact from climate change mitigation technology. However, the direct impacts of climate change to the industrial output and employment in the context of ASEAN are not sufficiently explored. Therefore, these issues need to be addressed first. In the future, the next step could be analysing the impacts of climate change adaptation and/or mitigation in the region.

## 3 Data

This study relies on yearly dataset of six ASEAN countries collected from two data sources from 1989 to 2016. We use two measures of climate change in this study: temperature and precipitation (millimetres). Precipitation is defined as any kind of water that falls from clouds as a liquid or solid. Temperature is average temperature in degrees Celsius. The climate change data on temperature and precipitation are taken from the World Bank's Climate Change Knowledge Portal (CCKP). The World Bank created the Climate Change Knowledge Portal, supported by the Global Facility for Disaster Reduction and Recovery and others. The data for output and employment are collected from Asian Development Bank (ADB). The output data are in year 2000 constant prices for all the countries. The six ASEAN developing countries are Cambodia, Indonesia, Malaysia, the Philippines, Thailand, and Vietnam. Appendix has provided the graphs for the climate change variables used.

# 4 Methodology

## *4.1 Autoregressive Distributed Lag (ARDL) Cointegration Method*

To empirically investigate the long-run relationships between climate change, industrial output and employment, the model has been estimated by using the bounds testing procedure proposed by Pesaran et al. (2001). The ARDL modelling approach was initially suggested by Pesaran and Shin (1999) and later extended by Pesaran et al. (2001). There are a number of methods for conducting cointegration analysis such as the residual-based EG approach developed by Engel and Granger (1987) and Phillips and Hansen's (1990) FMOLS, and the maximum likelihood-based approach (JML), proposed by Johansen and Juselius (1990) among others. The main advantage of the ARDL approach is that it does not generally require a knowledge of the order of integration of variables, which is necessary in the EG and FMOLS and also in the JML technique. These methods require that the variables have the same order of integration. However, a problem arises when the variables have different orders of integration; to overcome this problem, Pesaran et al. (2001) propose to use the ARDL technique that does not require the classification of variables into $I(0)$ or $I(1)$. Therefore, we are using ARDL approach in this study.

The ARDL approach to cointegration of Pesaran et al. (2001) involves estimating the conditional error correction (EC) version of the ARDL model for climate change, output, and mployment:

$$\begin{aligned}\Delta \ln(Ind\_Output)_t = \alpha_0 &+ \sum_{i=1}^{p} \varphi_i \Delta \ln(Ind\_Output)_{t-i} \\ &+ \sum_{i=0}^{p} \theta_i \Delta \ln(Temp)_{t-i} + \sum_{i=0}^{p} \lambda_i \Delta \ln(Precip)_{t-i} \\ &+ \delta_1 \ln(Ind\_Output)_{t-1} + \delta_2 \ln(Temp)_{t-1} \\ &+ \delta_3 \ln(Precip)_{t-1} + \upsilon_t\end{aligned} \tag{1a}$$

$$\begin{aligned}\Delta \ln(Employ)_t = \alpha_0 + \sum_{i=1}^{p}\varphi_i \Delta \ln(Employ)_{t-i} \\ + \sum_{i=0}^{p}\theta_i \Delta \ln(Temp)_{t-i} + \sum_{i=0}^{p}\lambda_i \Delta \ln(Precip)_{t-i} \\ + \delta_1 \ln(Employ)_{t-1} + \delta_2 \ln(Temp)_{t-1} \\ + \delta_3 \ln(Precip)_{t-1} + \upsilon_t\end{aligned} \tag{1b}$$

where temperature (*Temp*) and precipitation (*Precip*) are used as climate change proxies. The variables, ln(*Employ*) is the total employment, and ln(*Ind_Output*) is total industrial output, $\Delta$ is first-difference operator and $p$ is the optimal lag length.

The $F$-test is used in testing the existence of long-run relationship. When long-run relationship exists, $F$-test indicates which variable should be normalized. The null hypothesis for no cointegration among variables in Eqs. (1a, 1b) is $H_0$: $\delta_1 = \delta_2 = 0$ against the alternative hypothesis $H_1$: $\delta_1 \neq \delta_2 \neq 0$. The $F$-test has a nonstandard distribution which depends on (1) whether variables included in the model are $I(0)$ or $I(1)$, (2) the number of regressors, and (3) whether the model contains an intercept and/or a trend. Pesaran et al. (2001) test the critical values for the $I(1)$ series which are referred to as *upper* bound critical values, while the critical values for $I(0)$ series are referred to as the *lower* bound critical values. If the $F$-test statistic exceeds their respective upper critical values, we can conclude that there is evidence of long-run relationship between the variables. If the test statistic is below the upper critical value, we cannot reject the null hypothesis of no cointegration, and if it lies between the bounds, a conclusive inference cannot be made without knowing the order of integration of the underlying regressors.

In the next step, if there is evidence of long-run relationship (cointegration) of the variables, the following long-run model is estimated:

$$\begin{aligned}\ln(Ind\_Output)_t = \alpha_1 + \sum_{i=1}^{p}\varphi_{1i} \ln(Ind\_Output)_{t-i} \\ + \sum_{i=0}^{p}\beta_{1i} \ln(Temp)_{t-i} + \sum_{i=0}^{p}\theta_{1i} \ln(Precip)_{t-i} + \mu_t\end{aligned} \tag{2a}$$

$$\begin{aligned}\ln(Employ)_t = \alpha_1 + \sum_{i=1}^{p}\varphi_{1i} \ln(Employ)_{t-i} \\ + \sum_{i=0}^{p}\beta_{1i} \ln(Temp)_{t-i} + \sum_{i=0}^{p}\theta_{1i} \ln(Precip)_{t-i} + \mu_t\end{aligned} \tag{2b}$$

In this study, the orders of the lags in the ARDL model are selected by the Akaike information criterion (AIC), before the selected model is estimated by ordinary least squares (OLS).

The ARDL specification of the short-run dynamics can be derived by constructing an error correction model (ECM) of the following form:

$$\Delta \ln (Ind\_Output)_t = \alpha_2 + \sum_{i=1}^{p} \varphi_{2i} \Delta \ln (Ind\_Output)_{t-i} + \sum_{i=0}^{p} \theta_{2i} \Delta \ln (Temp)_{t-i} + \sum_{i=0}^{p} \lambda_{2i} \Delta \ln (Precip)_{t-i} + \psi \mathrm{ECM}_{t-1} + \vartheta_t \quad (3a)$$

$$\Delta \ln (Employ)_t = \alpha_2 + \sum_{i=1}^{p} \varphi_{2i} \Delta \ln (Employ)_{t-i} + \sum_{i=0}^{p} \theta_{2i} \Delta \ln (Temp)_{t-i} + \sum_{i=0}^{p} \lambda_{2i} \Delta \ln (Precip)_{t-i} + \psi \mathrm{ECM}_{t-1} + \vartheta_t \quad (3b)$$

where $\mathrm{ECM}_{t-1}$ is the error correction term, defined as,

$$\mathrm{ECM}_t = \ln (Ind\_Output)_t - \alpha_1 - \sum_{i=1}^{p} \varphi_{1i} \ln (Ind\_Output)_{t-i} - \sum_{i=0}^{p} \beta_{1i} \ln (Temp)_{t-i} - \sum_{i=0}^{p} \theta_{1i} \ln (Precip)_{t-i}$$

$$\mathrm{ECM}_t = \ln (Employ)_t - \alpha_1 - \sum_{i=1}^{p} \varphi_{1i} \ln (Employ)_{t-i} - \sum_{i=0}^{p} \beta_{1i} \ln (Temp)_{t-i} - \sum_{i=0}^{p} \theta_{1i} \ln (Precip)_{t-i}$$

All coefficients of short-run equation are coefficients relating to the short-run dynamics of the model's convergence to equilibrium and $\psi$ represents the speed of adjustment.

# 5 Results and Discussion

## 5.1 Descriptive Statistics

Table 1 summarizes the descriptive statistics for the series *Temp* and *Precip*. The minimum, maximum, and mean values of *Temp* for all the six ASEAN countries, Cambodia, Indonesia, Malaysia, the Philippines, Thailand, and Vietnam are close to each other. The minimum values are in the range of 24.243–27.212, while the maximum and mean values are within the range of 25.784–28.279 and 24.914–27.681, respectively. Small variations in the series *Temp* can be seen, implying that climate change has caused somewhat similar increases in temperature in the countries analysed. There are relatively larger variations in the series *Precip* as

**Table 1** Descriptive statistics of temperature and precipitation

| Series | Cambodia | Indonesia | Malaysia | Philippines | Thailand | Vietnam |
|---|---|---|---|---|---|---|
| *Temperature (Temp)* | | | | | | |
| Min. | 27.212 | 25.857 | 25.372 | 25.686 | 26.324 | 24.243 |
| Max. | 28.279 | 26.676 | 26.331 | 26.456 | 27.663 | 25.784 |
| Mean | 27.681 | 26.166 | 25.792 | 25.962 | 26.943 | 24.914 |
| Std. dev. | 0.3063 | 0.1729 | 0.2106 | 0.1600 | 0.3608 | 0.3734 |
| *Precipitation (Precip)* | | | | | | |
| Min. | 127.576 | 175.058 | 192.749 | 164.080 | 117.420 | 138.292 |
| Max. | 241.982 | 302.357 | 306.870 | 270.679 | 166.046 | 193.430 |
| Mean | 167.713 | 231.615 | 244.444 | 211.452 | 136.295 | 161.334 |
| Std. dev. | 23.8038 | 26.7527 | 29.9560 | 30.9238 | 13.7020 | 16.4253 |

Notes: Temperature is average degrees Celsius and precipitation is in millimeters

compared to the series *Temp*. For *Precip*, mean values are in the range of 136.295–244.444, minimum values between 117.420 and 192.749, and maximum values between 166.046 and 306.870. Nevertheless, the precipitation variability across these countries is not huge. The six ASEAN countries are in the regime of Southeast Asian summer monsoon. The countries in this regime are including Borneo, Cambodia, East India, Indonesia, Laos, Malaysia, South China, Myanmar, Singapore, Thailand, the Philippines, Portuguese Timor, Vietnam, Singapore, South China, Portuguese Timor, and western New Guinea. Monsoon is referred to as the large-scale seasonal reversals of the wind regime. The Southeast Asian summer monsoon is important in regulating rainfall in many countries within the tropical latitudes (Loo et al. 2015). The standard deviations of the series *Temp* for all the six ASEAN countries are in the range of 0.1600–0.3734 which is within the band of 1 standard deviation, Cambodia 0.3063, Indonesia 0.1729, Malaysia 0.2106, the Philippines 0.1600, Thailand 0.3608, and Vietnam 0.3734. This means the data points are close to the mean and not disperse from the mean. For the series *Precip*, the standard deviations are much higher than 1 standard deviation, Cambodia 23.8038, Indonesia 26.7527, Malaysia 29.9560, the Philippines 30.9238, Thailand 13.7020, and Vietnam 16.4253. This set of data values is widely spread out over the mean. The countries analysed exhibit consistency on low standard deviation of *Temp* and high standard deviation of *Precip*.

## 5.2 Augmented Dickey-Fuller Unit Root Test Results

It is necessary to test the stationarity of the series before proceeding with the ARDL bounds test. The stationarity status of all variables is important to determine their order of integration. In other words, this is to make sure the variables are not *I* (2) stationary so as to avoid spurious results. We apply a univariate Augmented Dickey-Fuller (ADF 1979) test for autoregressive unit root. Table 2 displays the test

**Table 2** ADF unit root test results

| Series | Cambodia *t*-stat. | Indonesia *t*-stat. | Malaysia *t*-stat. | Philippines *t*-stat. | Thailand *t*-stat. | Vietnam *t*-stat. |
|---|---|---|---|---|---|---|
| *Temp* | | | | | | |
| *I*(0) | −4.738* | −2.307 | −3.587** | −5.016* | −4.762* | −3.934* |
| *I*(1) | −7.539* | −5.841* | −6.217* | −7.213* | −6.946* | −8.063* |
| *Precip* | | | | | | |
| *I*(0) | −3.518** | −2.207 | −1.674 | −3.300** | −3.815* | −3.359** |
| *I*(1) | −4.145* | −3.275** | −5.225* | −5.799* | −7.575* | −4.646* |
| *Employ* | | | | | | |
| *I*(0) | −2.359 | −0.045 | −0.655 | −0.596 | −1.515 | −1.415 |
| *I*(1) | −2.768*** | −9.973* | −3.550** | −3.566** | −9.130* | 2.900*** |
| *Ind_Output* | | | | | | |
| *I*(0) | −3.920* | −0.408 | −0.764 | −10.130* | −3.047** | −0.546 |
| *I*(1) | 5.167* | −4.055* | −5.228* | −5.114* | −4.368* | −5.052* |

Notes: The optimal lags for the ADF tests are selected based on optimising AIC. *, **, *** significance at 1%, 5%, and 10% levels, respectively

results of the ADF test for *Temp*, *Precip*, *Employ*, and *Ind_Output*. All the variables are *I*(1) or become stationary after taking first-difference. We reject the null hypothesis of unit root process in all series based on the Akaike information criteria (AIC). It is, therefore, worth concluding that all the variables used in this study are not *I*(2). Having confirmed the variables are *I*(1), this suggests that OLS is not appropriate and the vector error correction model (VECM) approach is advisable. As the variables are changing in time and not constant, using OLS method to estimate linear regression model for these variables tends to produce spurious results. As a result, the model may have *R* square higher than the Durbin Watson statistic.

## 5.3 Bounds Tests for Cointegration

Having concluded the stationarity status of all the variables are *I*(1), the next step is to examine the long-run relationships between the variables. The optimal number of lags to be included in the conditional error correction model (ECM) is determined using the AIC, so that to ensure there is no serial correlation as emphasized by Pesaran et al. (2001). We find that the lag length that minimizes AIC is 2. Table 3 presents the computed *F*-statistics for the cointegration tests performed for each country.

The error correction term ($ECT_{t-1}$) indicates whether there is existence of long-run relationship between variables. In each model, almost all the coefficients of the lagged error terms have an expected negative sign and are significant at the 1% level, except the model *Ind_Output*|*Temp* and *Percip* for Vietnam in which coefficient with a positive sign is derived but it is not significant. Our results based on VECM provide strong evidence of cointegration relationships between the variables. We do

**Table 3** The ARDL results of cointegration relationships between climate change, output, and employment

| Countries | Model specification | *F*-stats | Lags | $ECT_{t-1}$ (*t*-stat.) | Decision |
|---|---|---|---|---|---|
| Cambodia | $(Ind\_Output \mid Temp, Precip)$ | 234.92* | 2 | −0.0157 (−34.27)* | Cointegration |
| | $(Employ \mid Temp, Precip)$ | 10.058* | 2 | −0.3884 (−8.023)* | Cointegration |
| Indonesia | $(Ind\_Output \mid Temp, Precip)$ | 4.684** | 2 | −0.5826 (−4.65)* | Cointegration |
| | $(Employ \mid Temp, Precip)$ | 5.571** | 2 | −0.0137 (−5.02)* | Cointegration |
| Malaysia | $(Ind\_Output \mid Temp, Precip)$ | 5.93** | 2 | −0.0498 (−5.22)* | Cointegration |
| | $(Employ \mid Temp, Precip)$ | 4.61** | 2 | −0.0440 (−3.85)* | Cointegration |
| Philippines | $(Ind\_Output \mid Temp, Precip)$ | 4.45** | 2 | −0.0724 (−3.31)* | Cointegration |
| | $(Employ \mid Temp, Precip)$ | 9.42* | 2 | −0.0194 (−5.63)* | Cointegration |
| Thailand | $(Ind\_Output \mid Temp, Precip)$ | 4.41** | 2 | −0.0568 (−4.32)* | Cointegration |
| | $(Employ \mid Temp, Precip)$ | 4.48** | 2 | −0.0190 (−4.54)* | Cointegration |
| Vietnam | $(Ind\_Output \mid Temp, Precip)$ | 1.712 | 2 | 0.0114 (1.574) | No |
| | $(Employ \mid Temp, Precip)$ | 60.94* | 2 | −0.0049 (−16.60)* | Cointegration |

Notes: *Temp* and *Precip* are temperature and precipitation variables. The critical values used are as reported by Narayan (2004). The critical values for $I(0)$ and $I(1)$ are (2.915, 3.695), (3.538, 4.428), and (5.155, 6.265) at 10%, 5%, and 1% respectively. *, **, *** significance at 1%, 5%, and 10% levels, respectively

**Table 4** ARDL long-run results of the relationship between climate change, output, and employment

| Countries | Dep. variable: *Output* | *Coeff. (std. error)* | Dep. variable: *Employment* | *Coeff. (std. error)* |
|---|---|---|---|---|
| Cambodia | *Temp* | −1.979 (0.550)* | *Temp* | −20.588 (7.321)** |
| | *Precip* | −0.058 (2.744) | *Precip* | 4.155 (0.746)* |
| Indonesia | *Temp* | −12.030 (16.429) | *Temp* | −0.079 (0.248) |
| | *Precip* | 3.501 (1.399)** | *Precip* | 0.0275 (0.014)*** |
| Malaysia | *Temp* | 117.41 (43.026)** | *Temp* | 8.033 (1497)* |
| | *Precip* | 4.012 (1.073)* | *Precip* | 6.467 (3.871) |
| Philippines | *Temp* | 26.342 (4.322)* | *Temp* | −0.473 (0.254)*** |
| | *Precip* | 0.5107 (0.238)** | *Precip* | −0.936 (0.620) |
| Thailand | *Temp* | −0.583 (0.285)*** | *Temp* | −0.369 (0.147)** |
| | *Precip* | −0.568 (0.513) | *Precip* | 0.109 (3.325) |
| Vietnam | *Temp* | −3.069 (4.152) | *Temp* | −0.286 (0.116)** |
| | *Precip* | −0.742 (0.585) | *Precip* | −0.004 (0.016) |

Note: Standard error in parentheses. *, **, *** significance at 1%, 5%, and 10% levels, respectively

not obtain sufficient evidence to establish a cointegration relationship for *Employ|Temp* and *Precip* in the case of Vietnam. The cointegrating equations serve as the negative feedback mechanisms. Each implies that departure from equilibrium in the previous period is reduced by certain percentage in the subsequent period. For instance, the coefficient −0.5826 indicates the highest rate of convergence to equilibrium with the model *Ind_Output|Temp* and *Precip* for Indonesia, and the coefficient −0.0049 indicates the slowest speed of convergence to equilibrium with the model *Employ|Temp* and *Precip* for Vietnam.

## 5.4 ARDL Long-Run Results of the Selected Models

This section further shows the ARDL long-run results for the estimated models. The long-run estimates of ARDL approach for the six ASEAN countries are presented in Table 4. Temperature negatively affects total industrial output in Cambodia (coefficient of −1.979) and in Thailand (coefficient of −0.583). The result of Cambodia is highly significant at the 1% level, and the result of Thailand is only weakly supported at the 10% level. Agriculture, forestry, and fishing play a vital role in the Cambodia economy besides the industry and service sectors. Over the last decade 2006–2016, the service sector has had a leading role in national output followed by the agriculture (agriculture, forestry, and fishing) and industry sectors. In recent years, the industry sector has become increasingly important. A large portion of Thailand's GDP is clearly driven by the service sector, followed by the industry and agriculture sectors in the last decade. For instance, in 2016, services contributed 55.8%, industry contributed 35.8%, and agriculture only contributed 8.3% of the country's GDP at current market prices. Considering the severe impact that temperature can have on agriculture, forestry, and fishing, therefore the strong evidence of temperature change impact on industrial output in Cambodia is expected. In Thailand, the impact is relatively less profound. In Malaysia and the Philippines, the total industrial output of each country seems to benefit greatly from rising temperature over long run, as indicated by the coefficients 117.41 and 26.342, respectively. The results are significant at least at the 5% level. The structures of national output in Malaysia and the Philippines are much similar to Thailand, where the services sector is leading and followed by the industry and agriculture sectors. The expansions in all the three sectors and the less dependence of economy on the agriculture sector may explain the positive impact of temperature on industrial output, in the cases of Malaysia and the Philippines. There is no evidence of any significant impact from temperature to industrial output in Indonesia and Vietnam. In addition, the industrial outputs in Indonesia, Malaysia, and the Philippines receive positive impacts from the increase in precipitation. These countries seem to have advantage from average weather and sunshine which attract visitors and smooth economic activities throughout the whole year, with the exception of irregular extreme rainfall and the resulted floods.

Turning to the long-run employment effect of climate change, the majority of the six ASEAN countries, including Cambodia, the Philippines, Thailand, and Vietnam, are negatively affected by rising temperature, most possibly because the agriculture sector is labour-intensive. The results for these countries are significant at the 5% or 10% levels. In the case of Malaysia, rising temperature seems to contribute towards higher level of industrial output. The coefficient is 8.033 and significant at the 1% level. In the last decade, the increase in employment in the agriculture sector of Malaysia had been slower. Labour supply had increased across other sectors. Meantime, Indonesia is not associated with the long-run employment effect from temperature change. Cambodia and Indonesia experienced increases in employment due to more precipitation, with the stronger impact observed in Cambodia (coefficient of 4.155) and relatively less impact in Indonesia (coefficient of 0.0275), significant at the 1% and 10% levels, respectively. Increase in annual rainfall is expected to enlarge the production of session paddy crop in Cambodia and Indonesia. The palm oil planting in Indonesia will maintain with reasonable rainfall. Thus, the employment in the agriculture sector is enhanced (the details on national output and employment of ASEAN countries as reported above are based on the Key Indicators for Asia and the Pacific 2017).

We apply a number of diagnostic tests to the error correction model. We find no evidence of serial correlation, heteroskedasticity, and ARCH (autoregressive conditional heteroskedasticity) effect in the disturbances. The model also passes the Jarque-Bera normality test which suggests that the errors are normally distributed.[1] Finally, the cumulative sums of squares of recursive residuals (CUSUMSQ) graphs are presented to show the stability of the ARDL estimations (see Brown et al. 1975).

[1]Results of these tests will be provided on request.

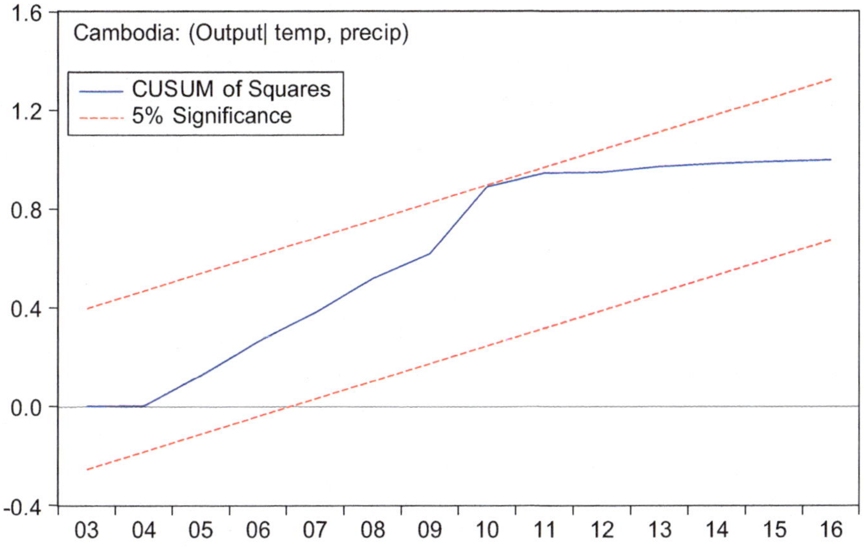


**Fig. 1** The plot of cumulative sum of squares of the recursive residuals of Cambodia

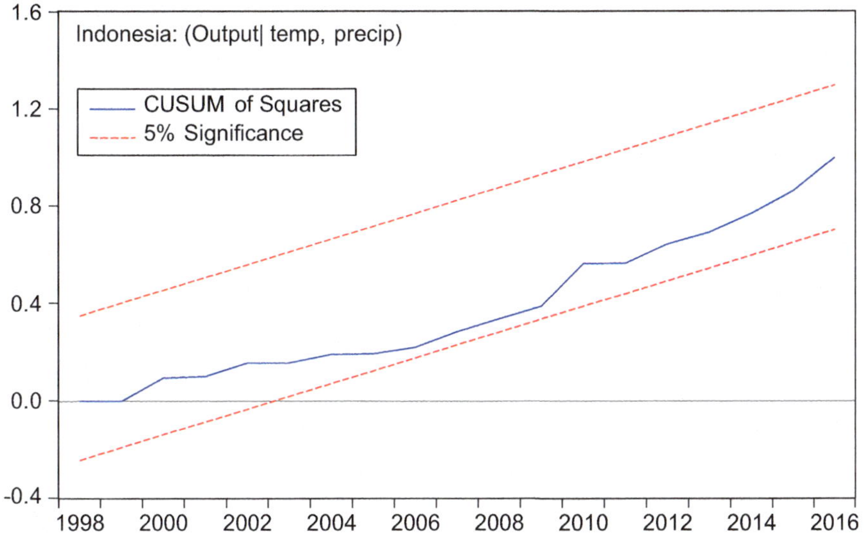


**Fig. 2** The plot of cumulative sum of squares of the recursive residuals of Indonesia

Figures 1, 2, 3, 4, 5, and 6 show the cumulative sum of squares of recursive residuals (CUSUMSQ) graphs which are resulted from the ARDL stability tests.

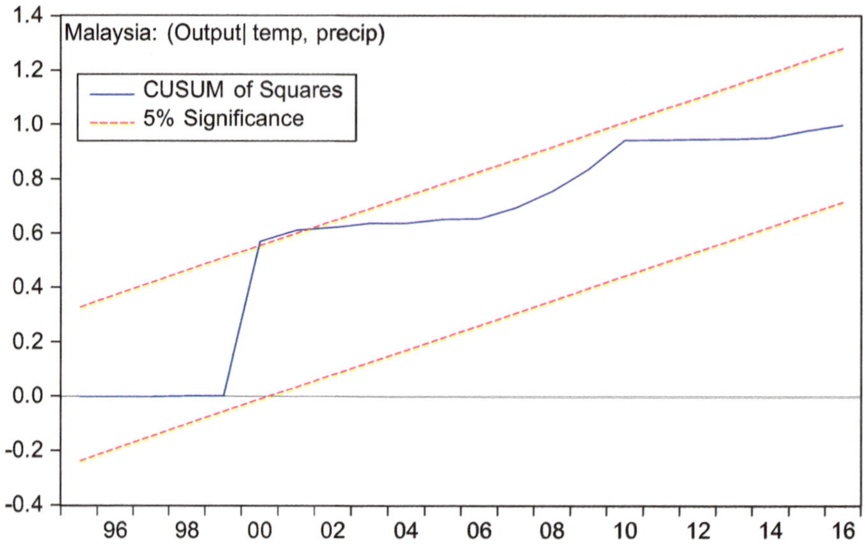


**Fig. 3** The plot of cumulative sum of squares of the recursive residuals of Malaysia

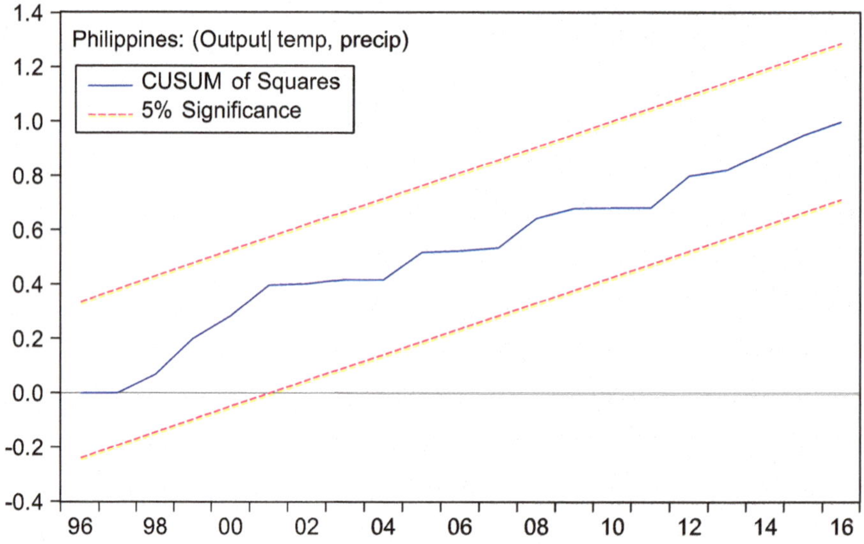


**Fig. 4** The plot of cumulative sum of squares of the recursive residuals of the Philippines

The null hypothesis of instability is rejected when the plots of the CUSUMSQ stay within the 5% significance level. However, the model is unstable when the plots

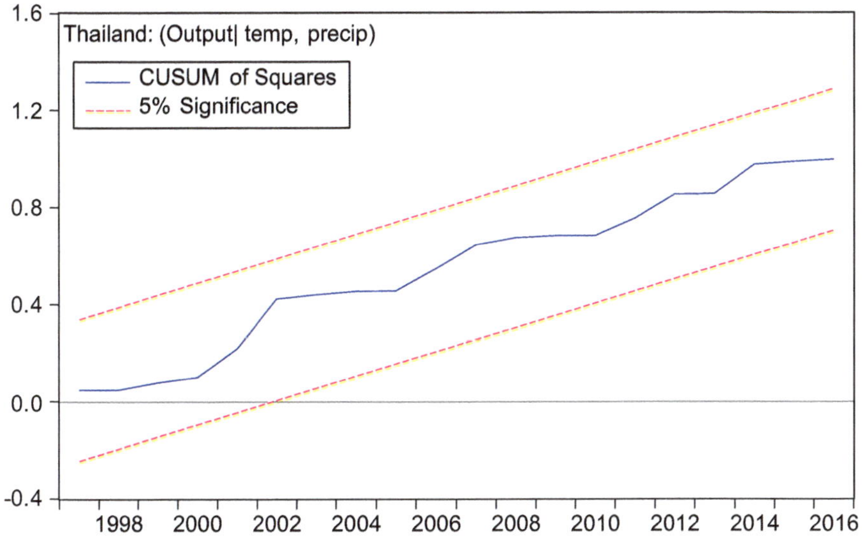


**Fig. 5** The plot of cumulative sum of squares of the recursive residuals of Thailand

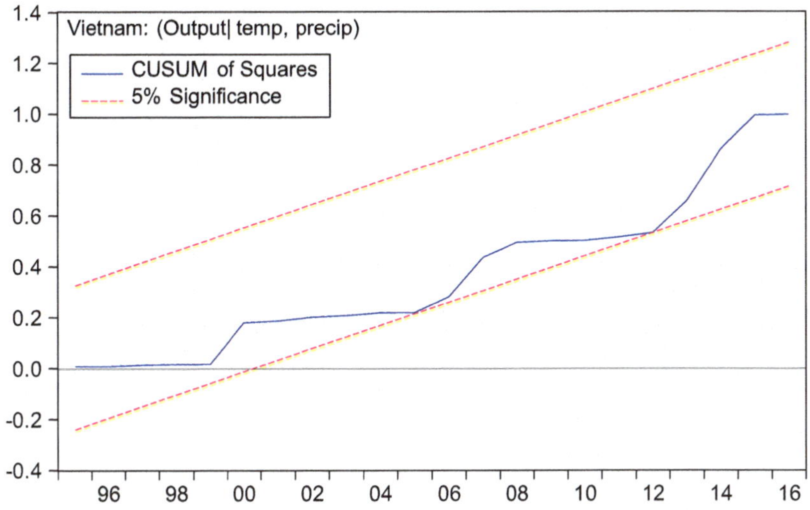


**Fig. 6** The plot of cumulative sum of squares of the recursive residuals of Vietnam

of the CUSUMSQ move outside the 5% critical lines. Our findings provide strong evidence for the stability of ARDL estimation results presented above.[2]

[2]In order to save the space, we only provide the plots of cumulative sum of squares of the recursive residuals of few models. However, the plots of cumulative sum of recursive residuals of other models will be provided on request.

**Table 5** ARDL short-run results of the relationship between climate change, output, and employment

| Countries | Dep. variable: *Output* | *Coeff. (std. error)* | Dep. variable: *Employment* | *Coeff. (std. error)* |
|---|---|---|---|---|
| Cambodia | *Δ(Temp)* | 0.137 (0.210) | *Δ(Temp)* | 4.761 (0.759)* |
| | *Δ(Precip)* | 0.037 (0.021)*** | *Δ(Precip)* | 0.066 (0.048) |
| Indonesia | *Δ(Temp)* | −1.251 (0.464)** | *Δ(Temp)* | 0.691 (0.473) |
| | *Δ(Precip)* | 0.578 (0.361) | *Δ(Precip)* | 0.041 (0.159) |
| Malaysia | *Δ(Temp)* | −1.639 (2.450) | *Δ(Temp)* | 0.403 (0.158)** |
| | *Δ(Precip)* | 0.135 (2.491) | *Δ(Precip)* | 0.053 (0.038) |
| Philippines | *Δ(Temp)* | 372.96 (489.154) | *Δ(Temp)* | −0.936 (0.524)*** |
| | *Δ(Precip)* | −6.323 (17.352) | *Δ(Precip)* | −0.624 (0.690) |
| Thailand | *Δ(Temp)* | 1.242 (1.454) | *Δ(Temp)* | −0.369 (0.126)* |
| | *Δ(Precip)* | 0.583 (0.234)** | *Δ(Precip)* | −0.052 (0.028)*** |
| Vietnam | *Δ(Temp)* | 13.914 (14.615) | *Δ(Temp)* | −0.291 (0.106)** |
| | *Δ(Precip)* | −0.011 (0.047) | *Δ(Precip)* | −0.004 (0.115) |

Note: Standard error in parentheses. *, **, *** significance at 1%, 5%, and 10% levels, respectively

### 5.5 *ARDL Short-Run Results of the Selected Models*

Table 5 reports the short-run results for different models. Some of the short-run results are consistent with the long-run results reported in Sect. 5.4 which include as follows: employment receives positive significant effect from temperature in Malaysia; employment is negatively and significantly affected by temperature in the Philippines, Thailand, and Vietnam. The results for short-run provide some new evidences. The employment in Cambodia is positively impacted by an increase in temperature in short run, but the impact becomes negative over long run. Furthermore, the outputs in Cambodia and Thailand only receive short-run positive impact from an increase precipitation, but there is no such impact in the long run. Lastly, Indonesia's national output reduces with higher temperature only in the short run, as the negative effect of temperature on output becomes insignificant over long run. The variations in the short-run and long-run results have implications for policy.

## 6 Summary and Conclusions

This chapter sheds light the climate change impacts to industrial output and employment in six ASEAN countries, namely, Cambodia, Indonesia, Malaysia, the Philippines, Thailand, and Vietnam over the period of 1989–2016. The industrial output and employment impacts are assessed based on temperature and precipitation as the proxies for climate change. The bounds testing procedure proposed by Pesaran et al. (2001) has been used to analyse the cointegration relationship between industrial output and employment with the climate change variables. Then, the autoregressive distributed lag (ARDL) approach of Pesaran and Shin (1999) and Pesaran et al. (2001) is used to model the long-run and short-run relationships between the variables. The majority of the variables exhibit cointegration relationships. Further, the evidences of long-run and short-run relationships are obtained from the models estimated. Our findings seem to suggest the impacts of climate change on industrial output and employment are largely through the agriculture sectors of the countries analysed, consistent with Chalise et al. (2017) and Wesseh and Lin (2017). We find that climate change can significantly affect employment, in that sense, this finding is consistent with Fankhauser et al. (2008).

The long-run and short-run results show some similarities and variations. In terms of similarities, the employment in Malaysia receives positive significant effect from temperature. Besides that, employment is negatively and significantly affected by temperature in the Philippines, Thailand, and Vietnam. The different results yield in comparison between long run and short run include as follows: the employment in Cambodia is positively impacted by an increase in temperature in short run, but the impact becomes negative over long run; the total industrial outputs in Cambodia and Thailand only receive a short-run positive impact from an increase precipitation, but there is no such impact in the long run; Indonesia's national output reduces with higher temperature only in the short run, as the negative effect of temperature on output becomes insignificant over long run (refer to Tables 4 and 5).

The long-run results also indicate the following: the industrial outputs in Cambodia and Thailand are adversely affected by temperature. The industrial outputs in Malaysia and the Philippines tend to increase with higher temperature. In Indonesia and the Philippines, the impact of precipitation to industrial output is positive and significant. Both Cambodia and Indonesia experienced higher employment with an increase in precipitation. It is important to stress that these are the long-run impacts which cannot be observed in short run (refer to Table 4).

In order to cope with the negative long-run impacts of climate change, it is generally accepted that the mitigation and adaptation approaches are crucial. In this regard, we note that the ASEAN economies have a direction towards regional cooperation to achieve an efficient use of resources. For instance, the ASEAN Socio-Cultural Community (ASCC) Blueprint 2009–2015 helps to promote environmental sound and clean technologies; and the ASEAN Plan of Action for Energy Cooperation (APAEC) is for the period of 2016–2025. In order to ensure the policies effectiveness, outreach of the institutional services is essential. Specifically, we suggest the policy-makers of ASEAN to enhance the climate-specific advisory services to benefit entrepreneurs, farmers, and fishermen who operate small businesses. In addition, the use of microfinance can be expanded to finance cleaner technology for energy supply and environmentally sound technology for equipment supply in the agriculture and fishing sectors. Meantime, the regional cooperation of ASEAN which aims for an efficient use of resources can be expanded and enhanced by the technologies from the advanced economies. As short-run negative climate change impact may disappear over long run, there is no urgent need for remedy using policy measures in the short run.

**Acknowledgement** This research was supported by a project code SBK0336-2017 from the SGPUMS, Universiti Malaysia Sabah.

# Appendix

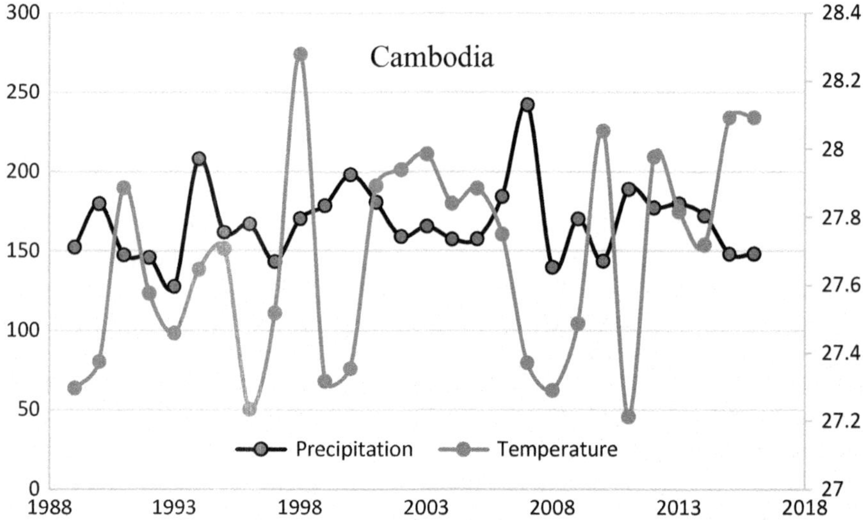

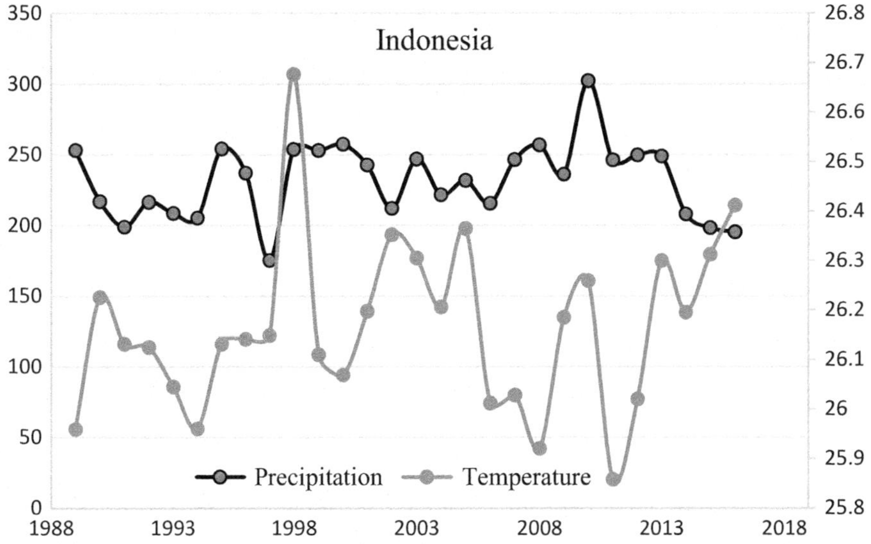
Indonesia
350
300
250
200
150
100
50
0
26.8
26.7
26.6
26.5
26.4
26.3
26.2
26.1
26
25.9
25.8
1988
1993
1998
2003
2008
2013
2018
Precipitation
Temperature

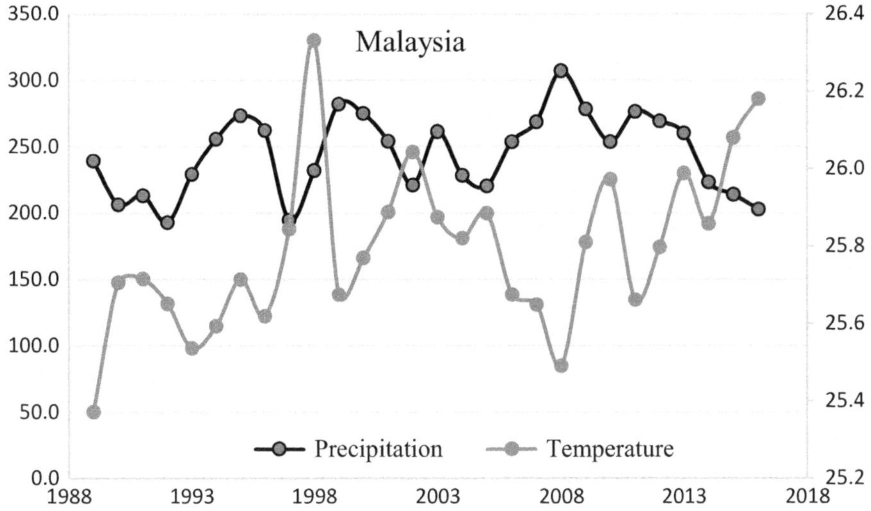
Malaysia
350.0
300.0
250.0
200.0
150.0
100.0
50.0
0.0
26.4
26.2
26.0
25.8
25.6
25.4
25.2
1988
1993
1998
2003
2008
2013
2018
Precipitation
Temperature

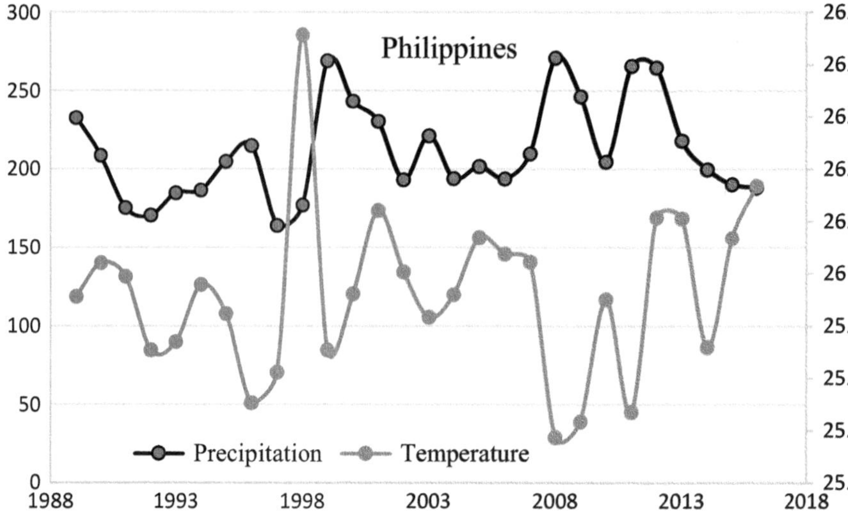
Philippines
300
250
200
150
100
50
0
26.5
26.4
26.3
26.2
26.1
26
25.9
25.8
25.7
25.6
1988
1993
1998
2003
2008
2013
2018
Precipitation
Temperature

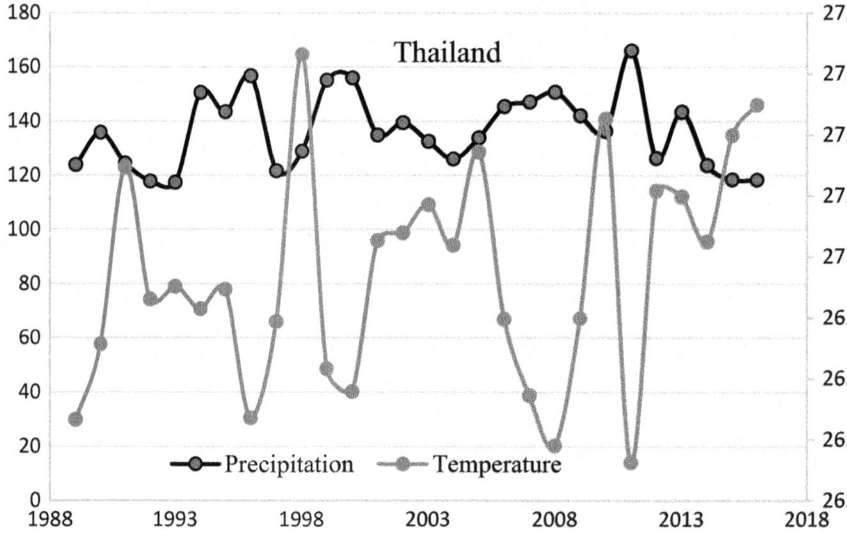
Thailand
180
160
140
120
100
80
60
40
20
0
27.8
27.6
27.4
27.2
27
26.8
26.6
26.4
26.2
1988
1993
1998
2003
2008
2013
2018
Precipitation
Temperature

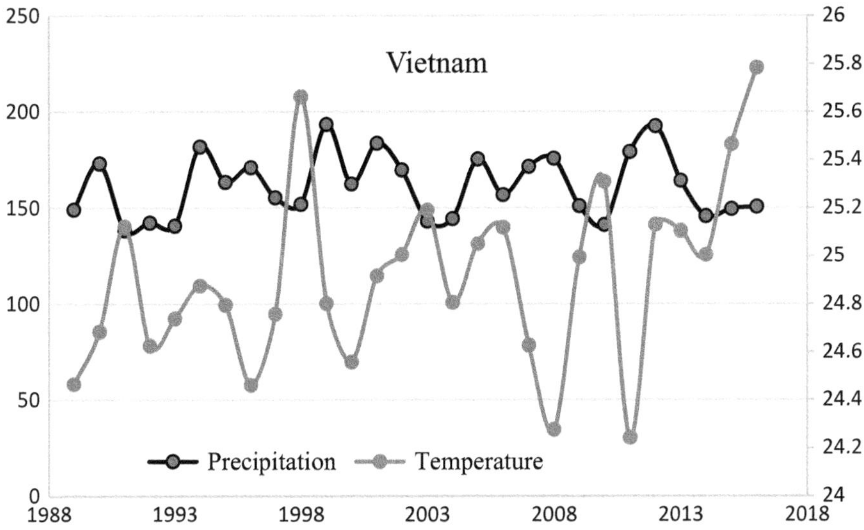


## References


Brown, R. L., Durbin, J., & Evans, J. M. (1975). Techniques for testing the constancy of regression relationships over time. *Journal of the Royal Statistical Society. Series B (Methodological), 37*, 149–163.

Chalise, S., Naranpanawa, A., Bandara, J. S., & Sarker, T. (2017). A general equilibrium assessment of climate change-induced loss of agricultural productivity in Nepal. *Econometric Modelling, 62*, 43–50.

Chen, K., Horton, R. M., Bader, D. A., Lesk, C., Jiang, L., Jones, B., et al. (2017). Impact of climate change on heat-related mortality in Jiangsu Province, China. *Environmental Pollution, 224*, 317–325.

Dey, M. M., Gosh, K., Valmonte-Santos, R., Rosegrant, M. W., & Chen, O. L. (2016). Economic impact of climate change and climate change adaptation strategies for fisheries sector in Solomon Islands: Implication for food security. *Marine Policy, 67*, 171–178.

Dickey, D., & Fuller, W. (1979). Distribution of the Estimators for Autoregressive Time Series with a Unit Root. *Journal of the American Statistical Association, 74*, 427–431.

Du, D., Zhao, X., & Huang, R. (2017). The impact of climate change on developed economies. *Economics Letters, 153*, 43–46.

Ebi, K. L., Frumkin, H., & Hess, J. J. (2017). Protecting and promoting population health in the context of climate and other global environmental changes. *Anthropocene, 19*, 1–12.

Elshennawy, A., Robinson, S., & Willenbockel, D. (2016). Climate change and economic growth: An intertemporal general equilibrium analysis for Egypt. *Economic Modelling, 52*, 681–689.

Engel, R. F., & Granger, C. W. J. (1987). Cointegration and error correction representation, estimation and testing. *Econometrica, 55*, 251–276.

Environmental Commissioner of Ontario. (2016). *Facing climate change* (Greenhouse Gas Progress Report 2016). Canada: Toronto.

Fankhauser, S., Sehlleier, F., & Stern, N. (2008). Climate change, innovation and jobs. *Climate Policy, 8*, 421–429.

Hewer, M. J., & Gough, W. A. (2017). Thirty years of assessing the impacts of climate change on outdoor recreation and tourism in Canada. *Tourism Management Perspectives, TMP-00349*, 14.

Hsiang, S. M. (2010). Temperatures and cyclones strongly associated with economic production in the Caribbean and Central America. *Proceedings of the National Academy of Sciences of the United States of America (PNAS), 107*(35), 15367–15372.

International Monetary Fund. (2017). *Regional economic outlook. Asia and Pacific: Preparing for choppy seas economic and financial surveys*. Washington, DC: International Monetary Fund.

Johansen, S., & Juselius, K. (1990). Maximum likelihood estimation and inference on cointegration—With applications to the demand for money. *Oxford Bulletin of Economics and Statistics, 52*, 169–210.

Lee, J. Y., & Kim, H. (2016). Projection of future temperature-related mortality due to climate and demographic changes. *Environment International, 94*, 489–494.

Loo, Y. Y., Billa, L., & Sigh, A. (2015). Effect of climate change on seasonal monsoon in Asia and its impact on the variability of monsoon rainfall in Southeast Asia. *Geoscience Frontiers, 6*(6), 817–823.

Lu, W., Huang, J., & Musshoff, O. (2015). Climate threshold, financial hoarding and economic growth. *Applied Economics, 47*(42), 4535–4548.

Luo, X.-S., Muleta, D., Hu, Z., Tang, H., Zhao, Z., Shen, S., et al. (2017). Inclusive development and agricultural adaptation to climate change. *Current Opinion in Environmental Sustainability, 24*, 78–83.

Nitschke, C. R., Nichols, S., Allen, K., Dobbs, C., Livesley, S. J., Baker, P. J., et al. (2017). The influence of climate and drought on urban tree growth in southeast Australia and the implications for future growth under climate change. *Landscape and Urban Planning, 167*, 275–287.

Narayan, P. K. (2004). *Reformulating critical values for the bounds F-statistics approach to cointegration: An application to the tourism demand model for Fiji* (Discussion Papers). Australia: Department of Economics, Monash University.

Pesaran, M., & Shin, Y. (1999). An autoregressive distributed lag modeling approach to cointegration analysis. In S. Strom (Ed.), *Econometrics and economic theory in the 20th century: The Ragnar Frisch centennial symposium*. Cambridge: Cambridge University Press.

Pesaran, M. H., Shin, Y., & Smith, R. J. (2001). Bounds testing approaches to the analysis of level relationships. *Journal of Applied Econometrics, 16*, 289–326.

Phillips, P., & Hansen, B. (1990). Statistical inferences in instrumental variables regression with I (1) processes. *Review of Economic Studies, 57*, 99–125.

Rodríguez-Rodríguez, G., & Ramudo, R. B. (2017). Market driven management of climate change impacts in the Spanish mussel sector. *Marine Policy, 83*, 230–235.

Rosselló, J., Riera, A., & Cardenas, V. (2011). The impact of weather variability on British outbound flows. *Climate Change, 105*, 281–292.

Rutty, M., & Scott, D. (2010). Will the Mediterranean become "too hot" for tourism? A re-assessment. *Tourism and Hospitality Planning and Development, 7*(3), 267–281.

Somanathan, E., Somanathan, R., Sudarshan, A., & Tewari, M. (2014). *The impact of temperature on productivity and labor supply: Evidence from Indian manufacturing* (Discussion Paper 14-10). New Delhi, India: Discussion Papers in Economics, Indian Statistical Institute, Economics and Planning Unit.

Susaeta, A., Adams, D. C., & Gonzalez-Benecke, C. (2017). Economic vulnerability of southern US slash pine forests to climate change. *Journal of Forest Economics, 28*, 18–32.

Wang, D. D. (2017). Do United States manufacturing companies benefit from climate change mitigation technologies? *Journal of Cleaner Production, 161*, 821–830. https://doi.org/10.1016/j.jclepro.2017.05.172

Weinberger, K. R., Haykin, L., Eliot, M. N., Schwartz, J. D., Gasparrini, A., & Wellenius, G. A. (2017). Projected temperature-related deaths in ten large U.S. metropolitan areas under different climate change scenarios. *Environment International, 107*, 196–204.

Weir, B. (2017). Climate change and tourism—Are we forgetting lessons from the past? *Journal of Hospitality and Tourism Management, 32*, 108–114.

Wesseh Jr., P. K., & Lin, B. (2017). Climate change and agriculture under CO2 fertilization effects and farm level adaptation: Where do the models meet? *Applied Energy, 195*, 556–571.

# The Role of Carbon Markets in the Paris Agreement: Mitigation and Development

**Henrique Schneider**

## Contents


1 The Paris Agreement ........ 110
  1.1 Development and Markets ........ 111
  1.2 Bottom-Up ........ 115
  1.3 Top-Down ........ 117
2 Carbon Markets and Sustainable Economic Development ........ 119
  2.1 Carbon Markets Under the Kyoto Protocol ........ 120
  2.2 A Review of the Clean Development Mechanism ........ 121
  2.3 Carbon Markets for Sustainable Economic Development ........ 126
3 What Follows Next? ........ 129
References ........ 131



**Abstract** In its Article 6, the Paris Agreement foresees international cooperation and tasks it with mitigating greenhouse gas emissions, guaranteeing environmental integrity and assuring sustainable development. In its paragraphs, the same article provides for three different systems of international cooperation. The so-called market mechanisms or carbon markets could play a role in two of them.

Article 6 makes it apparent that international cooperation and the use of markets are valid—but not exclusive—approaches in reconciling climate change and (sustainable) economic development. The question is how to align both through the operationalization of market systems. This paper provides first an overview of "carbon markets" within the body of the Paris Agreement. Then, it discusses why and how carbon markets reconcile climate change-related action with (sustainable) economic development. Third and lastly, the paper points at some safeguards in the operationalization of the international cooperation using markets as stipulated in Article 6 of the Paris Agreement. The most important safeguard regards to accounting principles.



H. Schneider (✉)
Swiss Federation of Small and Medium Enterprises, Bern, Switzerland

Nordakademie, Elmshorn, Germany




T. Sequeira, L. Reis (eds.), *Climate Change and Global Development*,
Contributions to Economics, https://doi.org/10.1007/978-3-030-02662-2_6

**Keywords** Paris Agreement · Carbon markets · Market mechanisms · Kyoto Protocol

At first sight, economic growth, sustainable development, and climate change do not seem harmonious. While sustainable development seems to slow economic growth, development per se and economic growth seem to be detrimental to climate change-related action (IPCC 2014), which is generally regarded as preservationist (for example, in Taylor 2008). This is one possible picture, but not by necessity the only one.

It is possible to conceive economic growth, climate change-related action, and sustainable development as complementing each other. In fact, most definitions of sustainable development make it clear that economic growth, development, and climate change-related action complement each other. It would be outside the scope of this paper to analyze if this more positive scenario is in principle, a matter of spontaneous order due to decisions taken by responsible and free agents (Kolk and Pinkse 2004), or of policy choice (Stern 2006). This paper will focus on policy choice since it is policy choice already. This choice happens at domestic and international level, and it is the latter that will be examined here. The Paris Agreement—best imagined as a global action plan on climate change—is committed toward reconciling sustainable (economic) development with climate change-related action. This relationship will be the object of this paper.

In its Article 6, the Paris Agreement foresees international cooperation and tasks it with mitigating greenhouse gas emissions, guaranteeing environmental integrity, and assuring sustainable development. In its paragraphs, the same article provides for two different systems of the so-called market-mechanisms or carbon markets. Article 6 makes it apparent that international cooperation and the use of markets are valid—but not exclusive—approaches in reconciling climate change and (sustainable) economic development. The question is how to align both through the operationalization of market systems.

This paper provides first an overview of "carbon markets" within the body of the Paris Agreement. Then, it discusses why and how carbon markets reconcile climate change-related action with (sustainable) economic development. Third and lastly, the paper points at some safeguards in the operationalization of the international cooperation using markets as stipulated in Article 6 of the Paris Agreement. The most important safeguard regards to accounting principles.

## 1 The Paris Agreement

This section explains the place of (sustainable) economic development and the so-called carbon markets within the Paris Agreement in three subsections. The first explores the overall context of the Paris Agreement as well as its Article 6; and the second and the third discuss the two types of the so-called carbon markets foreseen in the Paris Agreement, Article 6 paragraphs 2 and 3, respectively, Article 6 paragraphs 4–7.

## 1.1 Development and Markets

The Paris Agreement is an agreement—not an international treaty or an alliance, see Bodansky (2016)—within the United Nations Framework Convention on Climate Change (UNFCCC). It deals with the mitigation of greenhouse gas emissions, adaptation to the consequences of climate change, and financing mitigation and adaptation efforts. The Paris Agreement aims at including all state agents (called Parties to the Agreement) while allowing for some inner differentiation between the so-called developed and the other countries. Most of the Agreement's provisions become effective starting in 2020.

The language of the Agreement was negotiated by the representatives of 196 Parties at the 21st Conference of the Parties (COP) of the UNFCCC in Paris and adopted by consensus on December 12, 2015. The same conference also adopted a work program to operationalize the Paris Agreement in the form of a decision by the COP, i.e., a mandate directed to the conference itself to further elaborate the implementation of the Agreement. This work program is supposed to present rules, modalities, procedures, and processes until 2018. After that, individual countries—the signatory Parties to the Agreement—will each individually adapt their domestic laws and regulations to the provisions of the Paris Agreement whereby each country remains free to decide which provisions of the Agreement to pursue and use.

In a nutshell, the Paris Agreement builds upon the framework convention, the UNFCCC, and—for the first time—brings all nations into a common cause to undertake efforts related to climate change as well as to adapt to its effects, with enhanced support to assist developing countries to do so. The Paris Agreement's self-declared central aim is to strengthen the global response to climate change by keeping a global temperature rise in the twenty-first century "well"—a word that needs further interpretation—below 2 °C above preindustrial levels and to pursue efforts to limit the temperature increase even further to 1.5 °C. Additionally, the Agreement aims to strengthen the ability of countries to deal with the possible effects of climate change. To reach these goals, the Agreement continues to self-declare that appropriate financial flows, a technology framework, and an enhanced capacity building will be pursued, thus supporting action by developing countries and the most vulnerable countries, in line with their own national objectives. The Agreement also provides for enhanced transparency of action and support through a more robust transparency framework.[1]

[1] A guide to all relevant aspects of the UNFCCC, the Kyoto Protocol, and the Paris Agreement can be found here: http://bigpicture.unfccc.int.

A core issue for the Paris Agreement is its requirement of all Parties to put forward their best efforts through "nationally determined contributions" (NDCs)[2] and to upgrade these efforts in the years ahead. This includes requirements of all Parties to report regularly on their emissions and on their efforts in reaching their respective NDCs and implementing climate change-related policies. In 2018, Parties will take stock of the collective efforts in relation to progress toward the goal set in the Paris Agreement and to inform the preparation of NDCs. There will also be a global stock take every 5 years to assess the collective progress toward achieving the purpose of the Agreement and to inform further individual actions by Parties. A treatment of how to formulate NDCs is given by Levin et al. (2015); and one on how to track them is discussed in Boyd et al. (2015), for example. NDCs will be further addressed in Sects. 2 and 3 of this paper.

What role does (sustainable) economic development play in the Paris Agreement? At the level of NDCs, for example, the staff of the World Resources Institute analyzed in 2016 how they are aligned with the United Nations' goals for sustainable development finding out that climate actions communicated in NDCs align with at least 154 of the 169 sustainable development goals' targets (Northrop et al. 2016). They report:

> For example, Thailand's INDC identifies ambitious actions under its Environmentally Sustainable Transport System Plan, including extensions of mass rapid transit lines, construction of double-track railways and improvement of bus transit in the Bangkok Metro areas. Implementation of these actions has the potential to contribute to SDG target 11.2 to provide access to safe, affordable, accessible and sustainable transport systems for all while improving road safety. When it comes to actions aimed at increasing resilience and adaptive capacity, Mexico's INDC commits the country to improve watershed management and land conservation in order to guarantee food security and water access in the face of growing climate threats such as drought, floods and storms. This aligns strongly with SDG target 2.1 to end hunger and ensure all people's access—particularly the poor and vulnerable—to safe, nutritious and sufficient food year-round by 2030. Meanwhile, advancing action on the SDGs can in many cases contribute to mitigation and adaptation. For instance, Niger's INDC identifies the lack of literacy and numeracy in rural areas as a significant impediment to disseminating the technologies needed for climate-smart agriculture and land management. By targeting these rural communities in implementing SDG target 4.6 to improve literacy and numeracy by 2030, Niger can lay the necessary foundation for achieving its climate goals.

Even if sustainable development can be taken to permeate the Paris Agreement, in its text, explicit mentions of (sustainable) economic development are relatively scarce. Paragraph 8 of the preambular recital as well as Articles 2(paragraph 1), 4 (1), 6, 7(1), 8(1), and 10(5) foresee the necessity for a complementary view of climate change-related action and sustainable development—all economic development in the language of the Paris Agreement is sustainable development, but not necessarily, only interpretatively, aligned with the United Nations' sustainable

---

[2]At the time of negotiation and beyond, these contributions were called "intended nationally determined contributions" (INDCs). After the ratification of the Paris Agreement by a country-Party, the contribution is fixed, thus becoming a "nationally determined contribution" (NDC).

development goals. Technology, another issue closely related to (sustainable) economic development, is mentioned in paragraph 6 of the Agreement's preambular recital as well as in Articles 6(8), 10, 11(1), and 13(9;10).

This overview shows that Articles 6 and 10 are especially interesting for the scope of this paper, since only in those, development and, respectively, technology are mentioned several times. While Article 6 is concerned with international cooperation, Article 10 institutes the technology mechanism envisaging the exchange of climate change-relevant technology between countries. Because of this narrow scope of Article 10, it is Article 6 that addresses (sustainable) economic development in broader terms. Article 6 itself consists of a general clause and three distinct parts. The general clause in paragraph 1 reads:

> 1. Parties recognize that some Parties choose to pursue voluntary cooperation in the implementation of their nationally determined contributions to allow for higher ambition in their mitigation and adaptation actions and to promote sustainable development and environmental integrity.

The three parts following it are the provisions for the use of "internationally transferred mitigation outcomes" ITMOs (paragraphs 2 and 3), the establishment of a "mechanism to contribute to the mitigation of greenhouse gas emissions and support sustainable development"—called by many observers the "Sustainable Development Mechanism" SDM—(paragraphs 4–7), and the recognition of the "importance of integrated, holistic and balanced non-market approaches being available to Parties to assist in the implementation of their nationally determined contributions, in the context of sustainable development and poverty eradication, in a coordinated and effective manner, including through, inter alia, mitigation, adaptation, finance, technology transfer and capacity-building, as appropriate," generally referred to as "nonmarket approaches" (paragraphs 8 and 9).

Article 6 can be generally subsumed under international cooperation. This cooperation can take several forms. For example, it could rely on or use so-called carbon markets under 6(2–3) or 6(4–7), or they can take the form of exchange without commodification, under 6(8), for example, in the realms of joint technology development, multi- and supranational coordination of policies, or additional financing, inter alia. As per the UNFCCC–COP/APA[3] negotiations of 2017, the introductory clause of the article is taken as valid for all three types of cooperation.[4] Elements of this clause are the voluntary nature of any cooperation, the idea that it can extend to mitigation or adaptation actions, or the combination of both, as well as the commitment toward higher ambition, sustainable development, and environmental integrity. These assertions are not as clear as they seem, and there are considerable ongoing debates on what the general clause of Article 6 means, how the three different types

[3]Ad hoc working group on the Paris Agreement (APA).

[4]Article 6 has a highly politicized genesis (see Marcu 2016b). While some Parties opposed the use of the word "market" in the Agreement, others threatened with developing markets outside the scope of the UNFCCC. Finally, it was agreed to create an overall article involving all forms of international cooperation, market-based and not market-based.

of cooperation are to be implemented, and how they relate to existing provisions—those under the Kyoto Protocol (see Sect. 2). While it is not useful in this context to delve in these open issues (for that, see Marcu 2016a), it is important to discuss those that affect the relationship of (sustainable) economic development and climate change-related action.

The first issue is the relationship between the general idea of a market, the specific conception of a "carbon market," and the clauses of 6(1), namely, higher ambition in mitigation and adaptation action, sustainable development, and environmental integrity. One reading of it is that so-called carbon markets lead per se, i.e., automatically, to higher ambition in climate change-related action as well as in environment integrity and development. It is questionable if this reading yields if past experiences with carbon markets (see Sect. 2) as well as the spirit of the negotiations leading to the Paris Agreement (Marcu 2016b) are taken into account. The other way of reading this provision is: Since markets alone cannot vouch for any or either of the desiderata in the clauses, specific guidance has to be given ensuring that they do. Again, this reading is not plausible. During the negotiations leading to the Paris Agreement, an overall majority of countries maintained their support for market instruments and even discussed setting up a parallel pro-market organization within the negotiations (Marcu 2016b; Falkner 2016).

Furthermore, the more recent experience with so-called carbon markets shows that maintaining them to be detrimental to ambition, development, and environmental integrity does not hold (Spalding-Fecher et al. 2012). Via the World Bank, different countries such as the United Kingdom, Sweden, or Switzerland increased their respective levels of ambition using "carbon markets" (CI-Dev 2016).

There is a modified reading of the overall relationship of markets, environmental integrity, and development, as they are expressed in Article 6. This reading takes the first paragraph of Article 6 as a utilitarian task: Since there will be international cooperation and since this cooperation might take the form of "carbon markets" among other forms, the framework convention is being tasked with giving these markets certain goals beyond the goals more typical to the market mechanism—the allocation of prices and quantities. These further goals are sustainable economic development and environmental integrity. Here, immediately two questions arise. First: How are these additional goals given to be markets to be built into the market principle? The most likely answer is: By the design of its "products," i.e., ITMOs and/or other tradable units. These should yield to enhanced specifications to conform with the desiderata. The specifications are made by the framework convention in its deliberations on how to implement the Paris Agreement. This answer might seem straightforward, but the next subsection shows that it is not.

The second question that arises from the utilitarian view of paragraph 1 is: What is the status of higher ambition? The most likely answer takes the statement about higher ambition in the NDCs as a matter of fact. International cooperation allows countries to be more ambitious in their climate change-related actions,[5] and since

[5]The Paris Agreement makes it clear that there is no conditionality in the nations' levels of ambition.

this is the case, it applies equally to mitigation of greenhouse gas emissions as well as to adaptation to climate change.

But this matter-of-fact statement is not without importance, since the dynamics of the Paris Agreement foresees multiple stocktakes during or after which countries can adjust their NDCs. The level of commitment to international cooperation will be assessed by peers in this global stocktake. Those Parties that cooperate internationally, using so-called carbon markets among other instruments, will face an easier task if upgrading their respective NDCs making them more ambitious.

While these are the main tenets of the first paragraph of Article 6, the two provisions about so-called carbon markets follow in Paragraphs 2–3 and 4–7. They will be reviewed in the following sections.

## *1.2 Bottom-Up*

Internationally transferred mitigation outcomes (ITMOs) are at the center of paragraphs 2 and 3, which, together, form the first and wider understanding of so-called carbon markets by the Paris Agreement. This first type of carbon market is a bottom-up approach which 6(2–3) allude only to framework conditions. They read:

> 2. Parties shall, where engaging on a voluntary basis in cooperative approaches that involve the use of internationally transferred mitigation outcomes towards nationally determined contributions, promote sustainable development and ensure environmental integrity and transparency, including in governance, and shall apply robust accounting to ensure, inter alia, the avoidance of double counting, consistent with guidance adopted by the Conference of the Parties serving as the meeting of the Parties to the Paris Agreement.
>
> 3. The use of internationally transferred mitigation outcomes to achieve nationally determined contributions under this Agreement shall be voluntary and authorized by participating Parties.

Additionally, as the COP negotiated the Paris Agreement, it took a decision on how to work on the details of 6(2) UNFCCC (2015, paragraph 36 of decision 1/CP.21). It reads:

> Requests the SBSTA [Subsidiary Body of Scientific and Technological Advice] to develop and recommend the guidance referred to under Article 6, paragraph 2, of the Agreement for adoption by the CMA [Conference of the Members of the (Paris) Agreement] at its first session, including guidance to ensure that double counting is avoided on the basis of a corresponding adjustment by Parties for both anthropogenic emissions by sources and removals by sinks covered by their NDCs under the Agreement.

Paragraph 2 accepts international cooperation on a bi- or multilateral way. By accepting that there will be cooperation, it includes the possibilities of cooperation, in various forms, in the Agreement. Therefore, paragraph 2 does not endeavor in defining, designing, or measuring the forms or scope of this type(s) of cooperation but into accepting whichever possible forms of cooperation that will be established as international cooperation under the Agreement. However, paragraph 2 does not accept it passively. This acceptance goes hand in hand with guidance on this bi- or multilateral form of cooperation.

Paragraph 2 states that if such cooperation is the case and if it involves the transfers of mitigation outcomes, this corporation shall safeguard different desiderata, among others sustainable development. Paragraph 3 readdresses this point making it explicit that this form of international cooperation is voluntary and solely dependent on the participating countries. In other words, whatever the mode of cooperation is, is to be determined only by the countries participating in it. Furthermore, it is outside the regulatory scope of the framework convention. As per beginning of 2018, the only involvement of the latter with bi- or multilateral cooperation is the guidance it gives on technical matters of transparency and accountancy (including the reporting of corresponding adjustments) as well as on environmental integrity and sustainable development. This question, however, will continue to be discussed, since different Parties to the Agreement have different interpretations and desiderata. The technical nature of the guidance[6] is also recalled in the decision and narrowed down to two elements, accountancy and the reporting of corresponding adjustments in the overall "bookkeeping" of greenhouse gas emissions.

As seen before, here too, different readings compete. There is considerable discussion regarding how far this guidance given by the framework convention should go and how mandatory their content will be for the mode(s) of cooperation. One position would like a fine-grained guidance that remains, however, to large extent voluntary to countries engaged in cooperation to apply it.

Another position is to call for mandatory fine-grained guidance, replicating much of the rules to be established for 6(4–7). It would remain voluntary to cooperating countries to abide by the guidance, but in order for their emission outcomes to be counted as valid toward the implementation of NDCs, they would have to subscribe to the mandatory fine-grained guidance.

The third position endorses this idea of a mandatory implementation of guidance but asks for its content to be less fine-grained in order to make cooperation more broadly acceptable and more flexible. Finally, the fourth position wants voluntary and open guidance.

The probable conversion point of this discussion, as it unfolds toward the implementation of the Paris Agreement, is on the one hand, the framework convention will elaborate fine-grained guidance with a semi-mandatory character regarding technical issues such as robust accounting and transparency; on the other hand, it will issue very general guidance on governance, environmental integrity, and (sustainable) economic development. There are two reasons for this outcome. First, technical guidance is needed to make the use of ITMOs compatible with the system under the Paris Agreement and easily exchangeable along (linked) markets. Second, it is unlikely that Parties will allow for sovereign issues to be subjected to a standard under the framework convention. Finding consensus in matters concerning

[6]A guidance is less technical and less compulsory that rules, modalities, and procedures are.

environmental integrity and (sustainable) economic development is difficult and unnecessary for the instruments under Article 6(2–3) to develop. Yet another view is that these elements will develop anyway, and the guidance will only be able to catch up with whatever has emerged bottom-up.

This last remark also acknowledges that it is very likely that at least at the beginning, several different systems and instruments of international cooperation will arise. If these systems will be interlinked or will allow for trade or exchange among them, remains to be seen. It is even possible that competition between these international "carbon markets" could unfold. After all, in a bottom-up world, much is possible. Many argue, therefore, that "one-size-fits-all" approaches are detrimental to Article 6(2–3).

## *1.3 Top-Down*

The other type of market instrument conceived by the Paris Agreement is specified in paragraphs 4–7 of Article 6. This is a market mechanism under the centralized auspices of the framework convention. This market type will have rules, modalities, and procedures—a sharp contrast to the general framework conditions mentioned above—centrally instituted by the Parties of UNFCCC. The mechanism under Article 6(4–7) is top-down; these provisions read:

> 4. A mechanism to contribute to the mitigation of greenhouse gas emissions and support sustainable development is hereby established under the authority and guidance of the Conference of the Parties serving as the meeting of the Parties to the Paris Agreement for use by Parties on a voluntary basis. It shall be supervised by a body designated by the Conference of the Parties serving as the meeting of the Parties to the Paris Agreement, and shall aim:
>
> (a) To promote the mitigation of greenhouse gas emissions while fostering sustainable development;
> (b) To incentivize and facilitate participation in the mitigation of greenhouse gas emissions by public and private entities authorized by a Party;
> (c) To contribute to the reduction of emission levels in the host Party, which will benefit from mitigation activities resulting in emission reductions that can also be used by another Party to fulfil its nationally determined contribution; and
> (d) To deliver an overall mitigation in global emissions.
>
> 5. Emission reductions resulting from the mechanism referred to in paragraph 4 of this Article shall not be used to demonstrate achievement of the host Party's nationally determined contribution if used by another Party to demonstrate achievement of its nationally determined contribution.
>
> 6. The Conference of the Parties serving as the meeting of the Parties to the Paris Agreement shall ensure that a share of the proceeds from activities under the mechanism referred to in paragraph 4 of this Article is used to cover administrative expenses as well as to assist developing country Parties that are particularly vulnerable to the adverse effects of climate change to meet the costs of adaptation.

> 7. The Conference of the Parties serving as the meeting of the Parties to the Paris Agreement shall adopt rules, modalities and procedures for the mechanism referred to in paragraph 4 of this Article at its first session.

The decision by the COP that negotiated the Paris Agreement has further material concerning Article 6(4–7) (UNFCCC 2015, paragraphs 37 and 38 of decision 1/CP.21). It reads:

> 37. Recommends that the CMA adopt rules, modalities and procedures for the mechanism established by Article 6, paragraph 4, of the Agreement on the basis of: (a) Voluntary participation authorized by each Party involved; (b) Real, measurable, and long-term benefits related to the mitigation of climate change; (c) Specific scopes of activities; (d) Reductions in emissions that are additional to any that would otherwise occur; (e) Verification and certification of emission mitigation activities by designated operational entities; (f) Experience gained with and lessons learned from existing mechanisms and approaches adopted under the Convention and its related legal instruments.
>
> 38. Requests the SBSTA to develop and recommend rules, modalities and procedures for the mechanism referred to in paragraph 37 above for consideration and adoption by the CMA at its first session.

These paragraphs read differently from Article 6(2–3). They are about—possibly only—one mechanism under the authority of the framework convention. Of course, participation is voluntary, and countries still have responsibilities, even veto-rights, but the overall design—via rules, modalities, and procedures—governance, and administration of the mechanism are under the authority of the UNFCCC. For participating countries, all those provisions envisaged by the decision will be mandatory.

Paragraph 6 states that it is the market that is going to pay for its own administration and governance under the framework convention. Also, it is the market that is going to reserve some of its capital to aid the most vulnerable countries. So, there are two different types of transaction costs built into in this mechanism. The first is the administration fee and the second is development aid. These components are also mandatory.

While the goals of cooperation under Article 6(2–3) remain open, since the cooperating Parties or the cooperation entity sets its goals independently from the UNFCCC, under Article 6(4–7), these goals are given by Article 6. But even if the goals are given in the wording of the paragraphs, there is still controversy in their interpretation.

The mechanism under Article 6(4–7) serves toward multiple goals: mitigating greenhouse gas emissions, fostering (sustainable) economic development; delivering overall mitigation in global emissions—meaning that in utilizing this mechanism, cooperation must result in a lower emission of greenhouse gases and not just in an offsetting of one's emissions through another's mitigation; involving public and private entities; and allowing for transferal/trades of emission outcomes between countries, whereby paragraph 5 makes sure that each mitigation outcome only counts once, even when transferred. The discussion about these goals involves questions concerning the priorities among them, their granularity, the relationship between them, as well as possible conflicts within them.

These, and other, questions could be clarified in a specific guidance on goals of the mechanism under Article 6(4–7). This guidance would most probably not insist on the achievement of all the goals simultaneously and additionally but allow for various trade-offs between them. However, it is much more likely that the relationships between goals will be implicitly solved when developing rules, modalities, and procedures. This process will be a challenge of its own.

Summing up: This first section introduced the Paris Agreement and especially its Article 6. This article is the only one to mention repeatedly a commitment toward sustainable development. From the point of view of economics, it is no surprise that the Paris Agreement's provisions on so-called carbon markets—or, international cooperation, in the terms of the Agreement—are also those that address sustainable development. It was further shown that Article 6 foresees two different types of "carbon markets": a system of instruments emerging bottom-up and one mechanism designed, governed, and administrated top-down. It was also explained that there are different questions relating to each of these. Sustainable economic development has a place in the overall conception of the Agreement, its Article 6 as well as in each market type. However, it isn't clear at all how they interplay. In order to find this out, the next section reviews market instruments that already exist as well as those that are newly developing in order to map the relationship of carbon markets and sustainable economic development. The next section will make a case for why and how carbon markets reconcile climate change action with (sustainable) economic development.

## 2 Carbon Markets and Sustainable Economic Development

In a three-stepped approach, this section first introduces so-called carbon markets as they actually operate and then asks how the experience with these market mechanisms could so far inform (sustainable) economic development[7]; and finally, it makes the case for why and how carbon markets reconcile climate change-related action with (sustainable) economic development. While the section above was about outlying "carbon markets" in the Paris Agreement, this section aims at showing how economic development has profited from actual carbon markets and how it can continue to do so. It will be the last section that drawing on this experience will go back to the Paris Agreement and stipulate which market designs foment sustainable economic development.

[7]Remembering that it is each Party to the Paris Agreement that defines what its understanding of and policies regarding sustainable development are.

## 2.1 Carbon Markets Under the Kyoto Protocol

Up to this subsection, this paper has been using the formula "so-called" to refer to "carbon markets" or keeping them in quotation marks. After the explication of the term in this subsection, this paper will employ it without a caveat marker. The core of "carbon markets" is not, as the name might misleadingly make believe, trading of physical greenhouse gasses, but the trading of the right to emit these gases whereby the unit of account for these trades is a ton of carbon dioxide equivalent (tCO2e).

Market-based approaches are not referred to in the founding 1992 UNFCCC. It was in 1997 during the negotiations that led to the Kyoto Protocol that the idea, which has been discussed academically since the 1980, found its implementation (MacKenzie 2009; the following characterization is based on Yamin 2012).[8] In the Kyoto Protocol, several developed countries agreed to targets for their emissions of the six major greenhouse gases. Countries that opted to include themselves in an annex to the protocol can participate in different "flexible mechanisms" that are designed to allow these countries to meet their emission reduction commitments with reduced economic impact. Among these flexible mechanisms are international emissions trading (IET) and the sponsorship of activities that reduce greenhouse gas emissions in other countries so that the reduction can be used by the sponsor as an offset to its own emissions. These activities generate tradable credits that can be used by countries in meeting their emission reduction goals. The activities-based Kyoto Mechanisms are the Clean Development Mechanism (CDM) and Joint Implementation (JI). The CDM covers projects taking place in countries that are not in the annex to the Protocol, while JI covers projects taking place among countries in the annex.[9]

Emissions trading under the Kyoto Protocol rely on international multilateral oversight. All transfers are tracked using a registry called the international transaction log (ITL). A common accounting standard applies to all countries with emission targets. An executive board must approve the methodology CDM projects propose using a supervisory body as in charge of JI. Finally, under the protocol, only the international transfers it sanctions are considered legitimate to fulfill a country's emission cutting obligations. The Kyoto model provides important infrastructure for an international carbon market. Common accounting procedures ensure that any transfer meets an internationally agreed level of environmental integrity. Common

[8]There are several other "carbon markets" in place; but those under the Kyoto Protocol are the first—and remained for a long time the sole—international carbon markets.

[9]Each of these "carbon markets" has its own unit denomination; but all unit denominations go back to the tCO2e. The units are international emissions trading (IET) that allows countries that have reduced emissions below their targets to sell excess allowances to countries whose emissions exceed their targets. This is also known as a "cap and trade" system. The allowances are known as assigned amount units (AAUs). Joint Implementation (JI) allows countries in the annex to the Protocol to earn emission reduction units (ERUs) through emission reduction or removal projects in other countries. The Clean Development Mechanism (CDM) allows countries to earn certified emission reduction (CER) credits through emissions reduction projects in developing countries.

offset methodologies give a blueprint to replicate in projects across the globe. Moreover, when countries submit their national greenhouse gas inventories, any recorded transfers can be verified by checking the international registry thereby reducing the potential for emissions double counting.

The Kyoto Protocol's market mechanisms have, however, encountered shrinking participation. One reason has been a reliance on the emissions trading scheme of the European Union, a form of IET, as a source of demand. But the Union's low economic growth and restrictions placed on the accepted types of credits have created a generous oversupply of CDM credits. On the other hand, the overly complicated and controversial negotiations before and after the Paris Agreement made many participants of these markets nervous, leading to their withdrawal from the market or to price discounts (Schneider 2017). Independently from the actual situation of the mechanisms under the Kyoto Protocol, evaluating how they, especially the CDM, did in aligning (sustainable) economic development and climate change-related actions might prove informative for the way ahead under the Paris Agreement.

## 2.2 A Review of the Clean Development Mechanism

The CDM has a dual mandate to deliver climate mitigation and sustainability benefits. CDM projects are supposed to contribute to sustainable development in developing countries and also generate "real" and "additional" emission savings, i.e., savings that only occur thanks to the CDM project in question. The CDM has been able to issue 1.85 billion credits (UNFCCC 2017)—each representing a metric ton of avoided emissions—and mobilize over 400 billion US dollars in investment (Platonova-Oquab et al. 2012). Despite these impressive numbers, there is much criticism of the CDM. This critique will guide this subsection, not because it is always right but because it can be instructive for the further development of carbon markets and because it points at important aspects when thinking about how carbon markets and (sustainable) economic development interact. Three different "batches" of criticism will be addressed.

The first "batch" is the claim that the mechanism has failed on its dual mandate (e.g., Castro 2014; Schade and Obergassel 2014; Sterk et al. 2015). While most concede that it was able to deliver its mitigation, or offset, obligations, most equally state this as critique rather than praise. Delivering overly well on offsets is interpreted by critics as a sign of weak governance. From the actual point of view, it is also seen as the proof of low mitigation ambition impacting negatively on carbon prices and on climate change-related finance actions. The authors mentioned above find different reasons explaining the supposed failure of CDM, especially for its alleged failure of delivering benefits for (sustainable) economic development. One reason is that countries hosting the activities define their own sustainability criteria—sustainability being a national prerogative.

Developing countries rejected attempts to establish an international sustainability assessment process, arguing that it would violate their national sovereignty. Additionally, it is in the interest of the host country to secure as many CDM activities as possible because of the investment they bring. This means, according to the critics, that host countries have little or no incentive to require strong sustainability criteria that could dampen investment. The sustainability criteria therefore usually lack specificity, transparency, and stringency. Also, the assessment process that is performed by the host country Designated National Authorities (DNAs) is usually perfunctory. Even if the sustainability requirements are well developed, the requirements are often undermined by the lack of follow-up or verification of the sustainability criteria. Furthermore, critics point out that sustainability benefits have no financial value in the current system, as only greenhouse gas benefits result in monetary compensation (through the generation of credits). This results in the majority of offsets coming from projects with arguably little or no sustainable benefits (such as industrial gases and large hydropower). Finally, some critics point out that some activities caused significant harm to the local population such as displacement, job loss, and even increased pollution.

The critique expressed above has some validity. Under the CDM, the assessment of sustainable development issues is the prerogative of host countries. When approving projects, the DNA of the host country has to confirm that the projects assist the country in achieving sustainable development. Some host countries have developed and published criteria or guidelines to assess the sustainable development benefits of projects and programs, while most countries have not. Information on sustainable development impacts is self-declared by project developers in project design documents. In response to a request by the Conference of the Parties Serving as the Meeting of the Parties to the Kyoto Protocol (CMP), the CDM Executive Board has developed a tool that may be used on a voluntary basis to report on sustainable development co-benefits in a comparable and structured manner. The tool does not allow project developers to report on any negative effects. The CDM does not require third party verification or monitoring of sustainable development benefits. There are also no international environmental or social safeguards, except for Carbon Capture and Storage (CCS) projects.

On the other hand, there are plenty of private initiatives monitoring CDM activities and regularly publishing their findings as well as participating in climate negotiations under the framework conventions. As it occurs in every market, there is a process of differentiation between the investment options. Partly, this differentiation is driven by regulatory actions of countries buying CERs, for example, allowing only CERs fulfilling specific quality criteria to be bought or used. Partly, this differentiation is driven by the investors themselves searching for different levels of quality and demanding private organization to monitor and certify activities. One example for this is the development of labels, such as the "gold standard." The third driver for differentiation is civil society itself exercising influence over governments and investors. So, while the technical process shows gaps, in practice, all these gaps can be addressed by allowing for differentiation. This means, however, giving private sector, including civil society, a larger role to play, especially in monitoring and vouching for quality—and this inclusion of the private sector has been a contentious issue in past climate negotiations. The same logic of differentiation applies to the claim that CDM violate some of the peoples' and communities' rights. First, many private labels and government regulations make investors incorporate criteria regarding gender, indigenous rights, labor market, or environmental integrity in their choices. Then, all CDM projects qualify for the appeal processes regarding multinational enterprises foreseen by the Organization for Economic Co-operation and Development (OECD) (Reinert et al. 2016).

Claiming that the CDM does not build in the value of sustainable economic development as its co-benefit is correct. However, it remains doubtful if the problem supposed by the critique is well-addressed. The idea of the mechanism is to price the reduction of CO2 emissions, thus discovering lower abatement costs by allowing for separation of labor between emitter and host country of an offsetting activity. Pricing in other than mitigation outcomes would not only challenge the integrity of the priced unit (tCO2e) but also blur the line between a market instrument and development aid. The co-benefits of the CDM regarding (sustainable) economic development are meant to make the instrument more attractive for the host countries, for example, by allowing local entrepreneurship to develop around it, by making some of the project outcomes benefit the people directly, or by building knowledge and know-how in low-carbon activities, this leading to lower political obstacles or even endorsement. While the benefits of the mitigation aspect (offsets) are split between the emitter (credits) and the host country (financial inflows)—the emitter paying almost all transaction costs—the host country appropriates all the benefits of economic sustainable development of CDM activities. Countries hosting offsetting activities, therefore, could improve the materialization of development benefits by setting up good framework conditions, like incorporating the outcomes of activities in regional strategies or allowing for tax breaks for entrepreneurship developing around the activities.[10]

The second "batch" of critiques consists of two methodological issues. First, many CDM activities have been deemed un-additional. And second, some might lead to technological lock-in. To the first: to avoid giving credits to projects that would have happened anyway, "free riders," specified rules ensure the additionality of a proposed project, that is, ensure the project reduces emissions more than would have occurred in the absence of the use of CDM (Gillenwater 2012). At present, the

[10]Two anecdotal examples that this author assessed are: In Southeast Asia, a CDM financed project transforming methane from landfills (garbage) into electrical power was regarded skeptically by the local population. As the metropolitan government built net infrastructure making this electricity available for the local community, the attitude changed. The local political leader stated in a private conversation in 2015 "I don't care for the money, but electricity it's what we want. Since I have access to this electricity, I have more children in the school, less criminality and even less health-related problems." In Central Africa, a CDM-financed project substituted coal or wood stoves with lower carbon cook stoves. Around this project, local entrepreneurship developed, for example, selling utensils for the new cook stove or increasing the energy efficiency of other household components (cooling, water). These local entrepreneurs benefit from tax breaks.

CDM Executive Board deems a project additional if its proponents can document that realistic alternative scenarios to the proposed project would be more economically attractive or that the project faces barriers that CDM helps it overcome. Proving additionality requires some work. Zhang and Wang (2011) measured the large transaction costs of establishing additionality. Assessing additionality has created delays, or bottlenecks, in approving CDM projects, they claim.

On the other hand, it is not surprising that additionality is a difficult criterion to fulfill. Proving that a CDM activity is additional is complicated and creates transaction costs. But the definitive measurement if a project is additional or not only happens after implementation. After implementation, many initially additional projects might prove to be not additional. While there are guidelines of the Executive Board of the CDM for technologies and activities that are automatically considered additional, the technical idea behind the criterion is not to allow any activity to be financed by the mechanism. Balancing these two, contradictory, ideas and paying sufficient regard to the economic, social, and technological context of the individual activity is an admittedly difficult task. But at the same time, it is the safeguard addressing the critique voiced by the first "batch." Either the system is strict or taking time and effort to fulfill its strict criteria or the mechanism is permissive. The more permissive it is, the less environmentally integer it becomes, and the less probable it would lead to (sustainable) economic development of the host country. If there were no admittedly tortuous proof of additionality, the CDM would be a cheap offsetting tool, which it is not and should not be.

To the second: technology lock-in occurs, when the CDM finances investments that create a long-term path dependency on technologies that are deemed non-sustainable, for example, coal-based energy plants or, according to some, large hydropower. Also, some have criticized CDM activities to have locked-out smaller-scale projects, for example, solar energy for remote locations, biofuels, improved cooking stoves, or efficient lighting. Nonetheless, a review of costs for these technologies suggests that they could be cost-effective for developing countries. For projects requiring wide dispersal of household items, such as cooking stoves or lighting, the administrative burdens of CDM, especially the proof of additionality, provide a hurdle (Kim et al. 2013).

On the other hand, the differentiation between non-sustainable and sustainable investments is difficult to make. More often than not, they are a judgment call. All CDM activities must conform to specific technological guidance in addition to the national regulation on sustainable development. Several lock-in and lock-outs are prevented by this framework. Many critics that claim them nonetheless base their criticisms on counterfactual scenarios or knowledge ex post, i.e., after the activity has unfolded, which makes addressing the critique more difficult. Furthermore, the question of path dependency isn't developed along an academic benchmark but often recurs to the normative preferences of specific authors. For example, CDM has been criticized for technology lock-out of nuclear energy—by those that prefer it for being less invasive on the natural environment—and for technology lock-in in small hydropower, by those that prefer solar power. Nonetheless, this discussion points at the steady risk of technical system components being interpreted normatively and/or politically. In the CDM, technological questions have to be approved three times, making them the result of a wide consensus: first by the conventions under the UNFCCC, second by the Executive Board of the CDM, and third by the countries involved in a specific activity. There is no better process to scrutinize technology.

The third and last "batch" of critiques treated here addresses the economic logic of the CDM, more appropriately, the actual price levels of CERs (Rahman and Kirkman 2015; Newell et al. 2013). With low prices of CER, there is little incentive to invest in the CDM and its activities. This fact is aggravated by the timeline of the investment, since the investor and the project developer must inflow the whole investment sum at the beginning of the project but only receive the outflow of CERs with time, over the project cycle. The complete outflow will only be effective after many, up to 20, years—with the end of the predetermined cycle of a specific project. While this is common to all investments, falling unit prices and the expectation of falling prices diminish the expected economic benefit of the activity, for example, when activities are expected to over-deliver CERs, so that these CERs might become tradable. Additionally, the entity hosting activities depends on its operative revenues and carbon revenues. Because of the additionality criterion, without the carbon revenues, the operative revenues cannot cover economic costs. Low CER prices might drive the host to operate at a loss, leading to activities that become nonoperational or bankrupt.

The price configuration might indeed be a problem. However, the question should be posed, why the price configuration is such. Many would point at the lack of ambition by buyer countries that should have stepped-up their commitments, thus leading to higher demand. On the other side, these countries did not increase their commitments due to developing countries obstructing negotiations under the framework convention, namely, negotiations that could have led to clearer rules for the CDM addressing the questions mentioned above. Also, many developing nations refused committing to climate change-related goals which triggered developed Parties not to step up theirs. This had a negative effect on the CDM. Additionally, since 2012, there has been discussion about replacing the CDM with a "new market mechanism." The CDM is an artificial market as far as it was created by regulation. The ever-shifting discussion about its regulatory framework and the risk of it being replaced by other instruments could be argued to be the major driver of price decay (for an in-depth discussion, see Schneider 2017).

What can be learned after discussing these criticisms of the CDM? The mechanism strengthens the awareness and understanding about clean technologies, emission trading, and future action for climate change both in the private and public sector. It helps to attract financing for clean energy and energy efficiency activities. Where countries managed to set-up good economic conditions, a series of sustainable economic activities developed around CDM projects. It furthermore enables countries to gain first-hand experience and to enhance their local human capacity and institutions for managing and controlling greenhouse gas mitigation. It builds a significant carbon market infrastructure for project development, verifications, and finance services. And finally, it provides an opportunity for better understanding how to regulate and support carbon markets (UNFCCC 2012).

In addition, under the framework convention, many reforms have been undertaken: "The transparency of the framework has allowed identifying loopholes and spur reforms that have been ongoing since the inception of the CDM. Recent work led to the standardization of additionality demonstration and baseline setting as well as streamlining the procedures and giving more opportunities to underrepresented countries and sectors. On the other hand, more stringent baselines and performance benchmarks can help ensure net emissions reductions that could compensate for non-additional projects that manage to slip through validation. Some project types offer extremely high returns on investment which may encourage strategic behavior and rent seeking. This can be addressed through scrutinizing production technologies and introducing stringent benchmarks and/or crediting limits. The development of programs of activities as well as new sectoral crediting mechanisms that avoid project by project additionality demonstration may help achieve wider coverage. Positive lists and standardized baselines which are already being implemented within the CDM provide a good basis for further standardization, and hence scaling-up of the mechanism. Standardization also contributes to limiting the "judgment element" in project assessment (UNFCCC 2012)."

It is the task of the next subsection to make a case for carbon markets as complementing to (sustainable) economic development. The last section will, then, take these learnings from the CDM and incorporate them into the market mechanisms under the Paris Agreement.

## 2.3 *Carbon Markets for Sustainable Economic Development*

There are many ways of conceptualizing (sustainable) economic development and its interaction with carbon markets, as there are many ways of conceptualizing market instruments themselves. Sustainable economic development has different elements. First, there is the environmental element that addresses at least air, water, and soil as well as conservation; then, there are a wide array of social components such as health, welfare, or education; third, there is no development without macroeconomic components such as employment, growth, the quality and availability of energy, as well as imports and exports, for example; lastly, in order for development to be economically sustainable, there is a microeconomic component consisting of entrepreneurship and innovation, to mention some of its aspects.

It would be overbearing to expect a given activity under the premise of (sustainable) economic development to fulfill the totality of these components. It would be even wrong to expect an activity to address all these issues equally. But it can be expected for an activity to address more than one component and to balance several aspects of different elements not as trade-offs, but as complementary value drivers.

Carbon markets can yield to the expectation, not automatically, but depending on the design of policies and framework conditions. Let's go through the different components of (sustainable) economic development and asses how or under which conditions carbon markets can play the role as instigator, accelerator, trigger, driver, or catalyst—whatever metaphor suits best—in (sustainable) economic development.

Carbon markets naturally address the first element. By reducing emissions, carbon markets diminish the amount of greenhouse gases in the atmosphere. For the conservationist aim, the motives of the market agents are irrelevant. It is equally irrelevant where, why, and how the emission reduction occurs. From the conservationist point of view, it is only important to reduce greenhouse gas emissions.

There are, of course, different interpretations of the conservationist goal. The first is to accept as an emission reduction any unit of greenhouse gas that has not been released into the atmosphere. According to this version, all offsets by CDM count toward the conservationist goal. If greenhouse gas emissions occur in a given place and an emission reduction of the same magnitude occurs in a different place and these are used to offset the first, according to this first version, conservationist goals are fulfilled by the market instrument.

The second version of the conservationist goal is more ambitious claiming that market instruments have to provide for a net reduction of greenhouse gas emissions. This can happen either by discounting units credited by the markets or by setting emission reduction goals in both the emitting and the reducing countries. In the first variant, marketable projects would have to reduce greenhouse gas emissions by more than 1 tCO2e in order to credit and trade 1tCO2e. In the second variant, the unit of tCO2e remains undiscounted, but since both countries have to reduce emissions, the emitter will be able to use market credits only for those emissions it was not able to reduce instead of using them for all emissions, as in an offsetting system.

Carbon markets have to fulfill a goal, namely, to allow for price discrimination along the abatement cost. Since this goal is aligned with the Paris Agreement, it is difficult to make a case for the first version of the conservationist goal. Carbon markets are only able to play their role as an instrument in climate change-related action according to the Paris Agreement if they lead to a net decrease of greenhouse gas emissions. On the other hand, diluting the exchange unit of the market, tCO2e, is not a good idea. Any unit discount allows for the "politization of the unit itself"; with this "politization," the very base of the market can be called into question. This, in turn, diminishes the institutional stability of a market and discourages agents form operating in it. Furthermore, any discount in the unit leads to a price discount, which is one of the most important obstacles markets face already. On the other hand, setting greenhouse gas emission reduction goals for all participating entities is not only administratively easier than discounting units but also sends scarcity signal to markets. Markets work best in situations of scarcity; after all, it is their task to allocate scarce units based on productivity, and therefore, prices (for a more profound development of what the market is and how it deals with discounted units, refer to Kirzner 1963).

If markets operate clearing scarcity, they increase the productivity of all agents, supply and demand. It is because of this productivity effect that carbon markets address the third and fourth elements of (sustainable) economic development, macro- and microeconomics. Simply put, the entity that unfolds the activity related to carbon market is part of the economic components. But because of the additionality requirement, that entity cannot rely only on the prices of CERs but has to unfold an entrepreneurship of its own. Ideally, this entrepreneurship is not only directed toward carbon markets but especially aims at encompassing local, real markets. For example, the biogas activity will only have a small part of its revenues from selling CERs, the majoritarian share of revenues come from turning biogas into electricity and selling it to local electricity users. The same applies for activities increasing energy efficiency. By exchanging conventional lamps through LED lamps, the economic agent is benefitting from CERs but especially from those local customers buying the LED. A third example is the cooking stoves. Again, the CERs will only be a small part of the revenues, which are mainly made up of selling the stoves as well as appliances in order to use them.

Generally, carbon markets can serve as catalysts—or any other appropriate metaphor, see above—for the development of local entrepreneurship and economics. But carbon markets will never be the only part of entrepreneurship. Furthermore, whether they succeed in igniting entrepreneurial innovation depends, to a large extent, on the local framework conditions—not only in the economic but especially in the legal realm. Instead of seeing carbon market-related activities as stand alone, local governments could foment a network of different activities that can complement each other. Using the examples above, electricity generation from biogas, increasing energy efficiency, and substituting cooking stoves are activities that complement each other and benefit from network effects, if coordinated. They can, too, lead to as surge of entrepreneurship, once local infrastructure and revenue flows are established. With more electricity and more energy efficiency, local industrialization becomes achievable on scale. The revenues created by these activities plus the additional revenues from the trade in CERs can lead to the local accumulation and dispersion of capital, which, on its turn, leads to growing local markets. The aggregation of functioning entrepreneurship and local markets leads to a stable macroeconomic situation. Generally, this positive dynamics works well when the involved entities are privately owned. This diminishes the risk of market distortion and market power. In addition, the private entities involved in carbon markets and the general entrepreneurship described here should operate under full responsibility for their risks and full accountability for their success, i.e., without a state guarantee of any kind in case of failure and without a threat of expropriation or confiscation in case of success (Leeson and Boettke 2009).

Carbon markets cannot address any of these framework conditions on their own. They can only accelerate entrepreneurship and indirectly contribute toward positive economic development if these framework conditions are in place or guaranteed. While local governments can try to foment more marketable activities, for example, by forming clusters or providing for tax breaks, these instruments usually have little impact on markets and entrepreneurship. It is much more important to provide for good education and well-working framework conditions that set up instruments directed at "aiding" markets to establish themselves. On the contrary, it is under scarcity of means that market best work, provided that agents can unfold their activities freely.

Similarly, it is for the local level to create framework conditions for transforming the conservationist and economic entrepreneurial benefits of carbon markets into social benefits. As the anecdotal examples mentioned in the last subsection show, often, the social benefit created by carbon markets surpasses their economic effects. But, once again, carbon market activities only can allocate, i.e., create, benefits; it is a matter of local policy to distribute—i.e., create framework conditions for marketing—these benefits. Because of this characterization of markets, it would be detrimental to their effect making them credit "social benefit" or building "social benefits" into their unit of measurement, which is tCO2e.

To sum up the conclusions of this section, carbon markets surely fulfill the conservationist goals of (sustainable) economic development. But carbon markets are primarily catalysts or accelerators of (sustainable) economic development—expecting more from them would amount to overbear the economic logic of a market. It is a matter for local policy to develop favorable framework conditions for the establishment of markets, for the development of entrepreneurship, and for the distribution of social benefits. If it is the case, carbon markets surely work as catalysts, accelerating entrepreneurial and economic development as well as the allocation of "social benefits." Favorable framework conditions consist of a legal system that guarantees economic freedom as well as property that safeguards equality in the juridical procedures and that allows all interested agents participating in a market. They furthermore consist of low or no barriers to market entrance, of competition, and of good education. Especially, good framework conditions align risks and successes of market participants. But here, too, carbon markets can have a positive impact. If countries want to establish domestic or international carbon markets, they first need to reform their framework conditions, otherwise markets won't succeed.

## 3 What Follows Next?

In guise of a conclusion, this last section aggregates the learnings from the first and second sections into a list of desiderata for the further work under the Paris Agreement. This list aims at establishing carbon markets that serve as accelerators in (sustainable) economic development. This section is to be read as a summary of considerations that should inform the further work in implementing the Paris Agreement on international and domestic level.

As it stands at the beginning of the year 2018, the situation of these works looks like this:

The Paris Agreement establishes a fundamentally different framework from Kyoto. Rather than binding emission limits, which readily lend themselves to market approaches, the new climate regime requires all Parties to undertake nationally determined contributions of their own choosing. These contributions are not legally binding and come in many forms, ranging from absolute economy-wide targets to peaking years, carbon intensity reductions, and so on. A new transparency system will apply to all Parties but will be less prescriptive than the accounting that underpinned the Kyoto Protocol.[11] The future of the Kyoto flexibility mechanisms is also unclear, in particular whether the new mechanism will succeed the CDM and JI, or will sit alongside either of these. The Paris Agreement does not mention the CDM or JI, but notes that the new mechanism should draw on the experience gained from existing mechanisms. Similarly, it is unclear whether units generated under the Kyoto mechanisms will be eligible for compliance after 2020 and, if so, whether they will need to be converted to an alternative credit type to conform with credit issues under the new mechanism.

Over half of the Parties to the UNFCCC said that they plan to use or are considering market-based instruments to meet their national climate targets. But only half a dozen of them—Canada, Japan, New Zealand, South Korea, Switzerland, and possibly Norway—are likely buyers of international credits. The rest are developing countries hoping to be on the supply side. Major players—including the European Union and the United States—explicitly stated that they plan to meet their climate commitments without engaging in emission-reduction markets. But some experts think this might change as governments are faced with the economic realities of meeting their targets. But even countries that choose not to purchase emission reductions internationally may develop carbon markets at home in order to meet their targets, potentially creating significant new pockets of domestic demand. For example, China, the United States, and the European Union will face demand for credits but can probably meet that demand internally, as they have committed to do.

Taking into account the further work to be done, the following desiderata should be taken into account when developing guidance for 6(2–3) and rules, modalities, and procedures for 6(4–7):

Carbon markets should contribute to overall emission reduction.

- They must account for mitigation targets and their progression of all countries, i.e., all countries should have NDCs that can be compared.
- Carbon markets should be used by countries as accelerators in entrepreneurship, economic growth, and "social benefits" taking into account that the development of good framework conditions is a prerequisite for the successful functioning of markets; on the other hand, it makes no sense to expect from carbon markets more than markets can deliver, i.e., neither their units should be diluted nor should other consideration than the greenhouse gas abatement cost be factored in the prices; markets only work well if there is scarcity and the alignment of success and failure; additionality is a key element of carbon markets—strict additionality criteria make market environmentally feasible and prevent them from becoming instruments of blunt subsidization of an economy.

[11]All of the following are cooperative instruments enabling cooperating parties to increase their ambition, allowing for international transfer of units and allowing transferred ITMOs to adjust or to be counted toward an NDC: Linked ETS, Bilateral Offsetting, Government to Government transfers, Redd+, instruments under article 6. Subnational ETS and domestic policies are not cooperative approaches and do not allow for counting ITMOs toward national NDC.

## References

Bodansky, D. (2016). The legal character of the Paris Agreement. *Review of European, Comparative, and International Environmental Law, 25*(2), 142–150.

Boyd, R., Turner, J., & Ward, B. (2015). *Tracking intended nationally determined contributions: What are the implications for greenhouse gas emissions in 2030?* ESRC Centre for Climate Change Economics and Policy. Accessed August 31, 2017, from http://eprints.lse.ac.uk/64525/1/Boyd_Turner_and_Ward_policy_paper_October_2015.pdf

Castro, P. (2014). *Climate change mitigation in developing countries: A critical assessment of the clean development mechanism.* Cheltenham: Edward Elgar Publishing.

CI-Dev, Carbon Initiative for Development. (2016). *The impact of INDCs, NAMAs and LEDS on Ci-Dev operations and programs.* Washington, DC: World Bank Group.

Falkner, R. (2016). The Paris Agreement and the new logic of international climate politics. *International Affairs, 92*(5), 1107–1125.

Gillenwater, M. (2012). *What is additionality* (Discussion Paper). Greenhouse Gas Management Institute. Accessed August 31, 2017, from http://ghginstitute.org/2011/03/24/defining-additionality

Intergovernmental Panel on Climate Change. (2014). *Climate change 2014 – Impacts, adaptation and vulnerability: Regional aspects.* Cambridge: Cambridge University Press.

Kim, J. E., Popp, D., & Prag, A. (2013). The clean development mechanism and neglected environmental technologies. *Energy Policy, 55*, 165–179.

Kirzner, I. M. (1963). *Market theory and the price system.* New York: Van Nostrand.

Kolk, A., & Pinkse, J. (2004). Market strategies for climate change. *European Management Journal, 22*(3), 304–314.

Leeson, P. T., & Boettke, P. J. (2009). Two-tiered entrepreneurship and economic development. *International Review of Law and Economics, 29*(3), 252–259.

Levin, K., Rich, D., Bonduki, Y., Comstock, M., Tirpak, D., Mcgray, H., et al. (2015). *Designing and preparing intended nationally determined contributions (INDCs).* World Resources Institute and UNDP. Accessed August 31, 2017, from http://www.wri.org/publication/designing-andpreparing-indcs

MacKenzie, D. (2009). Making things the same: Gases, emission rights and the politics of carbon markets. *Accounting, Organizations and Society, 34*(3), 440–455.

Marcu, A. (2016a). *International cooperation under Article 6 of the Paris Agreement: Reflections before SB 44* (Background Paper). International Centre for Trade and Sustainable Development (ICTSD).

Marcu, A. (2016b). *Carbon market provisions in the Paris Agreement (Article 6)* (Working Paper No. 128). Centre for European Policy Studies.

Newell, R. G., Pizer, W. A., & Raimi, D. (2013). Carbon markets 15 years after Kyoto: Lessons learned, new challenges. *The Journal of Economic Perspectives, 27*(1), 123–146.

Northrop, E., Biru, H., Lima, S., Bouye, M., & Song, R. (2016). *Examining the alignment between the Intended Nationally Determined Contributions and Sustainable Development Goals* (Working Paper No. 43). World Resources Institute.

Platonova-Oquab, A., Spors, F., Gadde, H., Godin, J., Oppermann, K., & Bosi, M. (2012). *CDM reform: Improving the efficiency and outreach of the clean development mechanism through standardization.* Washington, DC: World Bank.

Rahman, S. M., & Kirkman, G. A. (2015). Costs of certified emission reductions under the clean development mechanism of the Kyoto Protocol. *Energy Economics, 47*, 129–141.

Reinert, K. A., Reinert, O. T., & Debebe, G. (2016). The new OECD Guidelines for Multinational Enterprises: Better but not enough. *Development in Practice, 26*(6), 816–823.

Schade, J., & Obergassel, W. (2014). Human rights and the clean development mechanism. *Cambridge Review of International Affairs, 27*(4), 717–735.

Schneider, H. (2017). After the Paris Agreement: A(ny) future for carbon markets? In S. Ismail & S. Ahmed (Eds.), *Energy, economy and sustainable development* (pp. 86–108). New Delhi: Bloomsbury.

Spalding-Fecher, R., Achanta, A. N., Erickson, P., Haites, E., Lazarus, M., Pahuja, N., et al. (2012). *Assessing the impact of the clean development mechanism* (Report commissioned by the High-Level Panel on the CDM Policy Dialogue). Accessed August 31, 2017, from http://www.cdmpolicydialogue.org/research/1030_impact.pdf

Sterk, W., Bolscher, H., Van Der Laan, J., Hoogzaad, J., & Sijm, J. (2015). Developing a sectoral new market mechanism: Insights from theoretical analysis and country showcases. *Climate Policy, 15*(4), 417–437.

Stern, N. H. (2006). The economics of climate change. *World Economics, 7*(2), 1–10.

Taylor, P. (2008). *An ecological approach to international law: Responding to the challenges of climate change*. Abingdon-on-Thames: Routledge.

UNFCCC. (2012). *Benefits of the clean development mechanism* (UNFCCC Report).

UNFCCC. (2015). *Adoption of the Paris Agreement* (Report No. FCCC/CP/2015/L.9/Rev.1). Accessed August 31, 2017, from http://unfccc.int/resource/docs/2015/cop21/eng/l09r01.pdf

UNFCCC. (2017). *CDM insights*. Accessed August 31, 2017, from https://cdm.unfccc.int/Statistics/Public/index.html

Yamin, F. (2012). *Climate change and carbon markets: A handbook of emissions reduction mechanisms*. Abingdon-on-Thames: Routledge.

Zhang, J., & Wang, C. (2011). Co-benefits and additionality of the clean development mechanism: An empirical analysis. *Journal of Environmental Economics and Management, 62*(2), 140–154.

# Part II
# The Role of Global Players

# The European Union as a Leading Environmental Player? A Critical Analysis on the Policy and Commitments Towards Global Development and Climate Change

**Ana Isabel Xavier**

## Contents

1 Introduction .......... 136
2 The European Union and the Environment: A Historical Background .......... 140
3 The EU as an Environmental Global Player? .......... 145
4 Conclusion .......... 149
References .......... 150

**Abstract** As soon as the environmental issues started to be recognised as a transnational cross- border and cross-cutting long-lasting challenge, the European Union (EU) has sought to become a relevant player in the regulation and fight against climate change. The 1986 Single European Act, the Europe 2020 Strategy, the Framework for Climate and Energy 2030 or the Kyoto Protocol (1997) and Paris Agreement (2015) clearly illustrates the policies and commitments that the EU is being pursuing towards a political climate model and a global environmental governance. But is the EU's authority in environmental issues externally recognised in the diplomatic arena? Moreover, has the environmental *acquis* led to more cohesion among Member States and a harmonisation of the EU policies for more autonomy? Caporaso and Jupille's four components to address an actor's capacity in global politics—recognition, authority, autonomy and cohesion—pave the way to the theoretical analysis and to identify some remaining shortfalls to succeed in the "green" roadmap targeted to the next decade.

**Keywords** European Union · Environment · Development · Climate change · Global player · Actorness

A. I. Xavier (✉)
CEI-IUL (ISCTE-IUL), Lisbon, Portugal
e-mail: ana.isabel.xavier@iscte-iul.pt; https://autonoma.pt/docentes/ana-isabel-xavier/

 
T. Sequeira, L. Reis (eds.), *Climate Change and Global Development*, Contributions to Economics, https://doi.org/10.1007/978-3-030-02662-2_7

## 1 Introduction

> Climate change is the poster child of global diplomacy today. In fact, it can easily be regarded as the most complex global policy problem.
>
> (Khan 2016: 14)

Global development and climate change are indeed one of the most challenging and pressing issues that international institutions, national governments and local communities face nowadays as a global phenomenon with strong transnational and cross border impact. As the world population continues to expand[1] and water (EUI 2017) and energy scarcities grow, the global environmental challenges to human development are becoming more pressing every day. Moreover,

> [c]limate-related risks, combined with socio-economic pressures generated by rapid industrialisation, urbanisation and shifting global economic power, have added another layer of uncertainty to national policy planning. The anticipated bottlenecks and constraints (in energy, water and other critical natural resources and in infrastructures)—together with these socio-economic shifts—will result in new instabilities that will be difficult to manage. (ESPAS 2011: 86)

In fact, it's quite unanimous that, nowadays, environmental problems are so "palpably linked" (Ferguson and Mansbach 2012: 96) that we cannot divide or label what is economic, social, military or political if we want to manage, regulate and address those phenomena. In other words, economic growth, social development and welfare cannot be dissociated from climate change policies and effects, both at the transnational, national and local level. Therefore, the awareness of both international organisations and national states paved the way as

> [b]y the early twenty-first century, environmental issues had been high on the international agenda for a whole generation of political leaders, government officials, scientists, industrialists, and concerned citizens. Since the 1960s, awareness of the risks and implications of a wide range of international environmental problems has increased greatly, and justifiably so. (Greene 2001: 387).

Within this framework, the EU has also sought to become a relevant player and even a prominent leader vis-a-vis the mitigation, regulation, fight and governance of environmental issues (Schreurs and Tiberghien 2007). Historically acknowledged as a project of peace for a Europe devastated by the Second World War (1939–1945), the EU's genesis was truly inspired by the ideals of peace, human rights, democracy and good governance, both within and beyond its borders. Although under a new and atypical formula, and despite not pursuing a conventional foreign policy, the EU has a strong international presence and is one of the main players on the contemporary international arena. The enlarged European Union covers a market of over 450 million people; the Euro is (despite the sovereign debts crisis of the last few

[1]The world population is expected to reach 8.5 billion by 2030, 9.7 billion in 2050 and 11.2 billion in 2100. By 2050, China, India, Indonesia, Nigeria, Pakistan and the USA are expected to exceed 300 million people, while those aged 60 or above will be more than double by 2050 and more than triple by 2100 (UNDESA 2015).

years in the southern countries) a strong and stable currency in international financial markets and is a major player in Africa, Latin America and competitive emerging economies like India and China.

In what concerns the aspirations to shape and lead the governance of Climate Change, Articles 11[2] and 191 to 193[3] of the Treaty on the Functioning of the European Union (TFEU) represent the legal basis, entitling the EU to act in all

---

[2]Environmental protection requirements must be integrated into the definition and implementation of the Union's policies and activities, in particular with a view to promoting sustainable development.

[3]*Article 191*

1. Union policy on the environment shall contribute to pursuit of the following objectives: preserving, protecting and improving the quality of the environment, protecting human health, prudent and rational utilisation of natural resources, promoting measures at international level to deal with regional or worldwide environmental problems, and in particular combating climate change.
2. Union policy on the environment shall aim at a high level of protection taking into account the diversity of situations in the various regions of the Union. It shall be based on the precautionary principle and on the principles that preventive action should be taken, that environmental damage should as a priority be rectified at source and that the polluter should pay.

In this context, harmonisation measures answering environmental protection requirements shall include, where appropriate, a safeguard clause allowing Member States to take provisional measures, for non-economic environmental reasons, subject to a procedure of inspection by the Union.

3. In preparing its policy on the environment, the Union shall take account of: available scientific and technical data, environmental conditions in the various regions of the Union, the potential benefits and costs of action or lack of action, the economic and social development of the Union as a whole and the balanced development of its regions.
4. Within their respective spheres of competence, the Union and the Member States shall cooperate with third countries and with the competent international organisations. The arrangements for Union cooperation may be the subject of agreements between the Union and the third parties concerned.

The previous subparagraph shall be without prejudice to Member States' competence to negotiate in international bodies and to conclude international agreements.

*Article 192*

1. The European Parliament and the Council, acting in accordance with the ordinary legislative procedure and after consulting the Economic and Social Committee and the Committee of the Regions, shall decide what action is to be taken by the Union in order to achieve the objectives referred to in Article 191.
2. By way of derogation from the decision-making procedure provided for in paragraph 1 and without prejudice to Article 114, the Council acting unanimously in accordance with a special legislative procedure and after consulting the European Parliament, the Economic and Social Committee and the Committee of the Regions, shall adopt: (a) provisions primarily of a fiscal nature; (b) measures affecting: town and country planning, quantitative management of water resources or affecting, directly or indirectly, the availability of those resources, land use, with the exception of waste management; (c) measures significantly affecting a Member State's choice between different energy sources and the general structure of its energy supply.

The Council, acting unanimously on a proposal from the Commission and after consulting the European Parliament, the Economic and Social Committee and the Committee of the Regions,

areas of environment policy, only limited by the principle of subsidiarity[4] and the requirement for unanimity in the Council in some specific fields.[5] In addition, the EU environment policy has been laying on the principles of precaution,[6] prevention and rectifying pollution at source and on the polluter pay[7] principles.

Furthermore, since the 1998 Cardiff process, integrating environmental concerns into other EU policy areas has become an important pillar in European politics. In fact, the Cardiff European Council has raised the political profile of integration of all environmental considerations into their respective activities, putting Article 6[8] of the EC Treaty into practice.

In addition, EU's decision to move forward with the ratification of the 1998 Kyoto Protocol, despite the US withdrawal, enhanced the key role and commitment towards "[e]nvironmental security discussions [that] are now changing into climate security discussions as the focus shifts to global warming and the effects it may have

---

may make the ordinary legislative procedure applicable to the matters referred to in the first subparagraph.

3. General action programmes setting out priority objectives to be attained shall be adopted by the European Parliament and the Council, acting in accordance with the ordinary legislative procedure and after consulting the Economic and Social Committee and the Committee of the Regions.

The measures necessary for the implementation of these programmes shall be adopted under the terms of paragraph 1 or 2, as the case may be.

4. Without prejudice to certain measures adopted by the Union, the Member States shall finance and implement the environment policy.
5. Without prejudice to the principle that the polluter should pay, if a measure based on the provisions of paragraph 1 involves costs deemed disproportionate for the public authorities of a Member State, such measure shall lay down appropriate provisions in the form of: temporary derogations, and/or financial support from the Cohesion Fund set up pursuant to Article 177.

*Article 193*

The protective measures adopted pursuant to Article 192 shall not prevent any Member State from maintaining or introducing more stringent protective measures. Such measures must be compatible with the Treaties. They shall be notified to the Commission.

[4]The principle of subsidiarity is defined in Article 5 of the Treaty on European Union whereby the EU does not take action (except in the areas that fall within its exclusive competence), unless it is more effective than action taken at national, regional or local level. It is closely bound up with the principle of proportionality, which requires that any action by the EU should not go beyond what is necessary to achieve the objectives of the Treaties.

[5]We are referring to fiscal matters, town and country planning, land use, quantitative water resource management, choice of energy sources and structure of energy supply.

[6]A risk management tool that may be invoked when there is scientific uncertainty about a suspected risk to human health or to the environment emanating from a certain action or policy.

[7]Implemented by the Environmental Liability Directive, aims to prevent or otherwise remedy environmental damage to protected species or to natural habitats, water and soil.

[8]The Union shall have competence to carry out actions to support, coordinate or supplement the actions of the Member States. The areas of such action shall, at European level, be: (a) protection and improvement of human health; (b) industry; (c) culture; (d) tourism; (e) education, vocational training, youth and sport; (f) civil protection; and (g) administrative cooperation.

**Table 1** Main features of the Kyoto Protocol and the Paris Agreement (Erbach 2016)

| | Kyoto Protocol (1998) | Paris Agreement (2015) |
|---|---|---|
| Scope | Mitigation | Mitigation, adaptation and finance |
| Duration | Phase 1: 2008–2012<br>Phase 2: 2013–2020 | Indefinite, with revision of NDCs every 5 years |
| Application | Only developed country parties have emission reduction targets | All parties must make (nationally determined) mitigation contributions |
| Coverage of global emissions | 14% in phase 2 | 99% of emissions are covered by already submitted INDCs |
| Mechanism | Emissions targets for developed countries, market-based mechanisms | Nationally determined contributions, voluntary cooperation between Parties |
| Compliance | Enforcement through suspension from emissions trading and additional emissions reductions in second commitment period | Expert-based and facilitative mechanism that is transparent, non-adversarial and non-punitive |
| Transparency | Different reporting requirements for developed and developing countries | Similar reporting requirements for all parties |

in coming decades" (Dalby 2013: 313). The 2015 Paris Agreement also illustrates the fact that the EU has been at the forefront of international efforts towards a global climate deal, both in nourishing a broad coalition of developed and developing countries in favour of high ambition that shaped the successful outcome of the Paris conference and in taking concrete steps to implement its target to reduce emissions by at least 40% by 2030.

Table 1 aims to illustrate those efforts and ambitions, showing how Kyoto Protocol features evolved to the Paris Agreement in 2015 in terms of scope, duration, application, coverage of global emissions, mechanism, compliance and transparency.

In fact, within the Paris Agreement, governments committed to a long-term goal of keeping the increase in global average temperature to well below 2°C above pre-industrial levels; to aim to limit the increase to 1.5°C, since this would significantly reduce risks and the impacts of climate change; on the need for global emissions to peak as soon as possible, recognising that this will take longer for developing countries; and to undertake rapid reductions thereafter in accordance with the best available science.

Within the framework of the United Nations, it's also worth to note the international commitment of both European Institutions and Members States towards the Millennium Development Goals (2000–2015) and the Sustainable Development Goals (2015–2030). But if in the first 15 years' roadmap, there was only one goal clearly devoted to this issue—ensure environmental sustainability[9]—in the second

[9]GOAL 7—Target 7.A: Integrate the principles of sustainable development into country policies and programmes and reverse the loss of environmental resources; Target 7.B: Reduce biodiversity loss, achieving, by 2010, a significant reduction in the rate of loss; Target 7.C: Halve, by 2015, the

agenda, environment appears as a cross-cutting topic when targeting clean water and sanitation, clean energy, sustainable cities and communities, responsible consumption and protecting the planet, life below water or life on land.[10] More recently, environmental policy integration has made significant progress in the transition to a low-carbon economy by 2050 in the sectors of power generation, industry, transport, buildings and construction as well as agriculture.

Bearing all these in mind, this chapter's aim is twofold: first, to briefly analyse the historical background of the EU's environmental policies (since the Single European Act to the Europe 2020 Strategy and the Framework for Climate and Energy 2030) and external commitments (Kyoto Protocol and Paris Agreement); second, to critically assess how those policies and commitments can actually embody the European Union as an environmental player in the global politics. We will conclude that, notwithstanding the EU has one of the world's highest environmental standards with significant impact in a greener economy, a more sustainable nature and a healthier quality of life of EU citizens, there are still some shortfalls to accomplish the targets' roadmap in the next decade. Jean Claude Juncker's scenarios on the future of Europe will also be helpful to trace the way Member States want to pursue in shaping EU's leadership and governance in environmental issues.

## 2 The European Union and the Environment: A Historical Background

Article 2 of the Treaty of Rome (1957) already highlighted the promotion of a harmonious development of economic activities as a key objective of the community's role, meaning that the economic expansion cannot be an end per se. But it is only in the 1970s that the European Economic Community (EEC) has been truly engaged in environmental issues.

In fact, in October 1972, a conference of Heads of State and Government held in Paris, in the aftermath of the first UN conference on the environment, invited the community institutions to establish, before July 31, 1973, an action programme towards an environment policy. Also, since 1973, the Commission has issued

---

proportion of the population without sustainable access to safe drinking water and basic sanitation; and Target 7.D: Achieve, by 2020, a significant improvement in the lives of at least 100 million slum dwellers.

[10]In terms of specific goals, the sustainable development agenda includes: ensure availability and sustainable management of water and sanitation for all (Goal 6); ensure access to affordable, reliable, sustainable and modern energy for all (Goal 7); ensure sustainable consumption and production patterns (Goal 12); take urgent action to combat climate change and its impacts (Goal 13); conserve and sustainably use the oceans, seas and marine resources for sustainable development (Goal 14); protect, restore and promote sustainable use of terrestrial ecosystems, sustainably manage forests, combat desertification and halt and reverse land degradation and halt biodiversity loss (Goal 15).

multiannual Environment Action Programmes (EAPs) setting out forthcoming legislative proposals and goals for EU environment policy. These multiannual environmental action programmes aim to set the framework for future action in all areas of environment policy, strategies and international environmental negotiations. Indeed, the current EU dual responsibility of the EU institutions and national governments up to 2020 is based on the 7th Environment Action Programme (2011–2020) and is entitled "Living well, within the limits of our planet" and sets the following ambitions for the next 30 years:

> In 2050, we live well, within the planet's ecological limits. Our prosperity and healthy environment stem from an innovative, circular economy where nothing is wasted and where natural resources are managed sustainably, and biodiversity is protected, valued and restored in ways that enhance our society's resilience. Our low-carbon growth has long been decoupled from resource use, setting the pace for a safe and sustainable global society.[11]

Later, in 1986, the Single European Act added a new title—Environment—providing the first legal basis of the community's policy on preserving the quality of the environment, protecting human health and ensuring rational use of natural resources. The Single European Act-added value in environment issues is twofold: first, the community's competences were explicitly extended to environmental issues calling for action to be based on the principles that preventive actions should be taken, that environmental damage should be rectified at source and that the polluter should pay; second, it specifies that the community can only intervene in environmental matters when this action can be attained better at community level than at the level of the individual Member States (subsidiarity principle). For that purpose, the Single European Act adds the Articles130R, 130S and 130T of the EEC Treaty which allow the community to preserve, protect and improve the quality of the environment, to contribute towards protecting human health and to ensure a prudent and rational utilisation of natural resources.

From 1990 to 2000, two major milestones should be emphasised as concrete steps of EU's role in environmental issues. First, the adoption of the 1992 Framework Convention on Climate Change (UNFCCC) that sets an overall framework for intergovernmental efforts to tackle the challenges posed by climate change, recognizing that the climate system is a shared resource whose stability can be affected by industrial and other emissions of carbon dioxide and other greenhouse gases; second, the ratification of the Kyoto Protocol,[12] an international agreement under the

---

[11]Please see http://ec.europa.eu/environment/action-programme/ and http://eur-lex.europa.eu/legal-content/EN/TXT/PDF/?uri=CELEX:32013D1386&from=EN. The programme sets out nine priority objectives—the protection of nature; stronger ecological resilience; sustainable, resource-efficient and low-carbon growth; and the fight against environment-related threats to health—and stresses the need for better implementation of EU environment law, state-of-the-art science, investment, and integration of environmental aspects into other policies.

[12]Despite President George W. Bush intention in 2001 to withdraw the USA from the Protocol, the European Council formally agreed to the Kyoto Protocol on 25 April 2002, and the European Commission, represented by Margot Wallstrom, jointly presented their instruments of ratification to the United Nations on 31 May 2002.

UNFCCC which aims to reduce emissions of greenhouse gases in an approximate percentage of at least 5% (compared to emissions to 1990) from 2008 to 2012, establishing binding reduction targets for industrialised countries.

In addition, several European countries established voluntary domestic emission reduction targets which ultimately led to an overall 8% reduction target adopted by the EU-15, 3% higher than that of other developed countries average. In 2007, the European Commission confirmed Europe's commitment in cutting its CO2 emissions by 20 percent of 1990 level by 2020, increasing the targets to 30 percent.

In turn, the 1992 Treaty of Maastricht succeeded in making the environment an official EU policy area, introduced the co-decision procedure and made qualified majority voting in the Council the general rule. In detail, Article 2 of the 1992 Maastricht Treaty recalls that one of the EU's mission is to promote a harmonious and balanced development of economic activities, sustainable and noninflationary growth which respects the environment. Also, Article 130r assumes that environmental protection requirements shall be integrated into the definition and implementation of other community policies, giving the legal basis to the EU competence to intervene in all areas of environmental policy.

Consequently, the Maastricht Treaty

> went a step further making the environment an explicit policy responsibility of the Community, giving the Commission greater powers to represent Member States in international organizations and with third parties, and calling upon it to promote measures to deal with regional and worldwide environmental problems. While the subsidiarity principle assures that many environmental decisions remain at the local and national levels, there has been a steady strengthening of the Community's powers with time (Schreurs and Tiberghien 2007: 27)

In 1997, the Treaty of Amsterdam enhanced the duty to integrate environmental protection into all EU sectoral policies and accepted the possibility of the EU to conclude international agreements based on its legal personality. The Treaty of Amsterdam also highlights the importance of promoting a harmonious, balanced and sustainable development of economic activities, sustainable and noninflationary growth aiming at a high level of protection and improvement of the quality of the environment (Article 2). It's Article 174 that clearly elaborates on the community policy in the field of the environment:

> 1. Community policy on the environment shall contribute to pursuit of the following objectives: preserving, protecting and improving the quality of the environment; protecting human health; prudent and rational utilisation of natural resources; and promoting measures at international level to deal with regional or worldwide environmental problems.
>
> 2. Community policy on the environment shall aim at a high level of protection taking into account the diversity of situations in the various regions of the community. It shall be based on the precautionary principle and on the principles that preventive action should be taken, that environmental damage should as a priority be rectified at source and that the polluter should pay.In this context, harmonisation measures answering environmental protection requirements shall include, where appropriate, a safeguard clause allowing Member States to take provisional measures, for noneconomic environmental reasons, subject to a community inspection procedure.
>
> 3. In preparing its policy on the environment, the community shall take account of available scientific and technical data, environmental conditions in the various regions of the

> community, the potential benefits and costs of action or lack of action, the economic and social development of the community as a whole and the balanced development of its regions.
>
> 4. Within their respective spheres of competence, the community and the Member States shall cooperate with third countries and with the competent international organisations. The arrangements for community cooperation may be the subject of agreements between the community and the third parties concerned, which shall be negotiated and concluded in accordance with Article 300.

However, the scope of those competences is limited by the principle of subsidiarity (as mentioned before, it restricts EU action to areas where it can be more effective than action at national or regional level) and the rule of unanimity in the Council on issues such as fiscal, land use, water resources management, energy source choices and the structure of energy supply. In fact, Article 176 clearly says that

> [t]he protective measures adopted pursuant to Article 175 shall not prevent any Member State from maintaining or introducing more stringent protective measures. Such measures must be compatible with this Treaty. They shall be notified to the Commission.

In turn, the Lisbon Strategy (2000) acknowledges the importance of promoting growth and employment through an increasing EU competitiveness. In addition, it sets the way to the 2001 EU Sustainable Development Strategy (ESD) aimed to

> identify and develop actions to enable the EU to achieve a continuous long-term improvement of quality of life through the creation of sustainable communities able to manage and use resources efficiently, able to tap the ecological and social innovation potential of the economy and in the end able to ensure prosperity, environmental protection and social cohesion[13]

In October 2005, it was launched the second European Climate Change Programme (ECCPII) showing the continuous commitment to climate change leadership.

In what concerns the Treaty of Lisbon, signed on 13 December 2007 and entered into force on 1 December 2009, combating climate change has become a specific goal, as well as sustainable development in relations with third countries. In addition, the Lisbon Treaty gives legally binding force to the Charter of Fundamental Rights of the European Union that recognises rights, freedoms and principles also in the environmental field[14] that apply to the EU institutions and Member States when they implement EU law. But there are more innovative aspects of this treaty with significant impact in the environmental field: the recognition of the EU's legal personality, entitled to conclude international agreements and become a member of international organisations; the acknowledgement of sustainable development as

---

[13]Please see http://ec.europa.eu/environment/sustainable-development/strategy/index_en.htm (last accessed on 14 September 2017)

[14]Article 37 of the Charter of Fundamental Rights of the European Union states that: «A high level of environmental protection and the improvement of the quality of the environment must be integrated into the policies of the Union and ensured in accordance with the principle of sustainable development».

one of the specific policy goals of the EU in its external relations (paragraph 5 of Article 3); the integration of environmental protection requirements in the definition and implementation of other EU policies and activities framed with regard for the need to preserve and improve the environment (Article 6); the clarification that environment and energy competences are shared responsibilities (Articles 2 to 5); and the reference to climate change as a worldwide environmental problem requiring the adoption of international measures, in parallel to internal measures.

In January 2010, the Europe 2020 strategy presents three matching pillars: smart growth (aimed to promote knowledge, innovation, education and the digital society), sustainable growth (making our productive apparatus more resource efficient while strengthening our competitiveness) and inclusive growth (increased labour market participation, skill acquisition and poverty reduction). The general framework is set by the EU institutions, but it's up to the Member States to translate these parameters into national targets: 75% of the population between the ages of 20 and 64 should be employed; 3% of EU GDP should be invested in research and development; the 20/20/20[15] climate/energy targets must be met; the drop-out rate must be less than 10% and at least 40% of the younger generation must have a higher education diploma; and 20 million people should no longer be at risk of poverty. This strategy must be read along with the 2015 Paris Agreement that sets out a global action plan to put the world on track to avoid dangerous climate change by limiting global warming below 2°C.

Bearing all these in mind, it is not surprising that

> [e]nvironmental policy is typically identified as a success story; actors within the EU took the creation of the single market as an opportunity to introduce an impressive array of environmental regulation and to place the Union at the forefront of international efforts to combat global environmental problems. (Burns and Carter 2012: 511)

In brief (see Table 2),

But what were the main drivers of the transformation of the European environmental policy? Burns and Carter (2012: 514–515) highlight four main aspects: (1) the transnational and cross sectorial nature of the environment was easily legitimised as a policymaking area; (2) the economic rationale first shaped regulation and harmonisation to prevent unfair competition; (3) the role of the EU institutions emerged the environment both as a "potential legitimating cause for the European project and a source for growth and jobs"; and (4) the pioneering role of Germany, Denmark, the Netherlands and the Nordic countries in exporting their industry and ecology "in promoting progressive environmental policy at the European level".

Notwithstanding these four important drivers, is the EU's authority in environmental issues externally recognised in the diplomatic arena? Moreover, has the environmental acquis led to more cohesion among Member States and a harmonisation of the EU policies for more autonomy? Caporaso and Jupille's four components to address an actor's capacity in global politics—recognition, authority,

[15]20% cut in greenhouse gas emissions (from 1990 levels); 20% of EU energy from renewables. 20% improvement in energy efficiency.

**Table 2** Chronological table of the EU's environmental framework (developed by the author)

| | 1970s | 1980s | 1990s | 2000s | 2010s |
|---|---|---|---|---|---|
| Treaties | – | Single European Act (1986)—Articles 130 R, S, T | Maastricht TFUE (1992)—Articles 11 and 191–193 | Lisbon Treaty (2007): Articles 2–5 | – |
| | | | Amsterdam Treaty (1997) (Articles 2, 174, 176) | | |
| European Council | – | – | Cardiff (1998) | – | – |
| Strategies and programmes | Multiannual environment action programmes (1973) | – | – | Lisbon Strategy (2000) | Europe 2020 (2010) |
| | | | | EU Sustainable Development Strategy (2001) | 7th Environment Action Programme (2011–2020) |
| | | | | Second European Climate change programme (2005) | |
| United Nations Framework | First UN Conference on the environment (1972) | – | 1992 Framework Convention on Climate Change | Millennium Development Goals (2000–2015) | Sustainable development Goals (2015–2030) |
| | | | Kyoto Protocol (1998) | | Paris Agreement (2015) |

autonomy and cohesion—pave the way to help us identify some remaining shortfalls for the EU to succeed in the "green" roadmap targeted to the next decade.

## 3 The EU as an Environmental Global Player?

Although under a *sui generis* umbrella, and despite not having a conventional foreign policy, the EU is currently recognised as one of the main players on the international arena and even a global player (Bretherton and Vogler 1999).

The EU seems able to act as a strategic player worldwide and as a community based on norms and values of stability and cooperation, within the EU, with its neighbourhood and with its strategic partners worldwide. We should question however if this is enough to project the EU's current external role towards a strategic power with effective presence as well as capacity to meet the natural and legitimate expectations on issues such as environmental governance.

In discussing the EU's activity in global politics, Caporaso and Jupille (1998: 214–220) suggest four requirements for the European Union to take over as a true global player: recognition ("acceptance of and interaction with the entity by others"); authority ("EU's legal competence in a given subject matter"); autonomy ("reflects the institutional distinctiveness and independence of an actor from other actors"); and cohesion ("degree to which an entity can formulate and articulate internally consistent policy preferences").

This chapter will now recall this theoretical framework and question (a) if the environmental *acquis* led to more cohesion among Member States and a harmonisation of the EU policies for more autonomy and (b) if the EU's authority in environmental issues is externally recognised in the diplomatic arena.

In what concerns the environmental issues, the EU has been playing an important role in international environmental negotiations, in a global, regional and subregional scale. In fact,

> (...) the EU has since the 1990s emerged as a key player in the field of international environmental politics: it is the signatory to more than forty multilateral environmental agreements and has sought to cast itself as a climate change leader. (Burns and Carter 2012: 519).

Just to mention some agreements on nature protection and biodiversity, climate change and transboundary air or water pollution where the EU's stance was crucial: in the 10th Conference of the Parties to the Convention on Biological Diversity held in Nagoya (Japan) in 2010, the EU contributed to achieve an agreement on a global strategy to halt the loss of biodiversity by 2020; the negotiation of the global Sustainable Development Goals (SDGs) in 2012; the United Nations Framework Convention on Climate Change (UNFCCC) on setting international climate standards; and the Convention on International Trade in Endangered Species (CITES) to pursue its fight against wildlife crime at the international level.

On the overall, the EU has one of the world's highest environmental standards in a greener economy, protecting the nature and safeguarding the health and quality of life of people living in the EU. Moreover, the EU has been striking to balance a more competitive economy that creates jobs and stimulates investment with a sustainable environmental framework focused primarily in improving people's quality of life. Though respecting the subsidiarity principle, the regulatory role still relies on the EU institutions with more than 500 directives, regulations and decisions being transposed or poured into national legislation in the last few years. But, at the same time, "[t]he failure to transpose and implement legislation properly has been a major concern in EU environmental policy since the 1980s" (Burns and Carter 2012: 516).

In fact, one key feature of authority surely relies on the effectiveness of the national, regional and local monitoring at the implementation level. And that is where cohesion gets tricky as there are still disparities in the level of implementation among Member States, although in accordance with the sayings of a top official of Directorate- General Environment, the environment is a great unifying issue for EU integration (*issue of preference*). Gupta and Grubb (2000) clearly identify the gaps: the complex multidimensional nature of the climate change that enfolds three key dimensions—structural, instrumental and directional; the still early stage of the development of effective and efficient instruments; the wide variety of complexity of the negotiation process; and the lack of effective international or even global governance and leadership. In addition, Vogler and Stephan (2007) highlight the need to make internal policy coherence a greater priority when faced with persistent diplomatic opposition from other coalitions (meaning China and the USA).

To overcome those kinds of structural problems, and despite its nonbinding nature, the European Parliament and the Council adopted in 2001 minimum standards for environmental inspections as well as effective, proportionate and dissuasive criminal sanctions for the most serious environmental offences. In addition, it's worth to mention other platforms that were shaped to give cohesion to the EU environmental policy: (1) the formation in 1990 of the European Environment Agency (EEA), based in Copenhagen and also open to non-EU members, with the mission to support the development, implementation and evaluation of environment policy; (2) the coordination of the European Earth Observation Programme (Copernicus), aiming to help policymakers take informed decisions and develop environmental legislation and policies; (3) the creation of the European Union Network for the Implementation and Enforcement of Environmental Law (IMPEL) providing a platform for policymakers, environmental inspectors and enforcement officers to exchange ideas and best practice; and (4) or the launching in May 2016 of the Environmental Implementation Review along with a regulatory fitness and performance programme, aiming to reach full implementation, monitoring and reporting obligations of EU environmental legislation.

In fact, these initiatives illustrate that the EU aims to be an international agenda settler and lead by example, driven by the pioneering role of the European Parliament, the European Commission and Member States. According to Schreurs and Tiberghien (2007), the European institutions have been striving to push forward an environmental agenda, namely, using the climate policy as a means to push EU integration forward and empower the European Commission with new regulatory tools; building an European foreign identity, especially in discrepancy with the US agenda; and sharing with the public opinion relevant outcomes and key environmental data such as the European Pollutant Release and Transfer Register (E-PRTR).

Therefore, authors such as Bäckstrand and Elgström don't hesitate to recognise the EU as a "leadiator" (2013) meaning a *leader-cum-mediato* between developing countries and major environmental players, such as China and the USA, to bridge legally binding agreements to put in practice the best environmental standards worldwide. The extension of the Kyoto Protocol towards the Paris Agreement as illustrated in Table 1 (page 6) clearly shows how the EU struggled for a binding framework for global actions to address climate change in the period after 2020.

But what about the future? How can we foresee the EU as an environmental global player? For sure,

> [f]ighting climate change successfully will certainly require sensible government policies to leave the economic playing field between clean and dirty energy, such as putting a price on carbon dioxide emissions. But it will also require policies that encourage investment in new clean energy, with even a level playing field may not generate in its own. (Sivaram and Norris 2016)

If we take in consideration the White Paper on the future of Europe (European Commission 2017), there are five possible scenarios of the potential future state of the union, depending on the choices Europe will make: (1) Carrying on (the EU27 focuses on delivering its positive reform agenda); (2) Nothing but the Single Market (the EU27 is gradually recentred on the single market); (3) Those Who Want More Do More (the EU27 allows willing Member States to do more together in specific areas); (4) Doing Less More Efficiently (the EU27 focuses on delivering more and faster in selected policy areas while doing less elsewhere); and (5) Doing Much More Together (Member States decide to do much more together across all policy areas).

Bearing those scenarios in mind and linking with the environmental issue, we believe that the near future can balance two scenarios. The first could be "Doing less more efficiently", enhancing the cohesion and consensus among the Member States in specific sectors but without major changes or developments in the existing *acquis*. That means that the EU would neither lose its legitimate action and concern for environment, development and sustainability and may even foster its global leadership (especially with the possible withdrawal from the USA of the Paris Agreement) nor turn back into their international commitments but within the existing framework. Other scenario could be "Those who want more do more," namely, accepting that the Nordic states might take the lead in priority sectors if the consensus among the 27 fails. But not only the Nordic states can take the lead. French President Emmanuel Macron held a summit in December 12, last year, 2 years after the adoption of the Paris Agreement on climate change and assembled more than 50 world leaders to fight against global warming and "make our planet great again". Despite US President Donald Trump's rejection of the Paris climate agreement and obvious unattendance in the summit, Macron's initiative (supported by President of the World Bank Group, Jim Yong Kim, and the Secretary-General of the United Nations, António Guterres) succeeded both in shifting the focus to the role of public and private finance to support and accelerate the fight against climate change and in launching 12 ambitious commitments[16] to be implemented until the second edition of the One Planet Summit.

Still, what the future will look like is still unknown but

> [i]n order to develop an adequate ethics of sustainability, it is necessary to understand the development and main themes of environmental ethics, just as it is necessary to have an overview of social and economic ethics (. . .). It is important to understand that environmental ethics encompasses a wide range of perspectives, with differing positions on many of the theoretical and practical issues involved in sustainability. (Kibert et al. 2009: 165)

[16]Please see https://www.oneplanetsummit.fr/en/ (last accessed in January 12, 2018).

However, by doing less more efficiently or doing more with who wants more, EU' s environmental *actorness* (Sjursen 2006; Hettne and Söderbaum 2005) surely requires a high level of internal and external credibility to pursue more coherent policies and investments for the European Union' citizens sake in the future to overcome the remaining shortfalls of EU's governance ambitions. In fact, coherence along with cohesiveness and effectiveness are crucial for a more reformist position in the environmental field if the EU succeeds in making that position externally feasible (Delreux 2014). In addition, when acknowledging the singular nature of EU's identity as an international actor, Pavese and Torney highlight that "EU actorness depends crucially on internal cohesion and the external opportunity structure" (2012: 140) which are of irregular pattern and limited degree. Therefore, those structural shortfalls must be overcome for the EU to succeed in the green roadmap that aims to trace and lead for the next decade.

## 4 Conclusion

This chapter's aim was twofold: first, to briefly analyse the historical background of the EU's environmental policies (from the Single European Act to the Europe 2020 Strategy and the Framework for Climate and Energy 2030) and external commitments (Kyoto Protocol and Paris Agreement); second, to critically assess how those policies and commitments can embody the European Union as an environmental player in the global politics based on Caporaso and Jupille's (1998) argument.

Regarding the first objective, this chapter intended to highlight the main background and historical milestones of the EU's environmental policy, showing that not only the EU has extensive environmental legislation (over 500 directives, regulations and decisions on energy, air quality, noise pollution, waste or water pollution), but it also has the ability to relate it with all the other EU's policies and consequently assure that Member States recognise environment as a crucial pillar of the communitarian action and their own. That leads us to our second question: critically assess how those policies and commitments can embody the European Union as an environmental player in the global politics.

In fact, as a global actor in environmental issues, the EU plays an important role to promote, manage and regulate climate change and sustainable development worldwide and does have a significant degree of actorness that favours its presence in the international climate and environmental agenda. The EU has one of the world's highest environmental standards with significant impact in a greener economy (the protection of environment and the implementation of innovations in this field can go hand to hand with a more competitive economy and investment), a more sustainable nature (legislation to assure that high environmental standards of water, air, natural habitats and species can coexist with human activities if they are sustainable) and a healthier quality of life of all European citizens (reduction or elimination of all the risks to health and wellbeing through a proper regulation of water, noise, air pollution and harmful chemicals). For that reason, both the environmental *acquis* led to more cohesion among Member States and a harmonisation of the EU policies for more autonomy, and the EU's authority in environmental issues is externally recognised and shapes the diplomatic arena.

That doesn't mean, however, that the work is done in what considers policy-making structures or Member States policy efforts. If we take in consideration Caporaso and Jupille's (1998) four components to address an actor's capacity in global politics—recognition, authority, autonomy and cohesion—it's quite clear that the EU meets those requirements in the environmental agenda but in different degrees, as internal cohesion has been evolving in the last few years but not always in the same extent than the other criteria. There's still indeed room for improvement in terms of cohesion, implementation, visibility and influence to succeed in the targets' roadmap set to the next decade and shape a new ethics of sustainability that is widely discussed in academia and politics all over the world. That is why the ongoing implementation of the 2015 Paris Agreement is so important (despite US alleged withdrawal), representing a truly global bridge between all governments of the world to become more engaged with climate action and more sensitive on how that severely affects global development and social economy.

In brief, to affirm its identity as an international climate actor, more action is needed to address current pressing environment threats and challenges and assure a truly sustainable and manageable climate change. More action is indeed needed to ensure that the existing programmes, strategies and roadmaps really meet the needs of current generations without compromising the ability of future generations to fulfil their own needs.

In the years to come, and notwithstanding the strongly intergovernmental nature of EU climate regime that still has a significant impact in terms of an autonomous, coherent and effective external diplomatic strategy, EU's actorness will then depend on the ability not only to contribute to global action to combat climate change but specially to be recognised worldwide as a leading and shaping authority in the environmental agenda.

## References


Bäckstrand, K., & Elgström, O. (2013). The EU's role in climate change negotiations: From leader to 'leadiator'. *Journal of European Public Policy, 20*(10), 1369–1386.

Bretherton, C., & Vogler, J. (1999). *Europe as a global actor*. London: Routledge.

Burns, C., & Carter, N. (2012). Environmental policy. In E. Jones & S. Weatherill (Eds.), *The Oxford handbook of the European Union* (pp. 511–525). Oxford: Oxford University Press.

Caporaso, J., & Jupille, J. (1998). States, agency, and rules: The EU in Global Environmental Politics. In C. Rhodes (Ed.), *The European Union in the World Community*. Boulder, CO: Lynne Rienner.

Dalby, S. (2013). Climate change and environmental security. In P. D. Williams (Ed.), *Security studies—An introduction* (pp. 311–323). New York: Routledge.

Delreux, T. (2014). EU actorness, cohesiveness and effectiveness in environmental affairs. *Journal of European Public Policy, 21*(7: Speaking with a single voice: The EU as an effective actor in global governance?), 1017–1032.

Erbach, G. (2016). *The Paris agreement: A new framework for global climate action*. Brussels: European Parliament http://www.europarl.europa.eu/RegData/etudes/BRIE/2016/573910/EPRS_BRI(2016)573910_EN.pdf
ESPAS. (2011). *Citizens in an interconnected and polycentric world—Global trends 2030*. Paris: European Union Institute of Security Studies.
EUI—The Economist Intelligence Unit. (2017). *Water security threats demand new collaborations: Lessons from the Mekong River Basin*. London.
European Commission. (2017). *White paper on the future of Europe*. Brussels.
Ferguson, Y. H., & Mansbach, R. W. (2012). *Globalization—The return of borders to a borderless world?* Abingdon: Routledge.
Greene, O. (2001). Environmental issues. In J. Baylis & S. Smith (Eds.), *The globalization of world politics—An introduction to international relations* (pp. 387–414). Oxford: Oxford University Press.
Gupta, J., & Grubb, M. J. (Eds.). (2000). *Climate change and European leadership—A sustainable role for Europe?* (Environment and Policy, Series No. 27). The Netherlands: Springer
Hettne, B., & Soderbaum, F. (2005). Civilian power or soft imperialism? *European Foreign Affairs Review, 10*(4), 535–552.
Khan, M. R. (2016). Climate change, adaptation and international relations theory. In G. Sosa-Nunez & E. Atkins (Eds.), *Environment, Climate Change and International Relations*. Bristol: E-International Relations Publishing.
Kibert, C. J., et al. (2009). *The ethics of sustainability*. Green Science Textbooks.
Pavese, C. B., & Torney, D. (2012). The contribution of the European Union to global climate change governance: Explaining the conditions for EU actorness. *Revista Brasileira Política Internacional, 55*(special edition), 125–143.
Schreurs, M. A., & Tiberghien, Y. (2007). Multi-level reinforcement: Explaining European Union leadership in climate change mitigation. *Global Environmental Politics, 7*(4), 19–46.
Sivaram, V., & Norris, T. (2016). The clean energy revolution—Fighting climate change with innovation. *Foreign Affairs. Council on Foreign Relations, 95*, 147.
Sjursen, H. (2006). The EU as a "normative" power: How can this be*? Journal of European Public Policy, 13*(2), 235–251.
UNDESA (United Nations Department of Economic and Social Affairs). (2015). *World population prospects: The 2015 revision*. New York: United Nations.
Vogler, J., & Stephan, H. R. S. (2007). The European Union in global environmental governance: Leadership in the making? *International Environmental Agreements: Politics, Law and Economics, 7(4)*, 389–413.

**Ana Isabel Xavier** is an assistant professor in UAL and ISCTE-IUL and a research fellow in OBSERVARE, also collaborating with CEI-IUL (Centre of International Studies) and at CISD (Research Centre in Security and defence of the Military University Institute). She is a PhD in International Relations/European Studies (University of Coimbra, Portugal, 2011). Currently, she is a postdoc researcher at the CEI-IUL (Centre of International Studies) on "The European Union as a global actor" and an associate research fellow at CISD (Research Centre in Security and Defence of the Military University Institute). As a guest assistant professor in NOVA University of Lisbon, she was responsible for teaching a course on "Environment, Development and Sustainability" for the master's students in International Relations and Political Science (2014–2017). Her main research interests, teaching and publications include the European Union, Security and Defence, Human Security, Human Rights and Globalization. From March 2015 to February 2017, she served as Defence Deputy Policy Director in the Portuguese MoD).

# The European Union Accession and Climate Change Policies in the Western Balkan Countries

**Teresa Maria Resende Cierco Gomes**

**Contents**

1 Introduction ........ 154
2 Climate Change, Institution Capacity and the Western Balkan Region ........ 155
3 The EU Environmental Policy and the Western Balkans ........ 158
4 Three Case Studies: Albania, Macedonia and Serbia ........ 164
4.1 Albania ........ 164
4.2 Macedonia ........ 166
4.3 Serbia ........ 167
5 Conclusion ........ 170
References ........ 171

**Abstract** Democracy, economic development and even statehood itself are all less than secure in the Balkans, a region with a history of conflict. Now, as the countries of the region (Serbia, Bosnia-Herzegovina, Macedonia, Albania, Montenegro and Kosovo) strive for individual and collective stability, they face the additional challenges brought by climate change and do so against a background of weak or recovering economies with limited budgets for addressing issues of environmental protection, insufficient environmental regulation or implementation, limited public participation and political tensions. Looking to the current legal and institutional framework on climate change in this group of countries, we try to identify the main obstacles to climate change policy and to access what has been the EU role at this level. We defend that although the several challenges that these states still must address on, such as corruption and weak governance, the EU accession process has been the main political driver of change in the region, providing opportunities for improving the environment in different ways. Using the comparative method, we argue that there is a correlation between institutional capacity and implementation of environmental laws and that governance aspects have a strong effect on environmental actions and outcomes.

T. M. R. Cierco Gomes (✉)
Faculty of Arts and Humanities, University of Porto, Porto, Portugal
e-mail: tcierco@letras.up.pt



T. Sequeira, L. Reis (eds.), *Climate Change and Global Development*,
Contributions to Economics, https://doi.org/10.1007/978-3-030-02662-2_8

**Keywords** Western Balkan region · Climate change · European Union · EU environmental acquis

## 1 Introduction

Climate change is a complex process (Gardiner 2006, p. 398; Devendra 2012, p. 2) that can exacerbate threats caused by poverty, weak institutions, broken communication channels and mistrust between communities and nations and by inadequate access to vital natural resources. Effective governance and political and economic stability are thus crucial to a country's or a region's capacity to cope with climate change (Wingqvist et al. 2012).

Characterised by weak governance, social and political tensions, limited public participation, economies with limited budgets for addressing issues of environmental protection, insufficient environmental regulation or implementation and heavy legacies from the past, such as corruption, the Western Balkan countries seem to have strong difficulties to react to the negative effects of climate change and to implement with success a climate change policy. Meanwhile, the Balkans are getting warmer and are projected to continue on this warming trend generally in proportion to the expected increase in global temperatures (Alfthan et al. 2015, p. 24). Similarly, the region is receiving less precipitation and is projected to experience further decreases, although precipitation patterns will continue to vary according to terrain, elevation and proximity to the sea. The effect of warmer temperatures on evaporation, together with the decline in precipitation, will make the region drier and highly vulnerable to climate change impacts (Idem).

Looking to the current legal and institutional framework on climate change in this group of countries, we try to identify the main obstacles to climate change policy and to access what has been the EU role at this level. The chapter is organised in three sections. In the first section, we provide an overview of research on concepts and linkages between climate change, institutions capacity and governance aspects, arguing that strong institutions can help to improve environmental outcomes. We then identify some of the environment challenges that the countries in the Western Balkan region lead with. In the second section, we argue that the EU accession process has been responsible for crucial developments in the protection of environment and the adoption of climate change policies in the region. Finally, in the third section, taking into account the EU accession process, we look to Albania, Macedonia and Serbia main problems in what concerns two governance aspects: rule of law and control of corruption. We defend that although there are challenges that these states still must address on, the EU accession process has been the main political driver of change, providing opportunities for improving the environment in different ways and increasing climate change policies. The main sources are the existing literature about climate change; the European Commission annual reports on Albania, Macedonia and Serbia; and interviews that were carried out with the heads of section for climate change in each of these countries. We then conclude that there is a correlation between institutional capacity and implementation of environmental laws and that governance aspects have a strong effect on environmental actions and outcomes.

## 2 Climate Change, Institution Capacity and the Western Balkan Region

United Nations Framework Convention on Climate Change (UNFCCC), in its Article 1, defines "climate change" as "a change of climate which is attributed directly or indirectly to human activity that alters the composition of the global atmosphere and which is in addition to natural climate variability observed over comparable time periods" (IPCC 2001). The UNFCCC thus makes a distinction between "climate change" attributable to human activities altering the atmospheric composition and "climate variability" attributable to natural causes. According to Hegerl et al. (2007), climate change refers to a change in the state of the climate measured by changes in the mean and in the variability of certain properties, such as temperature and precipitation that persist over decades. In other words, emissions of greenhouse gases and aerosols may cause a temperature change that leads to changes in precipitation patterns, rising sea levels and extreme weather events. These changes interact and have impact in water resources, natural ecosystems, agriculture and forestry.

In recent years climate change has been acknowledged as a challenge that carries potentially serious implications for international and regional peace and security. It has moved from being just an environmental problem and an energy challenge to being a matter of economic, social and political concern. Climate change will "affect the geopolitical landscapes as it brings direct and indirect risks related to energy security, food security, human health, water resources, patterns of weather and agriculture as well as disruptions to global markets" (ENVESEC 2017). States' failure to effectively adapt "to challenges such as floods, drought, coastal erosion, glacial melting and heat waves can result in disasters and societal instability. On the regional level, climate change will stress existing mechanisms for sharing resources like transboundary rivers and arable land" (Idem).[1]

In the Western Balkan region, climate change is expected to impact particularly agricultural regions, which will cause economic downturn and social consequences (ENVESEC 2017). Climate change produces serious threats to environment, ecological and socioeconomic systems. Increasing variability of precipitation coupled

[1]The agriculture, fisheries and tourism sectors will face considerable challenges in certain areas, and major investment will be required in adaptation activities to face drought, heat waves, forest fires, coastal erosion and flooding (EC 2009, p. 5). "Urban life will have to adapt to more extreme weather conditions which is likely to put a strain on the existing infrastructure in areas like water supply, drainage, health, energy and public transport. Adaptation will attempt to counteract the different impacts of climate change on regional conditions" (Idem).

with above-average temperatures can exacerbate existing stress factors and probably lead to a reduction in overall water availability, which in turn could affect the agricultural and energy security in this region, which is already characterised by political and social tensions over access to water and energy resources (Idem).

The Western Balkan region is not homogeneous. It is a concentration of different cultures, religions, ethnic groups, economies and geographies (Ek and Walter 2012, p. 40). This multicultural complexity added to a history of violent conflicts and to a demanding transition process that compromise this group of countries with profound reforms turns the adoption of climate change policies particularly difficult to achieve. In the last decade, the Western Balkan countries have been confronted with several challenges that contributed to environmental degradation and the pollution of the soil, air and water in the region. According to Ek and Walter (2012, pp. 2–3), "urbanisation, land abandonment, overexploitation of resources due to poverty, intensification of agricultural and forestry practices", the energy sector based on coal and fuel wood, inadequate local participation in establishing protected areas, insufficient or non-existing dialogue and lack of transparency and of investment in efficiency measures for power generation, transport, buildings and industry are just some of the problems. Despite considerable reductions in air pollutant emissions in some of these countries, "atmospheric pollution still poses a significant threat to human health and the environment". Ageing mining, energy and manufacturing plants generate high levels of air pollution, threatening human health (Idem, p. 5). "The environment legacy associated with extraction industries" and badly operated or abandoned mining sites has caused severe pollution, some with impacts reaching across national boundaries.[2] Moreover, one of the most difficult challenges regarding environment is to develop waste management systems which comply with EU legislation and, at the same time, are "affordable for consumers and businesses" (Idem, p. 8). The waste management services and infrastructure remain relatively underdeveloped in all these countries and generally lack advanced waste management practices such as separate waste collection (Macura et al. 2009).

In this setting, institutions play an important role and can have a key influence on climate change, based on how they structure it as a problem and how they devise responses to deal with its impacts. The most cited definition of institutions belongs to North (1991), who defines institutions as the rules of the game in a society or, in other words, institutions are humanly devised constraints that shape human interaction. Institutions from North's definition consist of both formal (i.e. rules, laws, constitutions) and informal (i.e. norms of behaviour, conventions, self-imposed codes of conduct) sets of rules, compliance procedures and moral and ethical behavioural norms designed to constrain individuals in the interest of maximising the wealth or utility of the principals. In terms of climate change, institutions are considered to be the multitude of means for holding society together, for giving it a sense of purpose and for enabling it to adapt (O'Riordan and Jordan 1999). It also helps "to define climate change both as a problem and a context, through such

[2]To see more about the environmental impact of the war in Yugoslavia, please see Clarke (2002).

socialized devices as scientific knowledge, culturally defined interpretation and politically tolerable adaptation policies" (Phong 2016, p. 83).

Central to our argument is also the concept of "institutional capacity" that has evolved over the years (Segnestam et al. 2003). Its meaning has expanded beyond individual organisations and their mandates, structures and processes. "Today, institutional capacity often implies a broader focus of empowerment, social capital and enabling environment, as well as the culture, values and power relations that influence us" (Idem). An IPCC (2007, p. 178) study argues that the existence of weak institutions in developing countries has implications for the capacity to adapt to or mitigate climate change. In general terms, capacity can be defined as "the ability to perform functions, solve problems and set and achieve objectives" (Fukuda-Parr et al. 2002). Skocpol (1985) defined capacity as "the ability of states to implement official goals, especially over the opposition of powerful social groups, or in the face of difficult economic circumstances". Linking the two concepts, Lindley (apud. Wickham et al. 2009) defined "institutional capacity" as the ability to "enhance the capability of local governments to perform intelligently and efficiently under their own direction" or "the ability of this broadly defined system of institutions, organizations, communities, relationships, rules, and values, within their different contexts, to achieve organizational and societal goals".

Institutional capacity is interrelated with the concept of governance. The term "governance" is broader in significance than "institutional capacity", which refers to institutional and decision-making structures. "Both are context dependent, relate to institutional environments through which citizens and government interact and are influenced by value systems and power relations" (Phong 2016, p. 84). "The conceptual scope of governance extends beyond that of institutional capacity to include and emphasise principles of accountability of stakeholders, equitable participation, transparency and predictable legal and regulatory frameworks" (ADB apud. Phong 2016, p. 84).

Climate change policies depend on the policy, institutional and legal framework related to environment as well as on the implementation capacity. "Although there is scope for improvement, the basic legal and policy framework are often in place in transitional countries. The major challenges are related to effective implementation of the existing framework. The gap between what is decided and actually implemented to improve environmental outcomes and climate change policies is called the implementation gap (or deficit)" (Wingqvist et al. 2012, p. 12). And, the implementation gap is particularly evident in the Western Balkan countries (EC 2016a).

There is now a growing consensus emphasising that governance has a strong effect on environmental actions and outcomes. Rule of law, citizens' rights of access to information, public participation and equal access to justice are bases for poverty reduction and sustainable development (UN 2012). According to Wingqvist et al. (2012, p. 12):

> weak governance is correlated with negative environmental outcomes and is closely associated with social ills such as corruption, social exclusion, and lack of trust in authorities. Good governance, on the other hand, has the potential to regulate and enforce environmentally sound policies and, as such, to steer individuals and societies into productive outcomes

> and sustainable use of the environment. Improved governance, combined with pro-poor legal frameworks and processes, may be powerful instruments contributing to poverty reduction and sustainable development.

Good governance aims at "ensuring inclusive participation and making governing institutions more effective, responsive and accountable and respectful of the rule of law and international norms and principles" (Idem, p. 13). "Participation of stakeholders including minority groups, access to information, adequate funding, transparency and accountability are crucial aspects to achieve good environmental governance" (Idem, p. 15).[3] Thus, the adoption of climate change policies is not sufficient to obtain good environmental governance. Strengthening the rule of law and controlling the corruption are equally or even more important than specific environmental policies to improve environmental actions and outcomes (Idem, p. 9). The rule of law implies legality, regularity and consistency in the enforcement of the democratic order. It is a necessary ingredient for the sustainability of climate change strategies and policies and fundamental to the effective implementation of EU environmental law. According to Esty and Porter (2005), developing the rule of law, eliminating corruption and strengthening governance structures are essential to successful environmental outcomes. In the same reasoning, Wingqvist et al. (2012, p. 10) stated that:

> efforts to improve environmental policies must go hand in hand with efforts to reduce corruption if they are to have the intended effects. Improved accountability, transparency, public participation and integrity can reduce the risk for corruption and create trust and legitimacy which facilitates implementation of different policy instruments.

Therefore, regarding the adoption and implementation of climate change policies, governance features become extremely important to achieve positive results.

In the next section, we are going to analyse the EU role regarding the environmental outcomes in the Western Balkan region and argue that these countries have a difficult and complex governance situation that hampers the implementation of the EU environmental legislation and the adoption of climate change policies.

## 3 The EU Environmental Policy and the Western Balkans

European environmental policy aims to preserve, protect and improve the quality of the environment, protecting human health, prudent and rational utilisation of natural resources, promoting measures at international level to deal with regional or worldwide environmental problems and in particular combating climate change (Article 191 Lisbon Treaty). It is based on "preventive action, the polluter pays principle,

[3]"Environmental governance is a specific form of the broader "governance" and refers to processes and institutions through which societies make decisions that affect the environment. It often includes a normative dimension of sustainability" (Wingqvist et al. 2012: 14).

fighting environmental damage at source, shared responsibility and the integration of environmental protection into other EU policies" (GRS 2017).

Protection of the environment is a well-established policy in the European Union since 1967 when the first environmental directive for harmonised classification and the labelling of dangerous chemicals was adopted. In November 1973, it was approved the First Environmental Action Programme, the framework for the EU's overall environmental policy development. The decision to do so was based on outcomes of the first UN Conference on Environment in Stockholm in 1972, addressing the public and scientific concerns about the "limits of growth". The Single European Act of 1987 introduced a new "Environment Title", "which provided the first legal basis for a common environment policy with the aims of preserving the quality of the environment, protecting human health, and ensuring rational use of natural resources" (Ohliger 2017). After that, all the treaty revisions strengthened the community's commitment to environmental protection: in Maastricht (1993) the environment was made an official EU policy area, and in Amsterdam (1999) environmental protection was integrated into all EU sectoral policies with a view to promoting sustainable development (Idem). In June 2013, it was adopted the 7th Environmental Action Programme that will run until 2020 and foresees the adoption of new targets for climate and energy, resource use and a wide range of new measures to make products more efficient, longer lasting and easier to repair and recycle, reduce waste production, protect citizens more effectively from hazardous chemicals including nonmaterials, restore Europe's biodiversity and protect oceans from marine litter.[4]

At the moment, the EU has some of the world's highest environmental standards, developed over the last decades; it is one of the most important forums for climate policy-making, with a far-reaching impact on the dynamics of international negotiations and national debates in other countries; it is thus an influential actor in environmental politics. One of its main concerns is to ensure that climate change policies are taken on board in other policy areas such as transport, energy and enlargement.

In their path towards accession into the EU, the Western Balkan countries must fulfil certain conditions—the Copenhagen Criteria: stable democratic institutions based on the rule of law protecting human rights, a functioning market economy and the alignment of national legislation with the legislation that is currently in force within the EU. This process of Europeanisation consists in "adopting new compliant laws, adapting domestic policies and ensuring that institutions are able to effectively implement these policies" (Borzel and Fagan 2015, p. 4). The adoption the EU laws and regulations in all the sectors, known as the acquis communautaire, is particularly

[4]To see more about the 7th Environmental Action Program: http://ec.europa.eu/environment/action-programme/

demanding concerning the environment.[5] Chapter 27 (Environment Chapter) is considered one of the most difficult and complex negotiation chapters (Börzel 2009, p. 40).

EU environmental legislation is detailed and specific, and it covers a wider scope of issues. It is increasingly interrelated, both between the various environmental sectors and with other EU policy areas such as energy, agriculture, transport and industry reflecting the EU overriding objective of promoting sustainable development. This approach is in accordance with specific treaty obligations (constitutional law) regarding the integration of environmental protection requirements into the definition and implementation of other EU policies, as well as with the inherent integrated nature of effective environmental legislation (ECRAN 2016, p. 16). The sheer volume of legislation and its cross-cutting character are not the only reasons that explain why the Environmental Chapter is considered one of the most difficult chapters of the acquis. Taking into consideration the extent of the environmental problems in the Western Balkan region and the state of environmental infrastructure, this arguably will be the most expensive part of the acquis. Moreover, to implement Chap. 27 has its politically sensitive aspects, such as the response to transboundary pollution, relations to the internal market rules and external trade issues.

According to Ek and Walter (2012, p. 14), EU environment acquis requires progress in three particular areas: legal, over 300 pieces of EU environmental legislation have to be transposed into national legislation in a short space of time; administrative, often weak and under-resourced administrations have to be significantly strengthened to be able to apply the environmental acquis; and financial, substantial investment in infrastructure and technology is needed to overcome years of insufficient funding and inadequate attention to environmental priorities.

The acquis comprises several major legal acts covering "horizontal legislation,[6] water and air quality, waste management, nature protection, industrial pollution control and risk management, chemicals and genetically modified organisms, noise and forestry" (GRS 2017). Thus, compliance with the environment acquis involves significant investment and a strong and well-equipped administration at the national and local level with the capacity to monitor and supervise the implementation and enforcement of the legislation. As Borzel and Fagan 2015, p 5, argues, "formal compliance refers to whether EU policies are fully and correctly transposed into

[5]The acquis has been divided into 35 different chapters, corresponding to different areas of Union policy. One of the chapters of the acquis is devoted to the Union's environmental rules and regulations.

[6]It is difficult to provide a substantive law for every specific environmental matter. To have effective and up-to-date instruments, which are able to regulate and to protect our environment, the EU decided to elaborate the so-called horizontal legislation. This category comprises several procedural pieces of law: the goal is to provide a strict procedural legislation to achieve a substantive protection. The application of this legislative framework leads to an improvement of the decision-making and legislative development through the regulation of the procedural aspect of the matter. Furthermore, the public have an essential role in guarantying an effective application of these rules: in terms of participation and as a supervisor (Env.Net 2017b).

national laws, conflicting domestic laws are repelled and administrative procedures are in place to ensure practical application. Practical application means the extent to which legally adopted EU policies are put into practice, i.e., rule targets bring their behaviour in line with the rule requirements and state actors monitor compliance with rule targets and impose sanctions in cases of violation". Considerable planning and management is also essential to ensure that the environmental benefits, such as the improvement of the ecosystem, public health and quality of life and the increased of tourism, one of the main sources of income in the Western Balkan countries, are obtained effectively and efficiently (Env.Net 2017a).

There are several EU strategies and instruments that aim to support the adoption of climate change policies in the Western Balkan region. The Stability Pact (launched in 1999), aimed at strengthening the Western Balkan countries' efforts in fostering peace, democracy, respect for human rights and economic prosperity, was responsible for the beginning of the adoption of environmental institutional structures in line with EU harmonisation requirements and has initiated legislative reforms. The Stabilisation and Association Process (launched in 1999 and strengthened at the Thessaloniki Summit in June 2003) established common political and economic goals, based on contractual relationships (the Stabilisation and Association Agreements), trade relations, financial assistance (the Instrument for Pre-Accession Assistance—IPA) and regional cooperation. In this EU-Western Balkan framework, the protection of the environment is always present and the object of great concern.[7]

IPA supports all the political, economic and social reforms, especially the adoption of climate change policies in the Western Balkan region. This is in fact the EU main instrument responsible for the adoption of the EU environmental principles and orientations that contributed to produce important environmental outcomes.

But, despite all the EU political and financial assistance, the governance in Western Balkan countries is still fragile. State institutions are still characterised as weak with high levels of bureaucracy, judicial system is still considered inefficient and not independent and corruption, lack of transparency and accountability of public institutions are still prevailing (Cierco 2016, p. 115; EC 2016a, p. 3).

Besides all the environmental problems that we have already identified in the previous section, Alfthan et al. (2015, p. 6) argue that the most "common gaps" in the Western Balkan countries "include inadequate policy coverage at different scales

---

[7]Several initiatives were taken to help this group of countries: the Regional Environmental Reconstruction Programme (2000–2009); the Regional Environmental Network for Accession (RENA) (2010–2013) that aimed to facilitate exchange of experience and best practice between the Member States, the candidate countries and the potential candidates; the Handbook for Implementation of EU Climate Legislation, which provides framework and step-by-step guidance, on the approaches and specific activities required to implement EC climate legislation; and, more recently, the Environment and Climate Regional Accession Network (ECRAN) project (2013–2016) that aimed to continue strengthening regional cooperation in the fields of environment and climate action and to assist the beneficiary countries on their way towards the transposition and implementation of the EU environmental and climate policies and instruments which is a key precondition for EU accession. For more information, please see http://www.ecranetwork.org

(e.g., regional, national and local); a lack of institutional coordination (including mechanisms) across sectors; a lack of or limited vertical integration from the EU to local administrations; and limited or low financial capacities to finance adaptation measures. In some cases, no policies exist to address existing or future risks". Another problem is corruption, a heavy legacy of the past. Corruption in this group of countries has high consequences in the environment. Every time that a project is approved with a bribe or corruption, it will have consequences to inhabitants by downgrading all factors of normal living socioeconomic, environmental, political, etc. According to UNODC-United Nations Office on Drugs and Crime, in the Western Balkans, every tenth company has paid bribe, being Kosovo, Serbia and Albania in the worst positions (Abedini 2016, p. 2). In these countries, corruption is deeply entrenched in the political culture; thus, reform programs that just borrow "best practices" from developed countries do not do enough to address the root causes of the problem. Corruption reforms need to be deep and comprehensive and have a long-term perspective; otherwise "the effects of the bad old days are likely to persist (Fredriksson and Neumayer 2014, p. 16). Fredriksson and Neumayer found that both corruption and democracy are statistically significant and substantively important determinants of climate change policies. Some of the most climate-vulnerable countries in the world fare worst on Transparency International Index (2017).[8] High levels of corruption, lack of transparency and low levels of participation may constrain the outcomes of the environmental efforts that are made.

The capacity of adopting and implementing climate change policies depends on the functioning of other checks and balances, including courts, law enforcement and a participative media and civil society. In this context, the efficiency and independence of key institutions in adopting climate change policies in accordance with the EU environmental acquis and promoting its implementation are serious compromise.

Candidate countries such as Albania, Macedonia and Serbia have limited institutional, financial and human capacity to effectively change, monitor and enforce environmental regulation or engage civil society, as they are building a market economy and a democratic system with all the political, economic and social transformation that this requires. Economic and social challenges include impacts of the recession with low levels of competitiveness, income and investment and of high and rising unemployment. There is awareness of the need for priority reforms and measures for growth, and there are opportunities to develop and implement green growth strategies. However, they face difficulties ranging from a lack of the necessary financial resources to an administrative structure that is not prepared to implement and enforce the EU legislation. The costs are high, especially in the areas of air pollution, water and wastewater management and solid waste management,

[8]To see more about this issue, see the 2017 Transparency International Report about *Global Corruption Report: Climate Change.*

besides the cost to set up the administrative structure that is needed to carry out the environmental acquis.[9]

Weak institutional capacity obstructs implementation of the legislation. Implementation is the process of having the national authorities that are in charge of the environment take EU environmental requirements into account when making individual decisions. Applicant countries will have to make improvements in many areas in order for successful implementation to occur. Among the necessary improvements are the creation of reliable data collection systems, effective systems of monitoring and reporting, increasing the awareness of industry and the public in environmental matters and facilitation of public participation in environmental issues. Perhaps the most pressing requirement for the applicant states, in order to successfully implement the environmental acquis, is the development of competent national authorities to carry out environmental administration. Therefore, it is necessary that the applicant countries have an administrative structure capable of ensuring compliance with the requirements of the acquis and the capacity to implement and enforce EU law. Unfortunately, this does not exist in most of the applicant countries. Most of the Western Balkan countries lack the qualified personnel necessary to implement and enforce the environmental acquis. These tasks belong to several different administrative agencies, resulting in a lack of coordination in ensuring compliance.

In the Western Balkan countries, there is an increased implementation gap between standards and the situation and practices on the ground. In the current institutional architecture of government, successful implementation will require significant enhancement of current intra- and inter-ministerial coordination and cooperation channels. And, as we are going to analyse in the next section, this is difficult to achieve in the Western Balkan countries. Thus, it is possible to affirm that in this region, the adoption of and adaptation to the acquis communautaire "has run into serious problems, with high costs meeting weak capacities" (Börzel and Buzogány 2010).

Nevertheless, EU acquis is driving Western Balkans environment institutions to adjust to a changing legal framework that has significant implications for the future scope and organisation of their work. Institutional changes are part of a government-wide adjustment that expands the role of environmental protection, increases accountability and demands improved communication on outputs and objectives of environment programs and investments (Env.Net 2017a). Consequently, EU support at this level to this group of countries creates an opportunity to adapt institutions systematically and access additional resources, guidance and technical assistance before completing the stabilisation and association process.

[9]Specifically related to environment, key challenges include air quality; waste management (including the Landfill Directive); water and wastewater management (including the Water Framework Directive); industrial pollution control and risk management (including Directive on Industrial Emissions); and nature protection (including Natura 2000 sites and lack of implementation of existing laws on nature protection).

## 4 Three Case Studies: Albania, Macedonia and Serbia

As we have already the opportunity to argue in the previous sections, governance is essential for environmental management. Strong and democratic institutions, such as environmental ministries, agencies and inspectorates, will guarantee the implementation of climate change policies and contribute to prepare and adapting the region to climate change impacts.

Taking into account our argument—the EU accession process has been the main political driver of change in the region, and there is a correlation between institutional capacity and implementation of environmental law—we are now going to analyse two governance indicators in Albania, Macedonia and Serbia, namely, the rule of law and the control of corruption. These two aspects are at the heart of the enlargement process and are considered both major problems in the region. Due to their importance in the EU accession, the "new" approach, endorsed by the Council in December 2011, requires that countries tackle judicial reform and the fight against corruption early in accession negotiations. This maximises the time to develop a solid track record of reform implementation.

Although with candidate status, Albania, Macedonia and Serbia are in different stages regarding the EU accession process. According to the last European Commission reports, Serbia is in the better position and Macedonia in the worst. Albania is still struggling with major difficulties regarding the two governance aspects that we have chosen: the rule of law and the control of corruption.

This study will be useful to access the EU influence on the improvement of environmental outcomes and to confirm the importance of institutional capacity in the implementation of climate change policies.

### *4.1 Albania*

The EU-Albania relation has been gradually developed in the last years. The Stabilisation and Association Agreement (SAA) was signed on June 2006 and entered into force on April 2007; the European Partnership, an instrument of the Stabilisation and Association Process, was established on February 2008; and finally, in June 2014, Albania got the status of candidate country for EU accession.

According to the last European Commission report (EC 2016a, p. 12), Albania has two major problems: the adoption and implementation of the judicial system reforms and the high level of corruption.

Looking to the rule of law, Albania adopted constitutional amendments related to the organisation and functioning of the justice system, including an "in-depth re-evaluation of judges and prosecutors" (EC 2016a, p. 12). Nevertheless, there are negative aspects that still persist, namely, the politicisation in the appointments process and the political interference in investigations and court cases (Idem, p. 15). Moreover, the "planning and distribution of administrative and financial resources within the judiciary" and the "efficiency of the court system" need to be improved (Idem, p. 16).

The country has continued other reforms, such as the public administration and the "establishment of solid track records of proactive investigations, prosecutions and convictions in the fight against corruption and organised crime", but institutions need to be strengthened, public service needs to be depoliticised and its efficiency and financial sustainability need to be increased (EC 2014, p. 10).

Looking now to the second indicator of our analysis, corruption, "Albania has one of the highest levels of perceived corruption in the region" (Abel apud. TI 2016). Crucial institutions for fighting corruption, such as the "Prosecutor's Office, the High Court or the Central Election Commission and the Inspectorate for the Declaration and Audit of Assets and Conflict of Interest are subject to political pressure and interference that undermines their ability to fulfil their role to stop corruption" (TI 2016). As Cornelia Abel (apud. TI 2016) argues, "it is troubling that institutions that are set up to ensure decision makers act in the interest of Albania citizens are not given the strength, impartiality and resources to do so". And, this situation has strong implications in the environmental outcomes.

Although the legal framework is almost in line with EU environmental acquis, its implementation is still lacking (EC 2016b, p. 4). The progress was mostly made in the sector of horizontal legislation, waste management, water quality and nature protection (Env.Net 2017c). Laws on Environmental Impact Assessment, Environmental Permits and Strategic Environmental Assessment entered into force, but "monitoring of implementation records are still lacking" (Idem).

In the field of climate change, there was very little progress (EC 2016b, p. 78). Some initial steps were taken to identify stationary installation for the purpose of future implementation of Emissions Trading System, but the country needs to strengthen its monitoring, reporting and verification capacities.

Albania, just like the other countries of this region, "lacks human, financial and administrative resources to successfully implement environmental legislation".[10] According to the European Commission report (2016b, p. 78), "industrial control and monitoring of emissions, water quality and waste management still remain poor. Resources and investments remain limited". The capacity to develop important projects in areas such as water and waste management infrastructure is weak, and insufficient resources are allocated to ensure sustainability of such projects.

Albania Environment Agency requires extensive investment in equipment and training to manage issues such as permit applications, permit monitoring, environmental monitoring and sampling and legislative enforcement. Local authorities have limited budgets and scarce administrative tools to develop the infrastructure and services required (Env.Net 2017c).

Despite the adoption of a national strategy for climate change, a national action plan for mitigation and a specific law on climate change, Albania still needs "more political commitment, investment and cooperation and coordination with and

[10] Interview with an Albanian Environment Ministry Official in September 2017.

between sectors", "to ensure coherence of the Climate Strategy with the EU 2030 Framework on Climate and Energy Policies" and increase public awareness and consultation on legislative initiatives or public investments, especially regarding climate change (EC 2016b, p. 78).

## 4.2 *Macedonia*

Macedonia was the first country of the Western Balkan region to sign a Stabilisation and Association Agreement with the EU, which entered into force in April 2004. A Joint Parliamentary Committee of the EU and Macedonia was established in 2004, to ensure the implementation of the SAA, and, in February 2008, it was adopted the Accession Partnership.

Although with candidate status since 2005, Macedonia is one of the Western Balkan countries with more difficulties to fulfil the set of reforms that are needed in its accession process towards the EU. Macedonia has been leading with a political crisis in the last years that has put into question its transition process and refrained the adoption of reforms in important sectors, such as the environment. Concerns about state capture affecting the functioning of democratic institutions and key areas of society persist, and "substantial progress is still needed in terms of concrete implementation of the Urgent Reform Priorities" (EC 2016a, p. 12).

Looking to the first governance indicator, the rule of law, political interference continues to "undermine the work and appointment of the judiciary" (EC 2016c, p. 13). There are "reports of selective justice in high-profile or politically sensitive court cases" and the political tension related to the wiretapping scandal continues (Idem). Deficiencies in the rule of law estimate a serious distrust on the citizens in the judiciary, as the institutions that should provide assurance in the rule of law, protection of the public interest and state prosperity are considered inefficient.

The country's capacity to effectively tackle corruption continued to be challenged by political interference. According to the European Commission report (2016c, p. 15), "corruption remains prevalent in many areas and continues to be a serious problem". The main institution responsible for prevention of corruption, the State Commission for Prevention of Corruption, "lacks functional independence: the selection and appointment process of its members is not transparent and seems to favour political loyalty over professionalism and integrity", and it also "suffers from certain structural shortcomings which the national authorities are not willing to address effectively" (Idem).

Just like in the Albania case, although there has been some progress in Macedonia alignment with the acquis in the field of environment, there was overall very little progress in the field of climate change. The "national strategy on environment and climate change has not yet been adopted", and the administrative capacity for implementing the Environmental Impact Assessment (EIA) and Strategic Environmental Assessment (SEA) directives, although has been improved, "remains insufficient, notably at local level" (Env.Net 2017a; EC 2016c, p. 73).

Meanwhile, the country has adopted the acquis related to the air quality, waste management and water quality. Most of the air quality monitoring system was made operational; plans for an integrated regional waste management system and a national water strategy were adopted; priority measures were identified to strengthen the implementation of environmental legislation at local level, and a coordination body has been set up for cooperation and dialogue with civil society. However, the access to information and public consultation "needs to be improved in connection with the implementation of the EIA and SEA directives" (EC 2016c, p. 74).

The complex political, economic and social situation in the country turns more difficult the adoption of new laws or their implementation and enforcement, especially in the environmental sector. It is still missing the law on environmental inspection; the existing law on inspection supervision is not in line with EU acquis and relevant best practice; important horizontal environmental directives were not still transferred and implemented in the country, such as the Environmental Liability Directive, INSPIRE and the Environmental Crime Directive (Idem).

Moreover, the administrative capacity for implementing and enforcing environmental and climate change legislation remains insufficient at both central and local levels.[11] According to a Macedonian Environment Ministry Official (Interview 2017), the main problems are the lack of political will and understanding of the complexity of the problem and the lack of institutional and technical capacities. Moreover, environmental protection and climate change requirements need to be better integrated into policy-making and policy implementation in other area, and the external funding requires strengthening of capacities and prioritisation of the sector (Idem).

According to the European Commission report (2016c, p. 75), the level of alignment with the climate change acquis is considered to be "at an early stage". So, the country still must "develop a comprehensive policy and strategy on climate action consistent with the EU 2030 framework" and "integrate climate change into other sectoral policies and strategies" (Env.Net 2017d).

## *4.3 Serbia*

EU-Serbia relations have developed fast since Kosovo declaration of independence in 2008. On that year, a European Partnership was adopted, setting out priorities for the country's membership application; in March 2012, the candidate status was received; and in September 2013, the Stabilisation and Association Agreement entered into force. Finally, on January 2014, EU-Serbia formal accession negotiations started.

Serbia is the only country of the three that has already opened EU accession negotiations with the EU. But, the overall pace of these negotiations will depend in

[11]Interview with a Macedonian Environment Ministry Official in September 2017.

particular on sustainable progress in two areas: the rule of law and the normalisation of its relations with Kosovo. The implementation of relevant strategies and action plans in the area of rule of law will be essential in the country's path towards the EU. Institutions must be strengthened; the institutional and administrative capacity for planning, monitoring and coordination needs to be enhanced; political and administrative reforms must proceed, especially in creating an accountable, independent and efficient judicial system (EC 2016d, p. 14).

According to the European Commission report (2016d, p. 4), "the judicial system has reached some level of preparation", but "further steps are needed to tackle political influence". The quality and efficiency of the judiciary and access to justice "remain undermined by an uneven distribution of workload, a burdensome case backlog and the lack of a free legal aid system", and the political influence in the recruitment and appointment of judges and prosecutors continues to obstruct judges' independence (Idem, p. 12).

Looking to the second governance indicator, corruption, we argue that this is still prevalent in many areas of public and economic life and continues to be a major concern (EC 2014, p. 17; EC 2016d, p. 4). The legal and institutional framework for the prevention and fight against corruption is largely in place, but "implementation is lagging behind" (EC 2016d, p. 16). The administrative capacity of relevant authorities and especially inter-institutional cooperation and coordination among relevant authorities such as judiciary, prosecution, police, customs and tax authorities remains insufficient for proper joint investigations (EC 2014, p. 18). Institutions must be substantially strengthened, both in terms of staff and in terms of respecting the full independence of law enforcement and judicial bodies dealing with investigations especially into high-level corruption.

In Serbia, important institutions such as the Anti-Corruption Agency benefits from some independence but still "faces obstacles to playing its role effectively, in the form of imprecise and unclear provisions of the law on the Anti-Corruption Agency and the lack of legal access to databases and records of other state bodies" (EC 2016d, p. 16). The "Anti-Corruption Council does not have the necessary resources and the government does not follow up and act on its recommendations systematically" (Idem).

In terms of the degree of compliance with the EU environmental acquis, intense legislative activity of the ministry in charge of environment and other institutions with responsibility in this area has led to a significant level of transposition of the relevant EU regulations. Serbia adopted its National Environmental Approximation Strategy (NEAS) in 2011 and a post-screening action plan called "Transposition and Implementation of Environmental and Climate Acquis—Chapter 27: Status and Plans" in September 2015 (EC 2016e, p. 2). In March 2016, Serbia set up a new financing facility (Green Fund) in the ministry. Thus, according to the European Commission report (EC 2016a, p. 74), "Serbia has achieved a high level of alignment with the acquis".

The majority of EU directives relating to the so-called horizontal sector (nature protection, management of chemicals and noise) are completely or almost completely transposed into national legislation (Env.Net 2017e). In other areas (air quality, waste management, water protection, industrial pollution control), the bulk of EU legislation has largely been transposed into national legislation.

There was progress in transposing EU directives on public access to environmental information and public participation in environmental decision-making: Serbia has opened a fifth Aarhus centre in 2015 in Belgrade,[12] what provides a platform to engage citizens, governments and the private sector in a dialogue on environmental challenges. But it is still required "to increase the effectiveness of public consultations" (EC 2016d, p. 76).

Given the institutional complexity and multidisciplinary nature of the field of environment, full transposition of the acquis may require a longer time frame, especially in the case of directives that require large financial investments. The existing administrative capacity, especially at the local level, is assessed as insufficient for adequate implementation.[13]

Regarding climate change, Serbia "has achieved some level of preparation but implementation is at a very early stage. There is a substantial amount of work to be undertaken as regards the implementation of legislation and the establishment of the necessary administrative and enforcement and control capacities required by the acquis" (EC 2016e, p. 17). It is still missing a national strategy on climate change, consistent with the EU 2030 framework on climate and energy policies; and, the National Climate Change Committee needs to integrate climate action into other sectors (EC 2016d, p. 77).

Similar to other countries of this region, Serbia also leads with problems regarding administrative and technical capacities and "lack of understanding of climate change issues in general in the country" (Interview 2017). The government itself stated that it accepted the acquis on environment and climate change, but it leads with difficulties regarding the implementation of some parts of the acquis due to the high costs of the required investments (EC 2016e, p. 2). According to the Screening Report on Serbia (Idem), "further institutional development and capacity building, staffing, training, financing as well as technical assistance" are required for the implementation of the EU acquis related to the environment and climate change policies. Moreover, the "inter-institutional cooperation and coordination at state and local level" must be developed and organised (Idem, p. 27). The institutions responsible for policy development, implementation and enforcement exist in the country,

[12]Access to information, public participation in environmental decision-making and access to justice in environmental matters are the three pillars of the 1998 Aarhus Convention. Serbia has a network of five Aarhus Centres. The first Aarhus Centre was inaugurated in 2010 in the city of Kragujevac, followed by the Aarhus Centres Novi Sad and Subotica in 2011. In 2012, an Aarhus Centre opened in the south-eastern city of Niš. In 2015, the Aarhus Centre Novi Beograd was inaugurated in the capital Belgrade. The Aarhus Centres of Serbia focus on building and facilitating participatory local governance processes and on promoting dialogue on environmental issues. The Aarhus Centres are also working on Disaster Risk Reduction (DRR), particularly flood risk reduction (OSCE 2017).

[13]Interview with a Serbian Environment Ministry Official in September 2017.

but they need to have the necessary resources (human, technical and financial) to ensure that "competences are matched by capacity" (EC 2016d, pp. 17–18).

## 5 Conclusion

Environmental problems in the Western Balkan region constitute a great challenge for both the countries within themselves and for their relationship with the EU. Undoubtedly, the EU enlargement process is a powerful driver of change in the region and provides opportunities for environmental improvements as well as (legal, administrative and financial) challenges. But, complying with EU environmental requirements may prove to be one of the most demanding tasks for these countries. While recent reform efforts include the important development of a new legal and policy framework for environmental management, effective implementation leading to measurable improvements remains a great concern. Thus, in the area of environment and climate change legal aligning, there is a growing implementation gap.

In this study, it is possible to confirm that there is a correlation between institutional capacity and the implementation of environmental laws. Institutional or governance indicators, such as the rule of law and the control of corruption, have a strong effect on environmental actions and outcomes. Weak governance is correlated with negative environmental outcomes and is closely associated with instability and political and social problems, such as corruption, and lack of trust in institutions and authorities (Wingqvist et al. 2012, p. 13).

Furthermore, the study also demonstrates that these governance indicators have an impact on the ability/capacity of the governments to perform or not the necessary legal changes and the institutional capacity needs associated with implementation and enforcement of environmental legislation. The strengthening of policies, institutions and capacities is needed in all of the Western Balkans countries as weak governance often hampers the adoption of climate change policies and the adaptation efforts to the EU environmental *acquis*.

The administrative challenges are becoming more obvious when it comes to applying and implementing the environmental legislation. There is a growing implementation gap, between what is agreed and decided and what is actually implemented. The lack of implementation and enforcement remains a concern and is related to weak administrative capacities, which have been identified as a major obstacle for EU accession process. All countries need to better integrate environment and climate change concerns into other policy areas. Strengthening of administrative capacity and inter-institutional cooperation and coordination among ministries of government institutions should remain a priority. Technical solutions to environmental problems are often not sufficient to obtain sustainable development. Improving governance is key and could provide important synergies with support to infrastructure investment. It provides an important vehicle to improve environmental performance in general as well as the relation with the EU.

Looking to the last European Commission reports, governance has been improving in the Western Balkan countries, but it is still plagued with several problems that contribute to the fragility of the rule of law and the weakness of institutions. Improving government effectiveness, supporting inclusion of private sector and civil society in policy-making and monitoring and improving the rule of law and control of corruption are thus important means to enhance implementation and enforcement of climate change policies.

But, although important, improving governance is not the sole solution. Environment and climate changes also need to be political and economic priorities given adequate budget allocations and incentives provided for environmental improvements. Areas identified as particularly important are improved integration of environment and climate change in development, enhanced capacities for environmental management and the transformation towards a greener economy.

The role of environmental authorities in the Western Balkan countries will continue to be important. However, in order to fully integrate environment into the core of development, also other ministries (like finance, economy, planning) play a key role.

## References


Abedini, B. (2016). *Corruption impacts climate change initiatives*. Macedonia: Center for Security Studies and Development, Flash Post.

Alfthan, B., Krilasevic, E., Venturini, S., Bajrovic, S., Jurek, M., Schoolmeester, T., et al. (2015). *Outlook on climate change adaptation in the Western Balkan Mountains*. Vienna: United Nations Environment Programme, GRID Arendal and Environmental Innovations Association.

Börzel, T. (2009). Environmental policy: The challenge of accession. In T. Börzel (Ed.), *Coping with accession to the European Union. New modes of environmental governance* (pp. 32–47). Houndmills: Palgrave Macmillan.

Börzel, T., & Buzogány, A. (2010). Governing EU accession in transition countries: The role of non-state actors. *Acta Politica, 45*(1), 158–182.

Börzel, T., & Fagan, A. (2015). Environmental governance in South East Europe/Western Balkans: Reassessing the transformative power of Europe. *Environment and Planning C: Government and Policy, 47*, 1–16.

Cierco, T. (2016). Bridging the gap: The Serbian struggle for good governance. *Journal of Contemporary Central and Eastern Europe, 24*(2), 113–129.

Clarke, R. (2002). Yugoslavia. In F. Carter & D. Turnock (Eds.), *Environmental problems in East Central Europe*. London: Routledge.

Devendra, C. (2012). *Climate change threats and effects: Challenges for agriculture and food security* (ASM Series on Climate Change). Malaysia: Academy of Sciences Malaysia.

ECRAN. (2016). *Handbook on the implementation of EU environmental legislation* (Hulla & Co Human Dynamics 2015–2016, Ed.). Accessed July 22, 2017, from http://www.ecranetwork.org/Files/Handbook_on_Implementation_of_Environmental_Legislation.pdf

Ek, G., & Walter, S. (2012). *Western Balkan – Environment and climate change policy brief*. Sida's Helpdesk for Environment and Climate Change. Accessed July 22, 2017, from http://sidaenvironmenthelpdesk.se/wordpress3/wp-content/uploads/2013/12/Regional-Wester-Balkan_EnvCC-Policy-Brief_Dec-2012.pdf

Env.Net. (2017a). *EU environmental policy*. Accessed July 27, 2017, from http://www.env-net.org/environmental-acquis/eu-env-policy/

Env.Net. (2017b). *Environmental acquis categories*. Accessed July 27, 2017, from http://www.env-net.org/environmental-acquis/eu-env-policy/environmental-acquis-categories/

Env.Net. (2017c). *Albania. Approximation process to the EU environment acquis in Albania*. Accessed July 27, 2017, from http://www.env-net.org/environmental-acquis/approximation-env-net-countries/albania/

Env.Net. (2017d). *The former Yugoslav Republic of Macedonia. Approximation process to the EU environment acquis in Macedonia*. Accessed July 27, 2017, from http://www.env-net.org/environmental-acquis/approximation-env-net-countries/republic-of-macedonia/

Env.Net. (2017e). *Serbia. Approximation process to the EU environment acquis in the Republic of Serbia*. Accessed July 27, 2017, from http://www.env-net.org/environmental-acquis/approximation-env-net-countries/serbia/

Environment and Security Initiative – ENVESEC. (2017). *Focus areas – Climate change*. Accessed July 27, 2017, from http://www.envsec.org/index.php?option=com_content&view=article&id=74&Itemid=92&lang=en

Esty, D., & Porter, M. (2005). National environmental performance: An empirical analysis of policy results and determinants. *Environment and Development Economics, 10*(4), 391–434.

European Commission. (2009). *Regions 2020. The climate change challenge for European regions*. Directorate General for Regional Policy, Brussels, March 2009. Accessed July 27, 2017, from http://ec.europa.eu/regional_policy/sources/docoffic/working/regions2020/pdf/regions2020_climat.pdf

European Commission. (2014). Indicative strategy paper for Albania (2014–2020). Instrument for pre-accession assistance. Accessed July 27, 2017, from https://ec.europa.eu/neighbourhood-enlargement/sites/near/files/pdf/key_documents/2014/20140919-csp-albania.pdf

European Commission. (2016a). *Communication on EU enlargement policy*. Brussels, 9/11/2016. COM(2016) 715 final. Accessed July 22, 2017, from https://ec.europa.eu/neighbourhood-enlargement/sites/near/files/pdf/key_documents/2016/20161109_strategy_paper_en.pdf

European Commission. (2016b). *Albania 2016 report*. Brussels, 9.11.2016 SWD(2016) 364 final. Accessed July 22, 2017, from https://ec.europa.eu/neighbourhood-enlargement/sites/near/files/pdf/key_documents/2016/20161109_report_albania.pdf

European Commission. (2016c). *The former Yugoslav Republic of Macedonia 2016 report*. Brussels, 9.11.2016 SWD(2016) 362 final. Accessed July 22, 2017, from https://ec.europa.eu/neighbourhood-enlargement/sites/near/files/pdf/key_documents/2016/20161109_report_the_former_yugoslav_republic_of_macedonia.pdf

European Commission. (2016d). *Serbia 2016 report*. Brussels, 9.11.2016 SWD(2016) 361 final. Accessed July 22, 2017, from https://ec.europa.eu/neighbourhood-enlargement/sites/near/files/pdf/key_documents/2016/20161109_report_serbia.pdf

European Commission. (2016e). *Screening report Serbia. Chapter 27 – Environment*. Accessed July 22, 2017, from https://ec.europa.eu/neighbourhoodenlargement/sites/near/files/screening_report_serbia_-_chapter_27_-_environment.pdf

Fredriksson, P., & Neumayer, E. (2014). Corruption and climate change policies: Do the bad old days matter? *Environmental and Resource Economics*, 1–19. http://eprints.lse.ac.uk/64180/1/Corruption%20and%20Climate%20Change%20Policies.pdf

Fukuda-Parr, S., Carlos, L., & Khalid, M. (Eds.). (2002). *Capacity for development: New solutions to old problems*. London: Earthscan.

Gardiner, S. (2006). A perfect moral storm: Climate change, intergenerational ethics and the problem of moral corruption. *Environmental Values, 15*(3), 397–413.

GRS – Government of the Republic of Serbia. (2017). *Negotiating team for accession of the Republic of Serbia to the European Union. Chapter 27: Environment*. Accessed July 27, 2017, from http://www.eu-pregovori.rs/eng/negotiating-chapters/chapter-27-environment/

Hegerl, G. C., Zwiers, F. W., Braconnot, P., Gillett, N. P., Luo, Y., Marengo Orsini, J. A., et al. (2007). Understanding and attributing climate change. In S. Solomon, D. Qin, M. Manning,

Z. Chen, M. Marquis, K. B. Averyt, et al. (Eds.), *Climate change 2007: The physical science basis* (pp. 663–745). Cambridge: Cambridge University Press.

IPCC – Intergovernmental Panel on Climate Change. (2001). *Climate change 2001: Synthesis report*. Accessed July 27, 2017, from http://www.ipcc.ch/ipccreports/tar/vol4/index.php?idp=204

IPCC – Intergovernmental Panel on Climate Change. (2007). Issues related to mitigation in the long term context. In B. Metz, O. R. Davidson, P. R. Bosch, R. Dave, & L. A. Meyer (Eds.), *Climate change 2007: Mitigation*. Cambridge: Cambridge University Press.

Lisbon Treaty. (2009). *Lisbon Treaty*. Accessed July 27, 2017, from http://www.lisbon-treaty.org/wcm/the-lisbon-treaty/treaty-on-the-functioning-of-the-european-union-and-comments/part-3-union-policies-and-internal-actions/title-xx-environment-climate-change/479-article-191.html

Macura, B., Bojovic, D., Tadić, M., Ćosić, N., Petrić, I., Jarić, I., et al. (2009). *A study of environmental conflicts and issues in South-Eastern Europe: Possible collaboration between CSOs and ecological economists* (A Report for the CEECEC Project). Accessed July 30, 2017, from http://www.ceecec.net/wp-content/uploads/2009/04/esee1.pdf

North, D. (1991). Institutions. *Journal of Economic Perspectives, 5*(1), 7–112.

O'Riordan, T., & Jordan, A. (1999). Institutions, climate change and cultural theory: Towards a common analytical framework. *Global Environment Change, 9*(2), 81–93.

Ohliger, T. (2017). *Environment policy: General principles and basic framework*. European Parliament. Accessed July 30, 2017, from http://www.europarl.europa.eu/atyourservice/en/displayFtu.html?ftuId=FTU_5.4.1.html

OSCE. (2017). *Aarhus Centres Serbia*. Accessed July 30, 2017, from https://aarhus.osce.org/serbia

Phong, P. (2016). *Climate change adaptation planning under uncertainty in Ho Chi Minh City, Vietnam*. Ph.D thesis, University of Westminster. Accessed July 30, 2017, from http://westminsterresearch.wmin.ac.uk/17373/1/Phong_Phu_thesis.pdf

Segnestam, L., Persson, A., Nilsson, M., Arvidsson, A., & Ijjasz-Vasquez, E. (2003). *Country environment analysis: A review of international experience* (Environment Strategy Papers, No. 8). Washington, DC: World Bank.

Skocpol, T. (1985). Bringing the State Back in. In P. B. Evans, D. Rueschemeyer, & T. Skocpol (Eds.), *Bringing the State Back in*. Cambridge: Cambridge University Press.

Transparency International. (2016). *Albania lacks independent oversight institutions leaving the door open to corruption*. Accessed July 30, 2017, from https://www.transparency.org/news/pressrelease/albania_lacks_independent_oversight_institutions_leaving_the_door_open_to_c

Transparency International Index. (2017). *How corruption affects climate change*. Accessed July 30, 2017, from https://www.transparency.org/news/feature/how_corruption_affects_climate_change

United Nations. (2012). *Resilient people, resilient planet: A future worth choosing* (UN Secretary-General's High-level Panel on Global Sustainability). New York: United Nations.

Wickham, F., et al. (2009). *Institutional capacity within Melanesian countries to effectively respond to climate change impacts, with a focus on Vanuatu and the Solomn Islands*. Honolulu: Bishop Museum and SPREP Report.

Wingqvist, G., Drakenberg, O., Slunge, D., Sjöstedt, M., & Ekbom, A. (2012). *The role of governance for improved environmental outcomes* (Report 6514, September 2012). Stockholm: Swedish Environmental Protection Agency.

# The BRICS Commitment on Climate Change: Process Towards an Effective Approach in the Path of Sustainable Development

**Marco António Baptista Martins**

**Contents**

1 Powers in Cooperation ... 176
2 Climate Change and Challenges ... 177
3 Polycentric Dynamics in the International Arena ... 182
4 Conclusions ... 185
References ... 186

**Abstract** Since 2008, the countries of the group known as the BRIC countries (Brazil, Russia, India and China) have acted as strategic partners with a twofold purpose: to change the architecture of international relations and to rearrange the balance of powers to soft balance world's largest military power, the United States. Countries that are today classified as 'growing economies' are undoubtedly the highest global consumers of energy, emitting polluting agents such as those from natural gas, oil and coal, which are all responsible for the greenhouse effect. The present chapter analyzes and examines whether, within the international context, the BRICS have sought to respect the United Nations Framework Convention on Climate Change (1992) (that is, the Doha Amendment to the Kyoto Protocol), and to establish new commitments for the period 2013–2020 under the Paris Agreement (2016). To that end, we note important political and economic aspects of the BRICS that converge—and sometimes diverge—as member countries pursue the climate change agenda, and we look at how these same aspects strengthen the decisions of the different governments of each BRICS. Of note in this examination is the role of the New Development Bank, known as the BRICS Bank, which is recognized as the

This study was conducted at Research Center in Political Science (UID/CPO/0758/2019), University of Minho/University of 'Evora and supported by the Portuguese Foundation for Science and Technology and the Portuguese Ministry of Education and Science through national funds.

M. A. B. Martins (✉)
International Relations, Evora University, CICP, Evora, Portugal
e-mail: mabm@uevora.pt



T. Sequeira, L. Reis (eds.), *Climate Change and Global Development*, Contributions to Economics, https://doi.org/10.1007/978-3-030-02662-2_9

driving force behind clean energy financing and for its contribution to sustainable development.

**Keywords** BRICS · Climate change · Sustainable development · Clean energy · Soft balance

## 1 Powers in Cooperation

The emergence of new players in international relations, known as the BRIC countries (Brazil, Russia, India, China), tended to combine member powers and minimize their differences in the face of the global challenges of climate change. In fact, these states are located in geopolitically strategic regions, where they claim a position of world leadership by which they may stem the continued hegemonic intentions and influence of the United States, as it transmits and applies unidirectional policies. The idea of 'leadership' connotes an actor's ability to influence the course of international relations. The actor may not only aspire to lead but must also be capable of enacting domestic governmental mechanisms, establishing institutions and a form of government, put forth an economic policy, practice cultural diplomacy and various policies for the energy sector (Langenhove et al. 2016). All this reflects the degree to which the actor is involved in the international arena and its interactions there.

The BRICS, as a group of regional powers, lies within a context and varying structural dimension, as it adapts to the international reality, as it attends to sensitive and problematic issues, such as security, respect for human rights, the environment and global warming and energy; not less important is the agenda related to financial and international trade sectors.

Each of these states develops its respective political agenda, taking into account the geopolitical space it occupies in a dynamic of participation at various levels. For example, China's macro-regional location allows it to participate in the Association of Southeast Asian Nations (ASEAN), the group Asia-Pacific Economic Cooperation (APEC), and the Shanghai Cooperation Organization (SCO). With Russia and the countries of Central Asia, China has a complex relation with Europe and its bordering countries, in addition to the United States and its spheres-of-influence, namely, the Commonwealth and its allies. South Africa, in its post-apartheid logic, affirms its regional identity, promotes sustainable development and aims to reduce poverty and guarantee the security of the region. To this end, this country reinforces the role of the Southern African Development Community (SADC), demonstrating evidence of its agenda for a greater multilateral diplomatic commitment in Africa to resolve the crises that plague the continent (Soko and Balchin 2016). For its part, Brazil relies on intra-regional cooperation through Mercosur and the Union of South American Nations (UNASUR) to boost its economy and promote its social and human development. As a partner and actor in bilateral relations with its neighbours and economic partners, Brazil strengthens political dialogue to achieve greater energy integration, supporting and respecting biodiversity and the ecosystem in the fight against the causes and consequences of climate change.

It should be noted that Brazil seeks to solidify an identity separate from other countries in the region regarding progressive recognition of the rights not only of its nationals but of residents of each member state of UNASUR. Brazil's aim is to rejoin, South-American association, and its policies are directed towards being interconnected regionally, but with a global focus. This is especially the case in Brazil's fight against terrorism and corruption, the trafficking of humans, arms, and drugs, or in countering any other transnational threat that might affect the internal security of each of the member states of UNASUR. In all these areas, Brazil's long view is to enhance the security of each citizen.

Also of note is the case of India within the framework of the South Asian Association for Regional Cooperation (SAARC), whose political and economic charter outlines the fundamental structure of the various areas of cooperation: agriculture, rural development, telecommunications, health, science, technological development, infrastructure and transport. In doing so, it aims to ensure the well-being of the population of that region.

The reinforcement and development of these intra- and extra-regional links, on the one hand, is intended to increase the degree of regional integration and, on the other hand, to enable small powers to play a strategic coalition role, enabling them to maneuver in the international system as they represent their regions at the level of a global community (Langenhove et al. 2016). Precisely in Quiliconi's opinion (2016), the rise of new players and powers in international relations is by its nature a challenge to the world order.

## 2 Climate Change and Challenges

Within the framework of the BRICs, it is critically important to emphasize that the countries comprise one of the largest areas of biodiversity richness in the world. In this sense, as emerging economies, they have high consumption and energy production indices, which increase the concentration of greenhouse gases in the atmosphere, having a direct impact on the ecosystem due to the excessive concentration of carbon dioxide, nitrous oxide, methane, chlorofluorocarbons, perfluorocarbons and sulfur hexafluoride. Thus, derived from the high consumption as referred to, these countries endanger their ecosystem.

In addition, problems related to the consequences of energy use are linked not only to economic issues, but also to society. Thus, climate change implies the transnationalization of the debate and a consensus on solutions that incorporate economic, financial, commercial and security considerations—among others. All these issues of major importance were highlighted in the United Nations Framework Convention on Climate Change (1992) and subsequently included in the Kyoto Protocol in 1997, where, for strategic reasons, the European Union advocated a strong and serious commitment to the countries involved; they were retaken at the

United Nations Conference on Climate Change (2015), which aimed at a universal agreement with the regency from 2020 to limit global warming to less than 2 °C by the year 2100.

The emergence of the BRICs, in focus with their agendas, in the framework of the geopolitical alliance of the BASIC (Brazil, South Africa, India and China), which does not include Russia, arose from an imperative necessity, the identification of attractive and alternative poles in the search for new markets and investments, in the international political system inserted in the dynamics of the global capitalist structure. The key question is this: What kind of order or disorder can these new actors generate or provoke?

South–South relations promote economic and trade links alongside specific coalitions in the trade blocs with the G20 or the World Trade Organization (WTO), which implies a convergence of the nature of the definition of the foreign policy of each of these states in a changing age so they do not become threats, but rather that they become cooperative powers, defying Western world domination. It is evident that one of the major convergences in this cooperation is evident in global economic governance, where the BRICS intend to demonstrate, even with the creation of the Bank for Development, that they can play a relevant role in the conduct of international politics together with the United States and the European Union.

In the second decade of the twenty-first century, the constant threat and specter of a new global financial crisis put the BRICS in a position of greater importance on this chessboard. Each of these states intends to become the preferred player, as was the case of Brazil during the presidency of Dilma Rousseff, or of Russia as a counterbalance to its eternal, ideological rival, the United States, during the Cold War.

However, from another perspective, China, through President Xi Jinping, presents itself not as a counter-power, but as a faithful player whose function lies in the constant rebalancing of the balance of power. However, the negative side of this common agenda, specifically, the limitations on any effective action the BRICS from imposing its policy reforms as an alternative to the dominance of the World Bank and its satellite institutions in financial and investment.

According to Quiliconi and Kingah (2016), related tensions exist in terms of inclusion or exclusion resulting from an unequal distribution of wealth within each of these countries, despite considerable efforts to reduce extreme poverty, in particular, in Brazil, India and China. All these elements are embedded in economic policies in the name of redistribution and social inclusion carried out by the leaders of these powers, with a greater or lesser degree of results in the various sectors of the population.

One of the characteristics referred to by O'Neill (2013) is the phenomenon of the rise of an unparalleled middle class compared to other Western countries, despite inequalities of social stratification, and a positive effect on the global transactional flows of capital, trade and labour.

Although the major urban centers of the BRICS readily allowed and contributed unequivocally to the emergence of this new middle class, this process did not bring about an eradication of poverty; on the contrary, inequalities persist alongside a reconfiguration of urbanization that, in a way, indicates and projects the image of two opposing worlds: those included in the process and those excluded from the system.

In addition, one of the major problems of and obstacles to the effective development of these countries in the name of the inclusion and eradication of human misery is the adoption of policies that promote the demolition of the wall or barriers to health, education, access to transport services and environmental practices, given that economic growth means high energy consumption, degradation of air and water quality and a disrespect of environmental policies, derived from the high rates of pollution. Incidentally, in the case of India, the government has been reluctant to accept the multinationals Walmart, Carrefour, Tesco and Metro because it considers them destructive to small businesses, and because they bring negative repercussions for local commerce, for both consumers and producers, problems that particularly affect the rural environment (Quiliconi and Kingah 2016).

As a result, the leadership of the BRICS starts at the domestic level and is reflected in the international arena, based on its behaviour regarding the reinforcement and containment of internal/international cleavages. Despite limitations, it is clear that the BRICs have been a force since 2009, from the moment they met at the first of several ministerial summits to address foreign affairs, and to affirm and develop their global leadership capacity. Since the 2011 BRICS summit, which involved meetings for the first time in the fields of education, health, agriculture and urbanization, it has been necessary to establish regular inter-ministerial meetings to follow up and monitor these areas of intervention. In summary, it is pointed out that the first summit, in Yekaterinburg, Russia, on June 16, 2009, before South Africa was a part of the group, debates converged on financial and economic issues, as it analyzed the impact of the 2008 crisis. The recovery of the world economy was part of the discussion regarding the need to reform the UN system.

The second summit in Brasília, Brazil, on April 16, 2010, served to strengthen political consensus among members of this group, announcing the intention to carry out and develop initiatives aimed at intra-BRIC cooperation, involving themes related to safety.

At the third summit, in Sanya, China, on April 14, 2011, South Africa was included in the group, taking into account its economic relevance and, above all, issues of geostrategic importance on the African continent as well as in constructive action in international politics, reaffirming the imperative to readapt the United Nations to contemporary reality, since the UN system did not then respond to or represent the new world order. Of the topics discussed, terrorism was condemned, the use of renewable energies was embraced and nuclear energy for peaceful and non-war purposes was supported, not as it was used during the Cold War. Equally, the relevance of the Millennium Development Goals, the fight against human misery and the eradication of hunger was equally emphasized.

The fourth summit, held in New Delhi, India, on 29 March 2012, consolidated the fundamental pillars of the BRICS, extending financial cooperation to non-member countries outside the group, through the decision to create the Development Bank in the name of financing infrastructure projects and sustainable development, in addition to helping and facilitating the granting of credits using local currencies.

The fifth summit in Durban, South Africa, on March 26–27, 2013, approved the report on the viability of the Development Bank and took advantage of the signing of two agreements: the establishment of the BRICs behalf Business Council and the establishment of the BRICS Think Tanks Council on of BRICS and Africa cooperation. With regard to Africa, the BRICS are the main partner; therefore, through the IBSA Dialogue Forum (India, Brazil, and South Africa) the volume of trade goes to around US $25 billion, offering and opening a new way to the achievement of new export, business and investment opportunities.

The BRICS have a particular interest in the exploitation of natural resources, in particular, energy, i.e., oil. Brazil intends to lead and create the new axis of the South, while Russia wishes to improve its image around a new perspective because of its action in previous times in Africa during the Cold War. For India, Africa is viewed politically as a strategic ally in seeking competitive advantages over its Chinese Asian competitor. However, for China, the countries of the African continent are considered long-term partners in the attempt to gain a foothold and, clearly, a certain global economic ascendancy. It is noted in passing that in just 20 years, from 1990 to 2010, China's trade grew from US $3.5 billion to US $100 billion, which reflects an exponential increase in Sino-African trade and its China's clear bet on the importance of international trade.

At the sixth summit held in Fortaleza, Brazil, on July 14–16, 2014, an agreement was signed by the Development Bank, the Reserve Contingent and the Memorandum of Understanding for Technical Cooperation between Credit Agencies and Guarantees to BRICS Exports. In addition, an agreement was implemented between the BRICS national banking sectors in the name of innovation cooperation.

At the seventh summit in Ufa, Russia, on July 8–9, 2015, the constituent agreements of the Development and Reserve Contingencies were ratified. at meetings of the Board of Governors. Furthermore, at this summit, the members decided not only to diversify but also to deepen trade to strengthen intra-BRICS investment.

At the eighth summit, in Panjim, India, on October 15–16, 2016, the BRICS decided to focus their attention on promoting the innovative global economy, linking global growth with sustainable development. From this summit, the role of the BRICS in the world was analyzed, and it was taken into account that the BRICS countries represented 43% of the world population and 30% of GDP and that they move around 17% of global trade, according to IMF the growth was on the order of 7.6% in 2016/2017. On the international agenda, climate change, regional security, the grave problem of international terrorism and the peace process in Syria were discussed. In the case of Brazil, Michel Temer called on remaining members of the group to aid and support Brazil in its exit from the deep recession.

The ninth summit, September 3-5, 2017, in Xiamen, China, was convened in the name of strengthening the intra-BRICS partnership, to promote solidarity and collaboration in order to improve global governance. It was deemed important to deepen cooperation and act pragmatically to achieve mutual benefit, increase interpersonal exchange, strengthen public support and create an institutional mechanism by which to improve the cooperation platform.

In this respect, China is expected to actively intervene, with President Xi Jinping the guarantor of success and fulfillment of such desires, in particular, to extend and deepen its influence on South–South cooperation, as it has already taken place to become a global platform of influence in international relations. Thus, the BRICS as a driving force at this summit had to sustain global economic growth accompanied by the promotion of democracy in international relations. According to Quiliconi and Kingah (2016), the BRICs involved the G20 in the process to "mask" their domestic differences, using G20 to follow up on their intentions to reform international financial institutions. However, the same author points out that the BRICS, despite their manifest desire to change the system, have focused more on increasing their participation in the global system than on properly proceeding with or contributing to their change.

Of the G20 issues in which the BRICS have shown interest, we highlight the following: (1) a better representation of the international financial institutions; (2) the Chinese devaluation resulting from increased imports of Chinese industrial products—an especial concern of Brazil, India and South Africa; (3) Brazil and China's criticism of the policy of US monetary expansion (India opted for the contrary position, demonstrating its support); (4) Russia's quest to restore confidence in the European investment market and to resolve the debt issue. From these common positions, Brazil continues to opt for cooperation to overcome the European crisis, in addition to requesting urgent reform of the IMF; the creation of intra-BRICS financial mechanisms, together with an increase in resources in the banking sector, namely the BRICS development banks, is the position assumed by Russia, India, China and South Africa (Quiliconi and Kingah 2016).

On the other hand, one of the major challenges of the BRICS in its claim to global leadership in place of the United States and the reform of the financial system dominated by the World Bank and its agencies is that each member country has to fight its isolationism in intra-BRICS logic by directing their bilateral relations and, clearly, in building a common global agenda. It is essential the BRICS develop policies of cooperation not only in matters of economic or finance, but in the large areas of employment, health, education, agriculture, environment, security and defense.

In this context, according to Käkönen (2013), each of the BRICS has its own internal vision of the group, being these: (1) Brazil, the connection of its interests with a South–South cooperation logic, specifically with economies such as India and South Africa, meaning an alternative channel of financing; (2) China, to share its approach to the concept of sovereignty, since it already acts on the world stage as a global power; (3) India, the possibility of exploring strategic options in the Indian Ocean region, beyond the feasibility of establishing multiple bilateral free trade areas with all countries currently contributing to global growth, adding Turkey, Indonesia and Nigeria, and it is of the utmost importance that China remain subject to a multilateral system and a rules-based international order.

India recognizes, at the global level, the relative decline of the alliance system around the United States is already evident; the US administration has been aware of this reality for more than a decade, presenting Trump with the challenge of finding a viable global strategy and the various regional levels (Khilnani et al. 2012); (4) Russia, in Lavrov's perspective (2012), taking the BRICS group as "a key piece of global chess" and supporting them as a new model of global relations. Replacing the former East–West and North–South relations is a key priority of Russian foreign policy; the Russia–India–China trilateral format is of equal importance, in addition to the demand for diplomatic support around Turkey, Egypt, Algeria, Iran, Saudi Arabia, Syria, Libya and Pakistan, among others; (5) South Africa, the continuation of the Bandung Conference and South–South relations, inserted in the African context of integration. Indeed, this group generates a sense of the search for an alternative order in the international community, in particular, the United States, which could put the dominant powers at risk and instigate violence and confrontation in international politics.

## 3 Polycentric Dynamics in the International Arena

BRICS individually or collectively may determine the health of the international market and the system of multilateral trade. In this sense, there are two ways of avoiding friction: on the one hand, it consists of a broad group of actors with new forms of thinking and acting on the social and economic consequences of BRICS globalization; and, on the other hand, it entails coming up with a strategy that involves the developed countries in their dialogue on the exploitation of the BRICS' place in a constantly changing world.

Under the chairmanship of Vladimir Putin (7 May 2000–7 May 2008), Russia becomes a world power and a counterbalance to Europe and plays a key role in the supply of hydrocarbons and geopolitical configuration energy. It should be emphasized that, according to Igor Sergeyevich Ivanov, Russia's former Minister of Foreign Affairs, Russia aims to enjoy full recognition by the European Union of the internal economic market, the marketing of Russian goods on EU soil, diversification of economic relations, exports to the APEC countries, the increase of foreign direct investment (FDI), and the representation of the national interest in the development of commercial relations, especially in legal matters. For Ivanov, external relations are part of a logic of expansion of multilateral and bilateral relations, where the intellectual sphere is properly present in initiatives of a scientific, cultural and educational nature (Ivanov 2002).

As far as foreign relations are concerned, Russia is moving forward in a dynamic and creative process, one oriented towards the individual, guaranteeing the interests of Russian citizens, actively supporting national entrepreneurs when penetrating foreign markets and, finally, supporting the development of international contacts in culture and science.

In general BRICS terms, let us note Brazil as a regional giant; Russia as a key factor in energy and within the framework of overall policy stability; India as having the largest global democracy and potential consumer market; and China being the largest authoritarian state with a capitalist economy as well as a huge consumer market. There is a positive development of the BRICS in terms of infrastructure development beyond its adoption of a new energy model: the Growth Acceleration Program (PAC) is accompanied by an increase in funds in the areas of social intervention, such as in the education system at child and middle/junior level.

However, in the economic context, we must bear in mind the volatility and instability of markets, which creates a considerable risk for both the BRICS and internationally. Within this context, it should be noted that the BRICS are associated in the international arena with poor countries, and therefore constitute not only the voice of the peoples of the 3A's (Africa, Asia, Latin America) but also form a kind of buffer zone.

Within the BRICS, we mention that Russia does not attach importance to the economic growth of Brazil, India and China due to these countries' poor contribution to Russian development. For the moment, the Russian government focuses its attention on the West in the context of technological imports. The biggest risk for Russia in terms of the global BRIC market is primarily its banking and financial sectors, in addition to the fact that it is characterized by its paternalistic government, its lack of transparency, its noncompliance and non-enforcement of environmental regulations.

Although we see a positive spillover of economic benefits within the population, an equitable distribution of these is lacking, which causes and generates disparity between social classes, between the richest and the poorest. However, if there is to be a positive effect on the world order, it is essential there be verifiable development in key sectors, such as those relating to economic and social issues. At the same time, countries such as Russia, India and China need to be strengthened in the technological area, and there needs to be a strategic implementation of the educational system.

Among the BRICS, Russia has positioned itself in the last 7 years with a GDP growth rate of around 7%, making it the tenth largest economy in the world, corresponding to GDP per capita of around US $10,000, building on a strong parallel economic and social growth during Vladimir Putin's presidential term. In comparative terms, since Russia is an economy based on the export of energy resources, it is closer to the Brazilian model than to the Indian or Chinese. It should be noted that although each BRICS country is quite different in overall, they are similar in social inequalities, especially in the distribution of wealth and the high poverty rate, in addition to how the government is involved in the functioning of the economy.

Russia has strengthened its economy to better meet the social conditions of the population through the systematic injection of capital in the short term. In addition, Russia sets up a stabilization fund from funds raised in energy exports through a balanced national budget. Indeed, one of the main problems for BRICS is the investment sector, specifically FDI, as these economies do not have sufficient funds to support the development of the global market. In the Russian context, the problem is at the governmental level due to the absence and insufficiency of policies aimed at attracting FDI; for example, a government monopoly in the natural gas sector and eventually in the oil sector.

In the case of China, export productive capacity is based on a mass-produced manufacturing industry, despite a diversified economy in which we find new technologies. However, in terms of sustainable development, if India and China follow similar paths, such as those taken by Japan and South Korea, they should certainly see substantial improvement in the areas of environment and income.

In fact, when addressing potential problems within the BRICS from a market and investment perspective, India and China are notorious examples of the poverty reduction effort, even as both economies gradually position themselves as potential centers of economic growth in the global economy. Of course, when we compare India and China to the United States, Japan and an organization like the European Union, we see different realities in both the cultural and social contexts: industrialized economies are guarantors of greater stability than emerging ones, owing to their immense monetary reserves.

At the global level, both India and China stand as the major economic forces. The main driver of economic growth was the opening of the global trading system, which has led to structural changes in the internal economies of these two countries, namely through the creation of private institutions to eliminate trade barriers. In addition, India and China capitalize on access to lower-cost labor and financing for technological development. BRICS economies choose increasing their economic levels over adopting preventive measures against environmental impact and global warming. If India and China maintain their current pace of economic growth using the same amount of energy, they will put enormous pressure over the next two decades on the world's energy resources.

We note that India's economic growth marks an average of 8%, despite no accumulation of capital or FDI. The areas of potential growth to attract FDI are those linked to the infrastructure sector and the retail industry. According to Goldman Sachs, India may be the long-term economic engine of the world, notably by building new rail networks in the country that can boost the transport sector, vehicle manufacturing and housing construction, creating a suburban corridor along this new network linking the entire country (O'Neill 2007).

The development of information technologies in India provokes an attractive offer for students seeking new areas of study in various knowledge-oriented professions. Alongside technological development, we have seen unparalleled urban growth, representing 10 of the 30 developing world. At the beginning of the decade of 2001, India had 35 cities with more than 1 million inhabitants.

In China, one problem of economic growth is seen in the aging of the population, brought on by the one-child policy. However, for Asian neighbors, China represents a center of investment opportunities in the industrial economy sector as well as a hub for multinationals, who can live there better off than they do in their countries of origin. It should be mentioned that consumption habits, with the evolution of time, will change as human capital is added, accompanied by urban growth and more luxury goods, travel, entertainment, media, fashion and acquisition industries.

On the other hand, in the Brazilian framework we have verified that savings and investment are not a political priority, since if they were, a fiscal adjustment and an improvement in the population's living conditions would have occurred to attract

FDI. To this end, the da Silva administration's policy (mandates: first, 2002–2006, second, 2006–2010) should focus on greater economic opening to substantially increase the investment to be applied in education, structural reforms to make productivity competitive. These measures would result in continued economic growth by 2050, bearing in mind that the main challenge is to restore global investor confidence so that emerging economies and industrialized countries can regain lost ground, in addition to a political circumvention of an international crisis.

In the energy and environmental framework, it is worth noting that Brazil in the last 6 years, despite its investment in energy infrastructures, abandons a clean energy policy and continues to burn forests, showing little respect for the Kyoto Protocol.

Similarly, we can mention that Russia is not concerned about the dangers of nuclear waste when it abandons its submarines in the North Seas and experiments with bacteriological weapons. It invests little in energy exploration, demonstrating an absence of orientation toward care and control of the environment. Members of the Russian mafia have tried to obtain nuclear missiles and atomic waste for ilegal trade, taking advantage of the government's policy of abandoning nuclear weapons and equipment.

China maintains the same position in the use of its energy sources, even provoking acid rain, despite sending a positive message to the international community. The industrialization of India and China demands energy resources, namely oil, and consequently puts pressure on crude oil prices and global demand/supply. It is estimated that India over the next 15 years will overtake China in this area and will need to consume increasing amounts of energy. In turn, both Brazil and Russia approach Japan's energy consumption habits, but will only reach Japan in this sense by the year 2025 (O'Neill 2006). In terms of world demand for oil, the value for Brazil, by 2025 is 4%; for India, by 2025, 7%, and in the case of Russia, 4%; in contrast, in the BRICS, China is highest at 16%. As a global reference, Japan will increase to 4% and the US to 19%, attesting to a decline in demand for oil, through implementation of renewable energy resources based on a long-term policy on environmental protection.

# 4 Conclusions

In the BRICS, poverty translates into a problem of the utmost gravity: education and the role of women. In India, someone with the most fortunate means, such as a driver, will tell his son to study rather than to pursue his father's line of work. This example reveals that it should be noted that, in the education sector, the BRICS has been developing rapidly and the sector has become more rigorous.

Taking into account cultural differences and social systems among the BRICS, the path of mutual understanding is the best way to prevent friction. Governments need to face the problems of misery and social contrasts in order to succeed.

Coming from the complexity of financial integration, no state even the most advanced will be able to guarantee their immunity from the impact of a new global economic crisis. The BRICS, clear examples of postcolonial formations, have been accelerating their economic growth, holding a regional market whose accessibility translates into a competitive advantage in productive expansion of their capacity, keeping them independent of the OECD club.

Russia might learn by following the example of China in business if it intends to increase investment and its position in the BRICS. It is noted that in an endogenous way, the market continues to operate according to the will of investors and analysts, both in gains and losses, inciting instability. In exogenous terms, the most diverse institutions for development—finance and universities—need to operate in a stable environment.

The emergence of BRICS in this twenty-first century is a testament to the fact that we are in an era of change, in a world where powers such as the United States and the EU continue to play a key role in reconfiguring international relations (Martins 2015). The emergence of new actors generates ambition and attempts to change the situation from geopolitics to the global economy in a world of marked contrasts. It thus increasingly appeals to a paradigm shift or a gradual transition to another system where the exercise of global leadership beyond American influence in politics and international relations, bearing in mind that the problems that humanity faces in your planet require solutions and initiatives that seek to promote multilateral cooperation.

In fact, the BRICS intend to assume a position in the international scene, even though domestically they face problems of the greatest dimension in terms of poverty, health, social inequalities and human rights and in judicial arena. In order to contribute to the transition or rebalancing of the international system as an alternative to US hegemony, it is difficult for the BRICS to impose their will or to replace this European Union, even in a marked crisis and an uncertain future, or if their role overlaps with that of the United Nations, ranging from humanitarian to guarantor of world security.

## References

Ivanov, I. S. (2002). *The new Russian diplomacy* (pp. 141–182). Washington, DC: The Nixon Center, Brookings Institution Press.

Käkönen, J. (2013). *BRICS as a new constellation in international relations?* IAMCR 2013 Conference Dublin, 25–29 June 2013. http://www.uta.fi/cmt/tutkimus/BRICS/materials/BRICS%20as%20a%20New%20Constellation%20in%20International%20Relations51.pdf

Khilnani, S., Kumar, R., Bhanu Mehta, P., Menon, P., Nilekani, N., Saran, S., et al. (2012). *Nonalignment 2.0. A foreign and strategic policy for India in the twenty first century*. New Delhi: National Defence College and Centre for Policy Research.

Langenhove, L. V., Zwartjes, M., & Papanagnou, G. (2016). Conceptualising regional leadership: The positioning theory angle. In S. Kingah & C. Quiliconi (Eds.), *Global and regional leadership of BRICS countries* (pp. 13–27). New York: Springer.

Lavrov, S. (2012). *BRICS: A new generation forum with a global reach*. Moscow: Ministry of Foreign Affairs.

Martins, M. A. (2015). The BRICS in the global order: A new political agenda? In M. Rewizorski (Ed.), *The European Union and the BRICS* (pp. 27–37). Switzerland: Springer. https://doi.org/10.1007/978-3-319-19099-0-3

O'Neill, J. (2013). *The BRIC road to growth (perspectives)*. London: Publishing Partnership.

O'Neill, J. (2007). *BRICs and beyond*. New York, NY: Goldman Sachs Group.

O'Neill, J. (2006). *The world and the BRICs dream* (pp. 37–54). New York, NY: Goldman Sachs Group.

Quiliconi, C., & Kingah, S. (2016). Conclusions: Leadership of the BRICS and implications for the European Union. In S. Kingah & C. Quiliconi (Eds.), *Global and regional leadership of BRICS countries* (pp. 243–253). New York: Springer.

Soko, M., & Balchin, N. (2016). South Africa's quest for leadership in Africa: Achievements, constraints and dilemmas. In S. Kingah & C. Quiliconi (Eds.), *Global and regional leadership of BRICS countries* (pp. 225–242). New York: Springer.

# Reducing Catastrophic Climate Risk by Revolutionizing the Amazon: Novel Pathways for Brazilian Diplomacy

**Joana Castro Pereira**

## Contents

1 Introduction ........ 190
2 The Amazon as a Global Natural Asset for Averting Catastrophic Climate Change ...... 192
2.1 The Need for Securing and Enhancing the Resilience of Forest Biomes ........ 192
2.2 The Key Role of the Amazon Rainforest: Risks and Opportunities ........ 194
3 Brazilian Governance of the Amazon (2005–2018): From Progress to Regression ....... 196
3.1 Curbing Deforestation (2005–2012): The Rise and Fall of Reformist Forces ....... 199
3.1.1 The Controversy Over Hydropower in the Amazon ........ 201
3.1.2 Biofuels: Ethanol and Biodiesel as Drivers of Amazonian Deforestation? .. 203
3.2 Reversing the Trend (2013–2018): The Dominance of Conservative Forces ....... 205
4 A New Development Paradigm for the Amazon and Novel Pathways for the Brazilian Diplomacy ........ 207
5 Conclusion ........ 211
References ........ 213

**Abstract** The Anthropocene poses profound challenges to the conservation of the Amazon rainforest, a global carbon reservoir. The forest's unsustainable model of development, based on severe land use changes, and the effects of anthropogenic climate change may result in irreversible damage. In addition, crossing tipping points for the survival of the Amazon rainforest may trigger catastrophic climate change (CCC), which reflects the urgent need for a new development paradigm for the region. However, the prospects of the dieback of the Amazon as well as of a CCC scenario are generally overlooked in the academic and political debates. This chapter aims to fill this gap by exploring the link between irreversible environmental changes in the forest and potential CCC and by analyzing the challenge of sustainable development in the Brazilian Amazon, demonstrating how close a climate catastrophe might be; it then presents a novel development paradigm for the region. The revolutionary technologies of the Fourth Industrial Revolution are proposed as central to this necessary paradigm shift. Because the Amazon rainforest is a global

J. C. Pereira (✉)
Portuguese Institute of International Relations (IPRI-NOVA), Lisbon, Portugal

Lusíada University, Porto, Portugal
e-mail: mail@joanacastropereira.com



T. Sequeira, L. Reis (eds.), *Climate Change and Global Development*,
Contributions to Economics, https://doi.org/10.1007/978-3-030-02662-2_10

natural asset for avoiding CCC, Brazil is also advised to follow a diplomatic plan that upholds fundamental ecological principles for creating a safe space for humanity, thus attracting international investments to manage and preserve the forest.

**Keywords** Amazon · Anthropocene · Brazil · Catastrophic climate change · Deforestation · Fourth Industrial Revolution

## 1 Introduction

Humanity has become a dominant force in driving global environmental change. Breaking with the relative environmental stability and predictability of the Holocene and profoundly changing the Earth system, human actions have become the drivers of the planet's transition to a new geological epoch, the Anthropocene (Crutzen and Stoermer 2000). Complex, uncertain, unstable, and dangerous, the Anthropocene threatens economies, societies, and ultimately the survival of the human species by challenging scientists, policymakers, and practitioners who address environmental conservation and management (Golladay et al. 2016; Zalasiewicz et al. 2010).

Under the Earth's new geological conditions, policymaking and academic and public discussions should consider the risk of transgressing thresholds that trigger abrupt, non-linear, and irreversible environmental changes. However, in a complex and uncertain environment, it is nearly impossible to define exactly where these thresholds are (Baum and Handoh 2014). Therefore, renowned scientists have established a set of planetary boundaries (PBs) that delimitate "a safe operating space for humanity"; if these boundaries are breached, humanity will probably risk catastrophic environmental events, since such breaches will bring humanity closer to crossing the expected Earth system's thresholds (Rockström et al. 2009). Alarmingly, scientists believe that humanity has already crossed four of the nine identified PBs, namely, altered biogeochemical cycles, land system change, biosphere integrity, and climate change. They suggest that crossing one boundary may cause others to be crossed, creating a "cascading collapse"; they also suggest that biosphere resilience and climate change are key boundaries that may drive the planet into a new state on their own if they are substantially and constantly crossed (Steffen et al. 2015). This is problematic, especially because the latest scientific research indicates that human actions are causing the climate to change 170 times faster than natural forces (Gaffney and Steffen 2017) and that the global rate of species extinction and population decline is more serious than previously thought, possibly leading to negative cascading effects on the functioning of ecosystems (Ceballos et al. 2017).

Political action has been profoundly disconnected from the gravity of planetary reality. Even if fully implemented, the pledges that countries have made under the Paris Climate Agreement have less than a 10% probability of keeping the planet below 2 °C above 1900 levels and a >50% chance of exceeding 3 °C.[1] Since Parties are not currently on track to meet their Nationally Determined Contributions (NDCs), the risk may be far greater than 50%. According to the Intergovernmental Panel on Climate Change (IPCC 2013), to stand a reasonable chance (>66%) of staying below 2 °C, humanity must not exceed a remaining global carbon budget of 225 GtC.

There is now a scientific consensus that self-reinforcing feedback loops are established once temperatures exceed a certain limit, potentially leading to a rapid increase in the concentration of greenhouse gases (GHG) in the atmosphere. Even a global warming increase of 2 °C may be sufficient to trigger tipping points in the biosphere that would push the planet into a completely new state (Abe et al. 2017). Considering that 2017 (+0.84 °C), 2016 (+0.94 °C), and 2015 (+0.90 °C) are the three warmest years since 1880[2]—and that 2017 experienced near-record temperatures even in the absence of El Niño events[3]—there are reasons for concern. Climate scientists believe that global temperatures may exceed 1.5 °C in the next one to two decades, placing the planet at a potentially dangerous level of climate change (Henley and King 2017). Considering this scenario, the prospect of a global environmental catastrophe should not be overlooked.

Nevertheless, academic, political, and public debates rarely acknowledge global catastrophic risks (Abe et al. 2017; Bostrom 2013; Rees 2013; Wiener 2016) such as catastrophic climate change (CCC). According to the definition of Bostrom and Cirkovic (2008)—commonly used in the still embryonic literature on global catastrophic risks—a global catastrophic event would potentially affect at least a large part of the current population or at most all future generations; it could cause significant harm while not destroying the quality of life completely, or, in the worst-case scenario, it could lead to "the premature extinction of Earth-originating intelligent life or the premature and drastic destruction of its potential for desirable future development" (Bostrom 2013, p. 15). To be considered catastrophic, an event would have to end the lives of no less than nearly 10% of the global population or have a comparable impact (Cotton-Barratt et al. 2016). CCC, one of the most pressing and challenging global catastrophic risks facing civilization, would probably cause this massive level of destruction. A CCC scenario would entail a global average temperature increase beyond 3 °C above preindustrial levels, a severe rise in sea levels, large-scale species extinction, substantial risks to regional and global food security, the combination of high temperature and humidity jeopardizing normal human activities, etc. (IPCC 2014a; King et al. 2015; Xu and Ramanathan 2017).

---

[1]Data from the Climate Action Tracker, available at http://climateactiontracker.org/global.html, accessed 24 January 2018.

[2]Data from the National Oceanic and Atmospheric Administration (NOAA), available at https://www.ncdc.noaa.gov/cag/time-series/global, accessed 15 June 2017.

[3]El Niño events usually magnify global warming; the hottest years on record correspond to the occurrence of these events. The first half of 2016 experienced a strong El Niño and was the warmest year ever recorded (NOAA 2017).

CCC would potentially be triggered by crossing one or more tipping points of the Earth's climate system.

The dieback of the Amazon rainforest is one of the tipping points in the Earth system[4] that may push the planet into a global climate catastrophe. Natural scientists suggest that the resilience of this vital biome may be deteriorating (Brienen et al. 2015; Fearnside 2012, 2016a; Rockström et al. 2016; Zemp et al. 2017). The forest's unsustainable model of development, based on severe land use changes, and the effects of anthropogenic climate change may result in irreversible damage; this reflects the urgent need for a new development paradigm for the region. Nevertheless, social scientists who study the Amazon as well as policymakers have generally not considered catastrophic climate risk and the link between the dieback of the forest and CCC thus far.[5]

This chapter aims to fill this gap by presenting the Amazon as a global natural asset in avoiding CCC (Sect. 2); analyzing the challenge of sustainable development in the Brazilian Amazon (which accounts for nearly 70% of the whole forest) and demonstrating the profound inconsistencies between the country's Amazonian politics and policies and the action needed to avert a climate catastrophe (Sect. 3); and suggesting a novel development paradigm for the region (Sect. 4). The revolutionary technologies of the Fourth Industrial Revolution are argued to be key for the necessary paradigm shift in the Amazon. Brazil is also briefly advised to follow a diplomatic plan that upholds the fundamental ecological principles for creating a safe space for humanity, thus attracting international investments to manage and preserve the forest.

## 2 The Amazon as a Global Natural Asset for Averting Catastrophic Climate Change

### *2.1 The Need for Securing and Enhancing the Resilience of Forest Biomes*

Global climate stability implies rapid and deep decarbonization, negative carbon emissions in the medium-term, and increasing biosphere resilience. During the next decade, Herculean efforts will be needed to decarbonize the world's economy and safeguard the Earth's climate. Abrupt carbon emission reductions within 40 years

[4]The others are the melting of Greenland's ice shelf, Arctic sea-ice loss, permafrost and tundra loss, boreal forest dieback, Indian monsoon chaotic multistability, Atlantic deep water formation, Sahara greening, a West African monsoon shift, a climate change-induced ozone hole, a change in the El Niño–Southern Oscillation amplitude of frequency, changes in Antarctic bottom water formations, and the instability of the West Antarctic ice shelf (Abe et al. 2017).

[5]Natural scientists widely recognize the gravity of the climate risk; however, CCC has been off of most social sciences' radar thus far.

starting in 2020 and a new carbon sink on the scale of the ocean sink are required if humanity wants to prevent dangerous climate change. Securing and enhancing the resilience of terrestrial biosphere carbon sinks and stocks are thus keys to meeting the 1.5 °C goal of the Paris Climate Agreement; hence, large forest biomes must be given top priority (Figueres et al. 2017; Rockström et al. 2016). Forests play an integral part in the carbon cycle. By absorbing and conserving carbon in soils and vegetation, they offset a large fraction of global anthropogenic emissions, making them the largest terrestrial stock by holding more than three-quarters of all above-ground terrestrial carbon (UNEP 2014). Some of the IPCC's Representative Concentration Pathways (RCPs) assume that carbon sinks will remain sufficiently robust to continue buffering anthropogenic interferences in the Earth's climate system (IPCC 2014b). However, this assumption is very uncertain.

First, it is worth noting that between 2013 and 2017, land temperatures persistently registered an annual average anomaly above 1 °C and 2016, the warmest year, reached +1.45 °C.[6] As global warming becomes more severe, forest droughts will certainly become more frequent, potentially reducing the resilience of terrestrial biosphere carbon sinks and stocks. Second, anthropogenic disturbances to forests (conversion to agricultural use, logging and other resource extraction, induced fires, hunting and wildlife trade, the spread of introduced species and pathogens, and human-induced changes to the climate) are seriously affecting the basic forces that determine the structure, functioning, and species composition of forests, possibly leading to ecological outcomes that we cannot foresee (Malhi et al. 2014). Under so many pressures, forest biomes may be pushed towards tipping points and alternate states of vegetation cover, releasing massive amounts of carbon (Reyer et al. 2015). However, the IPCC's carbon budget of 225 GtC does not consider carbon loss related to forest dieback, a self-reinforcing process that could potentially further destabilize the climate system (Abe et al. 2017).

In fact, as we shall see below, the resilience of vital biomes such as the Amazon rainforest is at risk. Rockström et al. (2016, pp. 468–469) assert that

> nations would be unwise to take for granted the robustness of the biosphere's capacity to buffer anthropogenic interferences. Carbon-cycle feedbacks may lead to long-term reservoir failure. (...) the restoration of ecosystems and large biome resilience must be on top of the agenda in the immediate future to ensure that critical sub-components of the Earth system continue to provide the services we take for granted.

Consequently, developing strong political coalitions to make significant public investments in conserving and restoring forests is fundamental. Urgent research and financial support are needed to promote reforestation projects and to establish carbon dioxide removal methods such as the afforestation of degraded land and the combination of second- and third-generation bioenergy production with carbon capture and storage (BECCS) (Rockström et al. 2017). The fact that terrestrial carbon stocks are vulnerable to human activities and natural events that release carbon back into

[6]Data from NOAA, available at https://www.ncdc.noaa.gov/cag/time-series/global, accessed 21 January 2018.

the atmosphere reinforces that need to promote their resilience and develop effective carbon stock management techniques and practices (Read 2008). This is particularly pressing in the case of the Amazon, a vital biome for the planet's climate stability that has been suffering from the effects of the Anthropocene.

### 2.2 *The Key Role of the Amazon Rainforest: Risks and Opportunities*

The Amazon holds the greatest continuous tropical forest in the world, covering nearly 5.5 million $km^2$. The forest provides a variety of key services at local, regional, and global levels, such as air filtration, water cycling, soil stabilization, the maintenance of biodiversity, and climate regulation. It is estimated that the forest stores approximately 50% of all tropical forest carbon, i.e. nearly 150–200 GtC (Brienen et al. 2015).

Some studies suggest a threshold for the dieback of the Amazon rainforest of 4 °C of global warming or deforestation exceeding 40% of the forest area (Nobre et al. 2016). Others indicate that the threshold may be closer, possibly at 20% deforestation (Fraser 2016). Crossing these tipping points would probably have massive consequences; the establishment of savanna-type vegetation would dry soils and carbon would be lost (Marengo et al. 2011). Since the Amazon is one of the Earth's climate system tipping points, the savannization of the forest could potentially trigger CCC. In the opinion of the IPCC experts, "the possibility of a critical threshold being crossed in precipitation volume and duration of dry seasons [in the Amazon] cannot be ruled out" (IPCC 2013, p. 71). In fact, the Amazon has warmed approximately 1 °C over the last 60 years and has already been deforested by almost 20% (Nobre et al. 2016). As we shall see in Sect. 3 in relation to the Brazilian Amazon, deforestation has been increasing again following a substantial reduction between 2005 and 2012.

According to Brienen et al. (2015), the Amazon acted as a strong carbon sink between 1990 and 2007, absorbing 0.42–0.65 GtC/year, which means that it was responsible for the uptake of 25% of the carbon absorbed by all terrestrial carbon sinks on Earth during that period. Due to the higher concentration of carbon dioxide in the atmosphere, which presumably increases the rate of photosynthesis, trees are believed to have been growing faster since the late 1980s—a phenomenon known as the carbon fertilization effect. As a result, the Amazon's vegetation has stored more carbon and helped to limit the effects of the increase in carbon dioxide concentrations. However, the authors' data also reveal that although the Amazon's mature forests acted as a biomass carbon sink between 1983 and 2011, there is a long-term declining trend of carbon accumulation (a decrease of 30% from the 1990s to the 2000s) because faster growing trees are dying sooner than expected. In other words, trees have been living faster and, consequently, dying younger. This phenomenon is also described by Allen et al. (2015), who warn about the underestimation of global vulnerability to tree mortality and forest die-off. In addition, as Monteiro et al. (2014, p. 693) stress, one cannot neglect that "numerical models suggest that vegetation loss leads to increased surface temperatures and loss in rainfall, which in turn can cause further losses in rainforest and a breakdown in the equilibrium between these parts". Actually, a recent study by Baccini et al. (2017) suggests that the world's tropical forests are already a net carbon source.

Droughts are integral to the natural climate variability of the Amazon; however, in 2005, 2010, and 2015, the Amazon experienced unusually extreme droughts that will probably have long-term effects (Nobre et al. 2016). Moreover, these droughts will certainly become more frequent, as temperatures are rising in the region. According to Feldpausch et al. (2016), during the 2005 and 2010 droughts, the Amazon temporarily lost its ability to absorb carbon dioxide. The 2015 drought may have had worse impacts. Jiménez-Muñoz et al. (2016) suggest that during the 2015–2016 course of El Niño, the forest faced record-breaking high temperatures; the affected forest area was approximately 20% greater than in past extreme El Niño events. The same authors also noted that during the 2015 and 2016 dry seasons, forest fires intensified significantly in Brazil, further contributing to tree mortality and biomass loss. As stated by FAO (2012) experts, fires, deforestation, and climate change form a vicious circle in the Amazon and have the potential to degrade up to 55% of the forest.

Furthermore, even the Amazon's indigenous territories and protected natural areas, generally considered free from risk, have surprisingly been experiencing growing threats related to growth in the agriculture, grazing, mining, petroleum, timber, and transportation sectors. A study by Walker et al. (2015) reveals that more than one-third of the above-ground carbon in those areas is currently at risk of being released; when the entire Amazon is considered, this number rises to nearly half of the carbon stored above ground in the forest. This is a conservative study that does not quantify illegal extractive activities or projected deforestation threats; consequently, the risk may be greater than the authors suggest.

The need to protect, conserve, and restore the region thus seems greater than ever. Maintaining the forest's carbon stocks is a key ecological service and essential for achieving global climate security. Action should be taken to reduce and prevent emissions into the atmosphere by conserving existing carbon pools in soils and vegetation and enhancing the uptake of carbon in forest reservoirs (Smith et al. 2014). Providentially, reducing emissions from deforestation and increasing forest restoration is one of the most cost-effective climate solutions available to policymakers today (Chazdon et al. 2016; Stanfurt et al. 2015). As reported by Aragão et al. (2014), doubling the current area of secondary forests and avoiding the additional removal of primary forests would potentially help the Amazon counteract approximately 42% of global emissions related to land use change. This is technically and economically feasible as well as socially desirable (Stanfurt et al. 2015). Restoration of the Amazon forest by reforesting part of the 20% that has already been converted or degraded is hence a powerful tool in helping to limit the risk of CCC (Ometto et al. 2011). Furthermore, experts believe that if the Amazon's biodiversity is protected, the forest may hold the best solutions to creating the disruptive technologies essential for developing the necessary zero and negative emission energy sources that will sustain the future of civilization.[7] Accordingly, the Amazon is a global natural asset for averting CCC. However, as Fearnside (2013, p. 18) has noted,

> not yet incorporated into policy is the value of the environmental services of the forest (. . .) and the nearness of the threat posed by climate change killing the forest itself. Unfortunately, science and technology are often totally ignored in Amazonian policy making when the results are inconvenient.

The quality of governance in the Amazonian region, which has unfortunately declined dramatically since 2014, requires substantial improvement in order to conserve the forest. The future does not look promising.

## 3 Brazilian Governance of the Amazon (2005–2018): From Progress to Regression

To understand the Brazilian governance of the Amazon and the challenge of sustainable development in the region, it is important to look first at the evolution of the country's GHG emissions profile. Between 1990 and 2015, Brazil's gross GHG emissions increased by 17%; however, considering carbon dioxide removal from the atmosphere by land use changes (for instance, pasture turning into secondary forest) and the maintenance of natural forests in indigenous territories and conservation units, the country's GHG emissions in 2015 are roughly the same as they were in 1990.[8] Throughout those 25 years, the trajectory of emissions had distinct stages of growth and reduction (Fig. 1). The non-linear variations over time in the country's GHG emissions profile have been intrinsically related to emissions from land use, land use change, and forestry (LULUCF), especially deforestation in the Amazon forest (Fig. 2).

Brazil's GHG emissions profile has historically differed from that of the major GHG-emitting countries due to Brazil's relatively low-carbon energy matrix. In 2016, renewable energy sources accounted for approximately 44% of the country's primary energy supply, with biomass from sugarcane (17.5%) and hydropower (12.6%) representing the biggest shares of that percentage (EPE 2017). Unlike many emerging countries whose electricity comes mostly from fossil fuels, Brazil's electricity is mainly hydroelectric; hydropower accounts for approximately 68% of the Brazilian electric energy matrix (EPE 2017). LULUCF has historically been the

[7]Interview with Professor Roberto Schaeffer (from the Federal University of Rio de Janeiro), 16 April 2017.

[8]Data from the Brazilian Greenhouse Gas Emission Estimate System (Sistema de Estimativas de Emissões de Gases de Efeito Estufa—SEEG in Portuguese), using the Global Warming Potential metric provided by the IPCC in its Fifth Assessment Report (GWP–AR5), available at http://plataforma.seeg.eco.br/total_emission, accessed 21 March 2017.

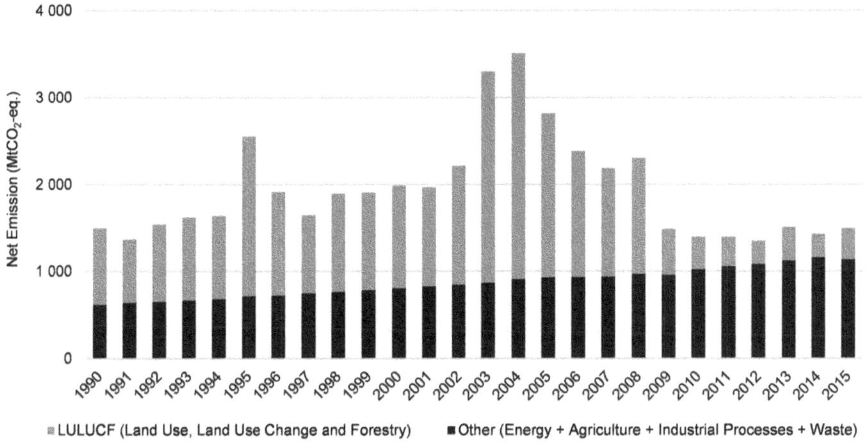


**Fig. 1** Brazil's net GHG emissions (MtCO2-eq.; 1990–2015). Source: Developed by the Author (2017) with data from SEEG (http://plataforma.seeg.eco.br/total_emission)

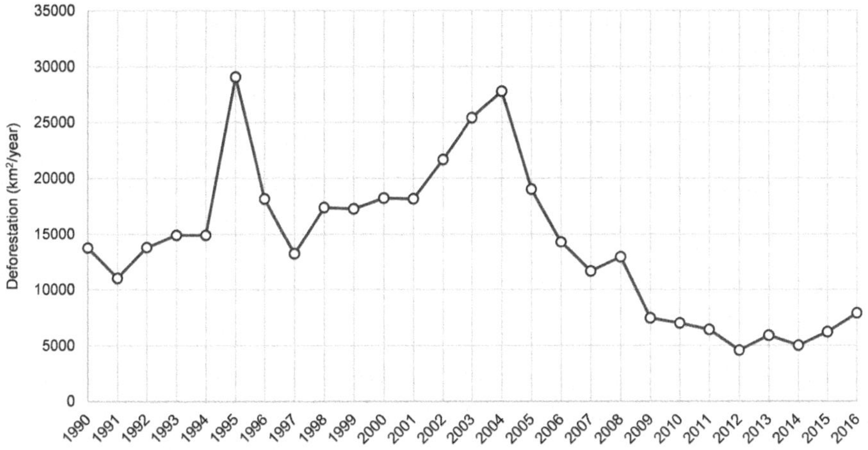


**Fig. 2** Annual deforestation in the Brazilian Legal Amazon (km$^2$; 1990–2016). Source: Developed by the author (2017) with data from INPE (http://www.obt.inpe.br/prodes/prodes_1988_2016.htm)

main source of the country's GHG emissions; in 2005, the sector accounted by about two-thirds of Brazil's net GHG emissions profile. However, since 2005, deforestation in the country has declined significantly to the point that between 2005 and 2012, Brazil reduced its overall gross emissions by more than 40%.[9] As reported in Brazil's third national communication to the United Nations Framework Convention on Climate Change (UNFCCC) (Brazil. Ministry of Science 2016), LULUCF's

[9] Data from SEEG, available at http://plataforma.seeg.eco.br/total_emission, accessed 21 March 2017.

share in Brazilian emissions dropped to approximately 26% in 2010; as a result, it was no longer the main driver of emissions, as it was overtaken by the agriculture (35%) and energy (27%) sectors.[10] In 2015, the emissions profile remained essentially the same.[11] Encouraged by these results, Brazil has stated in its NDC that it pledges to eliminate Amazonian deforestation by 2030 (Federative Republic of Brazil 2015); however, the current deforestation trend does not indicate that this goal will be achieved.

Deforestation is no longer the main source of Brazil's GHG emissions, but its volume is still substantial. Considering the deforestation rates in the Amazon since 2013, there are reasons for concern. Although deforestation has decreased by more than 70% since 2004 (when forest loss reached approximately 28,000 $km^2$), 2016 brought the largest reported forest loss in the region since 2008. Between 2015 and 2016, deforestation increased by nearly 29%; the Amazon has lost almost 8000 $km^2$ of forest (Fig. 2).[12] Illegal logging, cattle ranching, and soybean farming have been the main drivers of deforestation.

According to Viola and Franchini (2013), environmental governance is the result of the interactions between the reformist and conservative forces that constitute the government, the business sector, and the civil society. Reformist agents seek to push the system onto a low-carbon path, while the goal of conservative agents is to maintain the traditional carbon-intensive development paradigm. During the first decade of this century, Brazilian environmental movements contributed to and benefited from progress made regarding deforestation reduction and became more active; however, there have been several setbacks since 2010 affecting the reformist agenda (Schaeffer et al. 2015). Presently, the conservative forces prevail. The struggle between reformist and conservative forces is visible in Amazonian politics and policy.

As we shall see below, in addition to the dramatic decrease in deforestation rates, Amazonian politics and policy in Brazil from 2005 to the present were marked by (a) the slow development of hydropower projects in the region due to environmental restrictions and opposition by society; (b) the substantial encouragement of biofuels production caused by the emergence of flexible-fuel vehicles, which then faded after the discovery of the Brazilian pre-salt oil reserves (Viola and Basso 2014); and the recent (c) series of efforts to erode protected areas and weaken environmental licensing (Crouzeilles et al. 2017; Fearnside 2016b).

[10]Net emissions using the GWP–AR5 metric.

[11]Data from SEEG, available at http://plataforma.seeg.eco.br/total_emission, accessed 21 March 2017.

[12]Data from the Brazilian State Institute for Space Research (Instituto Nacional de Pesquisas Espaciais—INPE in Portuguese), available at http://www.obt.inpe.br/prodes/prodes_1988_2016.htm, accessed 8 February 2017.

### 3.1 *Curbing Deforestation (2005–2012): The Rise and Fall of Reformist Forces*

Deforestation became a concern in Brazil during the 1970s when the military government encouraged the occupation of the Amazon to ensure the national sovereignty of the territory, promoted migration to the region with the aim of avoiding land reform in highly populated areas of the country, and stimulated the exploitation of the land to produce commodities and improve the Brazilian balance of payments. During the late 1980s, when sustainable development concerns entered the international agenda and the Amazon started to attract drawing public attention, the Brazilian government took initial measures to combat deforestation (Carvalho 2012). However, as we have seen, deforestation rates were persistently high throughout the 1990s and the first half of the 2000s; 2005 was the turning point for the protection of the forest.

As stated by Viola (2013), reducing deforestation in the Amazon was at the core of the Brazilian government's agenda from 2005 to 2009. Between 2010 and 2012, the federal government's prime concern was no longer reducing deforestation but avoiding a new increase; deforestation continued to drop but at a slower rate. Deforestation reduction was the result of several factors: enhanced institutional capacity and stricter law enforcement; the creation of new and extensive protected areas; the formation of coalitions by various stakeholders to oppose the consumption of soybeans and beef produced in deforested areas; greater influence by national and international NGOs and the scientific community on the media; greater cooperation between the Amazon state governments and the federal government; and the cooperation agreement made between Norway and Brazil by which the Norwegian government pledged up to one billion dollars in results-based compensation through the Amazon Fund, the main fund for preventing, monitoring, and combating deforestation in the Amazon (Boucher et al. 2013; Viola and Basso 2014). The dramatic decrease in deforestation was accomplished with no apparent negative effects on the country's economic growth. In fact, despite the 2009 economic decline, Brazil's GDP has grown on average by 4% per year between 2005 and 2012 despite the abrupt fall in deforestation.[13]

We will now more closely examine the rise of reformist environmental forces in Brazil during the second half of the 2000s and their impact on Amazonian politics and policy. In 2003, Marina Silva, a renowned Brazilian environmentalist, was appointed to head the Ministry of the Environment (ME) during the first term of President Lula da Silva and made the conservation of the Amazon a priority in the federal government's agenda. In 2005, Marina finally gathered enough support to urge a profound shift in Amazonian politics and policy. The federal police and other federal agencies were ordered to cooperate closely with the ME to combat illegal

[13]Data from the Brazilian Institute of Geography and Statistics (Instituto Brasileiro de Geografia e Estatística—IBGE in Portuguese), available at https://www.ibge.gov.br/estatisticas-novoportal/economicas/contas-nacionais/9300-contas-nacionais-trimestrais.html, accessed 14 March 2017.

deforestation, which substantially increased law enforcement in the region. Simultaneously, more conservation units and constraints were created in the Amazon. Additionally, the Brazilian state's ability to monitor deforestation was significantly enhanced in 2007 with the substantial reinforcement of the scientific and technological capabilities of the Institute for Space Research (INPE). During the same period, NGOs such as Greenpeace and the World Wildlife Fund, supermarkets such as Carrefour, corporations such as Cargill, some local governments, members of the scientific community, and universities formed coalitions to draw attention to the irrational scourge of deforestation. These efforts had a considerable impact on the media and consequently on Brazilian public opinion. In 2008, Marina Silva was replaced by Carlos Minc, another renowned environmentalist, who was able to increase the influence of the ME in Brazilian politics (Viola and Franchini 2014).

The successful fight against deforestation weakened a belief that had long persisted in Brazil: the idea that deforestation was inevitable and the country would not be able to prevent it. Encouraged by the good results in terms of deforestation reduction, the Brazilian government started to shift away from some of the country's historical international positions regarding forests under the UNFCCC regime. Traditionally, Brazilian diplomacy has refused to recognize forests as subjects of international regulation to avoid external interference in the county's national sovereignty over the Amazonian region. During COP12 in Nairobi in 2006, Brazil proposed the creation of a global fund financed by the Annex I Parties to the UNFCCC and corporations to help curb deforestation. The fund's resources would be distributed according to performance in reducing deforestation. For the first time, Brazil was associating deforestation reduction with global financial tools; however, the country continued its traditional rejection of carbon markets (Viola 2013; Viola and Basso 2015).

In 2009, climate change was under scrutiny due to international expectations on the outcome of COP15, which was to be held in Copenhagen. In this context, the Amazon state governments, headed by the Amazon and Mato Grosso states, pressured the federal government into accepting the inclusion of a mechanism that had emerged within the UNFCCC in 2005, REDD+,[14] into the Clean Development Mechanism[15] or any other market mechanism, thus pressing Brazil to change its traditional position concerning forests in the international climate regime. Other requests, such as committing to reduce emissions until 2020 and negotiating avoided deforestation through market mechanisms, were made by corporate coalitions. In

[14]Reducing Emissions from Deforestation and Forest Degradation and the role of conservation, sustainable management of forests and enhancement of forest carbon stocks in developing countries. On REDD+ see, for example, Gupta (2012), Pistorius (2012), and Turnhout et al. (2017).

[15]The Clean Development Mechanism "allows a country with an emission-reduction or emission-limitation commitment under the Kyoto Protocol (Annex B Party) to implement an emission-reduction project in developing countries. Such projects can earn saleable certified emission reduction (CER) credits, each equivalent to one tonne of $CO_2$, which can be counted towards meeting Kyoto targets"—UNFCCC, available at http://unfccc.int/kyoto_protocol/mechanisms/clean_development_mechanism/items/2718.php, accessed 15 May 2017.

addition, Marina Silva announced that she was running for the Brazilian presidency, and she managed to bring the environmental and climate agenda into the spotlight during campaign discussions for the 2010 elections, forcing President Lula da Silva and his candidate, Dilma Rousseff, to address sustainability issues. These events were all very important in developing the Brazilian voluntary commitment to reduce GHG emissions by 36–39% from business as usual by 2020 presented at COP15 (Carvalho 2012). This pledge was included in the National Plan on Climate Change[16] in which measures for controlling deforestation in the Amazon strengthened previous policies for the region and whose results have met with considerable success. Hence, 2009 was marked by a substantial change in the Brazilian position within the UNFCCC negotiations. Traditionally conservative and unwilling to accept emission targets (with an attitude based on a North/South divide that placed all responsibility for global warming on developed countries), the country began to assume a more reformist position at that time (Pereira 2013a, b).

In 2010, Izabella Teixeira replaced Carlos Minc as Minister of the Environment, which marked the beginning of a conservative turn in the Brazilian ME. Between 2011 and 2012, the positive environmental trajectory in Brazil started to slow; conservative forces were progressively regaining ground, and the National Climate Change Policy became stagnant. The power of conservative forces, namely, the agribusiness lobby, was evident in 2011, when the Congress voted by a seven-to-one margin to largely reduce environmental protections in the Brazilian Forest Code, even though approximately 80% of the population was opposed to changes in the code (Fearnside 2016b; Pereira 2013b; Schaeffer et al. 2015). Even the ME supported the reform of the Forest Code—which paved the way for a new wave of deforestation in the Amazon and acted in synergy with the conservative economic growth paradigm of the new president at the time, Dilma Rousseff. The expansion of the oil sector, increases in gasoline consumption, the stagnation of the production of ethanol, and the continuous expansion of individual transport were other setbacks in this period that confirmed the fall of the reformist forces in Brazil. In fact, in 2012, the federal government bet on a carbon-intensive industrial impulse package to respond to the effects of the international crisis. In that year, environmental forces broke their alliance with the Workers' Party (Viola and Franchini 2014).

### 3.1.1 The Controversy Over Hydropower in the Amazon

Another significant development in this period strictly linked to Amazonian politics and policy was the resolution by some reformist forces to reinvest in hydroelectric power plants. Brazil holds one of the world's largest hydroelectric power potentials—in terms of volume of water, the Amazon basin is the largest in the world. The

[16]In Portuguese, Plano Nacional sobre Mudança do Clima (PNMC), available at http://www.mma.gov.br/estruturas/smcq_climaticas/_arquivos/plano_nacional_mudanca_clima.pdf, accessed 11 January 2017.

majority of the country's southern river basins have already been explored, and approximately 70% of the remaining Brazilian potential lies in the Amazon (Vieira and Dalgaard 2013). It is worth noting that most of the country's hydropower plants were built in the 1970s and 1980s when it was easier to obtain construction licenses; during the 1990s, higher environmental standards were in force, and licenses to build hydropower plants became much more difficult to obtain. Although hydropower expansion has positive effects in terms of GHG emission reductions, its negative environmental impacts—especially associated with the construction of big reservoirs—have attracted considerable resistance by some segments of the reformist sector of Brazilian society, especially within the environmental movement. The construction of hydroelectric reservoirs affects the ecological dynamics of ecosystems, altering hydrological, hydrosocial, and fundamental processes (Tundisi et al. 2014).

Although they usually have no concern for climate mitigation issues, conservative forces that were focused on traditional economic growth supported the move towards reintroducing hydroelectric power plants because of the relatively low price of hydroelectric energy (Viola and Franchini 2014). Hence, some segments of the reformist forces concerned with anthropogenic climate change agreed with conservative forces on reinvesting in hydropower in the Amazon, although for very different reasons. However, the segment of reformist environmental forces opposing such a move were able to press the government to opt for run-of-the-river hydroelectric power plants characterized by a low amount of water storage—which would presumably lower the environmental impacts—in the new hydropower plants built on the Amazon, namely, Jirau and Santo Antônio on the Madeira River. The direct consequences of the adoption of this type of technology have been an increase in the uncertainty regarding the real capacity of electricity production to match demands. With little or no stored water, the electric production of run-of-the-river hydropower plants is dependent on the river's run-off flow in a region with high rainfall variability and extreme vulnerability to droughts. To compensate for the intermittence of hydroelectric production, complementary energy sources need to be applied (Lucena et al. 2016; Schaeffer et al. 2013). For this purpose, the best option would be to couple run-of-the-river power plants with biomass; however, fossil fuel thermal power plants have played this role, exacerbating environmental impacts (Pereira 2013b).

On the Xingu River in the Amazon, the Belo Monte hydropower plant, whose project was redesigned during the second half of the 2000s to apply run-of-the-river technology, is under construction (Lucena et al. 2013). Once concluded, it will be the world's fourth largest hydroelectric dam in terms of installed capacity. This dam has been a controversial project, especially due to its effects on indigenous populations. Despite applying run-of-the-river technology, environmentalists claim that because this huge project floods swathes of forest, it will result in the release of significant amounts of methane, which would have a negative impact on climate change. These environmentalists also claim that the project will threaten ecosystems, including animal life, and displace indigenous communities (Vieira and Dalgaard 2013).

Amidst controversy in Brazil, alternatives have emerged, such as an ambitious project between Brazil and Peru regarding the construction of six hydroelectric power plants in the Peruvian Amazon; these plants would supply Peru, and the surplus would be exported to Brazil. Nevertheless, potential negative impacts on the environment and opposition by indigenous populations have also been an obstacle, and the project was blocked by local social movements; in addition, because of the huge corruption scandal engulfing Brazil's largest construction firm, Odebrecht,[17] which was involved in the project, it may be very difficult to make any progress over the next few years.

The future of hydropower in Brazil is unclear. The government has vast plans for the construction of dams and highways in the Amazon region that would probably have massive environmental consequences[18]; however, it is worth noting that the lack of production from some of Brazil's largest hydroelectric power plants due to recent droughts is increasing the awareness in the Brazilian energy sector that Brazil must reduce its dependence on hydropower. This need for change may open the country to the expertise of engineering firms that can increase efficiency and generation capacity in existing hydroelectric power plants through technological and engineering services; it may also encourage the expansion of wind and solar energy (Pereira 2017a). Furthermore, societal resistance to the construction of hydroelectric power plants in the Amazon remains very strong.

### 3.1.2 Biofuels: Ethanol and Biodiesel as Drivers of Amazonian Deforestation?

The exceptional encouragement to produce biofuels in Brazil in the second half of the 2000s also had implications for the Amazon. The country has traditionally been a producer of ethanol from sugarcane; as a result of the oil shocks of the 1970s, the use of ethanol as a substitute for petroleum derivatives, especially for gasoline, registered an exponential increase. At that time, Brazil was heavily dependent on crude oil imports, importing more than 80% of the oil it consumed, and the high oil prices in the international market led to the creation of the National Alcohol Program in 1975. Governmental incentives and subsidies promoted the establishment of large sugarcane plantations, especially in São Paulo and Pernambuco. In 1985, 85% of the cars sold in Brazil were powered by alcohol (EPE 2007). However, the fall in oil prices in the following years and a severe supply crisis led to a decline in the ethanol sector. Still, due to the compulsory use of anhydrous alcohol in the gasoline mixture, the existence of cars exclusively powered by hydrated alcohol, and the maintenance of the ethanol supply in more than 90% of the country's gas stations, ethanol has

[17]Now being sued in the ten Latin American countries in which it was operating.

[18]See Brazil's Growth Acceleration Plan (Programa de Aceleração do Crescimento—PAC in Portuguese), available at http://www.pac.gov.br/, accessed 24 May 2017.

survived (IPEA 2010). With the development of flexible-fuel vehicle technology—which allows consumers to alternate between ethanol and fossil fuel—at the beginning of the twenty-first century, ethanol was revived. In 2004, considering the international prices for oil, ethanol in Brazil became economically competitive with gasoline (Goldemberg 2008; Goldemberg et al. 2008). By 2010, flex-fuel cars accounted for more than 85% of light vehicle sales (EPE 2011).

In 2006, President Lula also launched the "ethanol diplomacy". Brazil's sugarcane-based ethanol production is far more efficient than the corn-based ethanol produced in the USA (Goldemberg 2008). Biofuels became an integral part of the international insertion strategy of the country; President Lula aimed to create the conditions for a global economy for Brazilian biofuels. Brazil has exported its know-how to several countries, such as some of its Central and South American neighbours and a number of African countries. The high point of "ethanol diplomacy" occurred in March 2007 when Brazil and the USA announced an agreement to develop this market. The dynamism of the Brazilian investment in biofuels was evident; in 2009, the country invested nearly 3.5 billion dollars, and between 1995 and 2009, ethanol production increased from nearly 13 million to 28 million cubic metres per year (Benavidez and Cadena 2011; Pinto et al. 2012).

Heated discussions occurred regarding the potential relationship between sugarcane plantations—which require large parcels of land—and deforestation in the Amazon; however, the dramatic decline in deforestation in the region revealed that such plantations were not the driver of this issue. Cultivation is currently concentrated in São Paulo and the surrounding areas. Still, as noted by Viola and Basso (2014), the expansion of sugarcane production in these areas has displaced cattle grazing and soybean farming to the Amazon region, which means that these plantations were partly and indirectly responsible for deforestation; however, it would be unreasonable to say that they were its main driver.

Additionally, "ethanol diplomacy" started to fade after the discovery of the pre-salt offshore oil reserves in 2007. In fact, President's Lula discourse shifted from the defence of ethanol as a substitute for fossil fuels to the promotion of an international image of Brazil as a future major exporter of oil (Vieira and Dalgaard 2013). Oil prices started to be used as a heterodox policy tool to artificially maintain greater economic growth rates, which translated into changes in the relative prices of gasoline/ethanol, compromising ethanol's competitiveness. Due to a number of obstacles, such as the postponement of the reserves' auctions due to disputes among state governments over the distribution of revenues, the shale gas revolution in the USA (which would probably be the largest market for Brazilian oil), and the loss of financial capacity by the semipublic Brazilian energy company, Petrobras, national euphoria regarding offshore oil reserves has declined substantially since 2013. However, this situation has led only to small changes in the relative prices of gasoline/ethanol, and the ethanol production has stagnated in recent years (Goldemberg et al. 2014; Viola and Basso 2015). However, at the time of the writing of this chapter, the Brazilian Congress passed a bill for a new federal programme for biofuels, RenovaBio,[19] whose aim is to expand the sustainable national production of ethanol, biodiesel, and other biofuels so that they account for 18% of the Brazilian energy matrix by 2030, helping the country to achieve its pledges under the Paris Climate Agreement.

This expansion plan may be particularly problematic in the case of biodiesel. Brazil is one of the world's largest producers of biodiesel. The Brazilian biodiesel market emerged in 2004 when the federal government launched the National Programme for the Production and Use of Biodiesel. Since January 2010, the diesel available in Brazil has contained a mixture of biodiesel (10% in 2018), thus ensuring a market for this fuel (La Rovere et al. 2013). The government plans to increase the percentage of biodiesel in diesel to 15% in 2023. Considering that soybeans are the main source of Brazilian biodiesel and that the country's freight sector uses mostly diesel, the production of biodiesel has contributed to deforestation in the Amazon in recent years. Nevertheless, it would also be unreasonable to say that biodiesel production is responsible for the deforestation rates of recent years, since the amount of soybeans exported for animal and human consumption is far greater than the share of soybeans used in biodiesel production (Viola and Basso 2014). However, governmental incentives and subsidies under the RenovaBio programme may change this scenario, especially considering that the current Brazilian government is profoundly conservative and environmental conservation is very low on its political agenda.

### *3.2 Reversing the Trend (2013–2018): The Dominance of Conservative Forces*

By 2013, Brazil was struggling with deteriorating public services, rising inflation, falling commodity prices, and a massive wave of street demonstrations. These protests triggered a chain of events that culminated in a serious economic recession and a cycle of political instability that continues to persist. These problems began as a complaint against the increase in public transportation fares, but they rapidly grew to include issues such as the quality of healthcare and education and corruption. In 2014, the country's GDP registered anaemic growth (+0.5%), followed by a profound recession in 2015 (−3.8%) and 2016 (−3.6%).[20] Additionally, operation *Lava Jato* (Car Wash), launched in the same year, exposed huge corruption scandals involving politicians, Petrobras, and several Brazilian infrastructure companies. In August 2016, President Rousseff was impeached.

---

[19]Available at http://www.mme.gov.br/web/guest/secretarias/petroleo-gas-natural-e-combustiveis-renovaveis/programas/renovabio/principal, accessed 27 December 2017.

[20]Data from the Brazilian Institute of Geography and Statistics (Instituto Brasileiro de Geografia e Estatística—IBGE in Portuguese), available at https://www.ibge.gov.br/estatisticas-novoportal/economicas/contas-nacionais/9300-contas-nacionais-trimestrais.html, accessed 14 March 2017.

Amidst a severe political, economic, and ethical crisis, environmental conservation in Brazil has suffered various setbacks. Public attention to environmental issues has seriously declined since 2014. Although Marina Silva has once again run for the Brazilian presidency, the environmental and climate agenda played a limited role during campaign discussions for the October 2014 elections. Instead, issues such as corruption, economic growth, public spending, inflation control, and employment prevailed during the 2014 presidential campaign and were Rousseff's top priorities during her second term (Schaeffer et al. 2015). The weak results of the Green Party in the municipal elections of 2016 are symptomatic of the reduced role that environmental issues currently play in the country's political agenda. Taking advantage of the country's chaotic situation, conservative forces have been trying to reverse environmental progress by reviving proposals that had been "dormant" for some time.

Between 2016 and 2018, the Brazilian government and its president, Michel Temer, who was supported by a coalition of conservative/particularistic interests, made efforts to soften environmental licensing and erode protected areas. Short-term economic plans to attract foreign investment, improve exports, and revive the country's recessive economy, as well as the political dominance of the Agribusiness Parliamentary Front—which accounted for 40% of the Brazilian Congress during the 2015-2018 legislature—have resulted in a number of anti-environmentalist bills and laws to weaken environmental licensing, suspend the ratification of indigenous lands, reduce the protection of 60,000 $km^2$ in the Amazon and Atlantic Forest, and allow those who have illicitly appropriated land to easily and cheaply legalize those holdings (Crouzeilles et al. 2017).

The country's environmental licensing system is fundamental for restraining infrastructure projects entailing very high impacts on the environment. PEC-65, a proposed constitutional amendment (pending since 2012), would allow any project to progress to completion as long as an environmental impact assessment was submitted, regardless of the assessment's content. Moreover, there is a bill (654/2015) also waiting for a full Senate vote that would allow any project considered "strategic" and of "national interest" to go through a simplified and accelerated environmental approval process—the usual sequence of 3 licenses (which commonly takes 4–5 years) would be reduced to 1 with a deadline of 8 months to approve the license; after the deadline, the project would automatically be authorized (Fearnside 2016b; Moutinho et al. 2016). As we have seen, Brazil has significant plans for building dams and highways in the Amazon. PL 1610/1996 is another anti-environmental bill currently being considered; it would authorize mining on indigenous lands (El Bizri et al. 2016).

It is worth noting that indigenous lands are particularly relevant since they have a positive record of resisting deforestation and protect a larger area than conservation units (Nepstad et al. 2006). PEC-215, another proposed constitutional amendment awaiting approval (pending since 2000), aims to delegate the authority to create indigenous lands as well as conservation units exclusively to the Congress; in essence, as long as ruralists control the legislature, new protected areas would not be created.[21]

[21] Available at http://www.camara.gov.br/proposicoesWeb/fichadetramitacao?idProposicao=14562, accessed 24 May 2017.

Another threat to the conservation of the Amazon forest lies in palm oil exploitation. Under the aegis of the Ministry of Agriculture, Livestock, and Food Supply, the Brazilian Agricultural Research Corporation has plans to transform Brazil into a "palm oil giant" able to compete with Indonesia and Malaysia, the world's largest producers of this commodity. The country's current production volumes are insignificant, but the sector is expanding and has substantial potential for massive future growth; because most of this potential is in the Amazon, environmentalists fear that this ambitious project in the palm oil sector will boost land grabbing, conflict, and deforestation (Levitt and Araujo 2017).

Finally, it is worth noting the severe budget cuts in the fields of science and technology in Brazil that profoundly affect research on biodiversity, a fundamental basis of public policies for nature conservation and sustainable development (Overbeck et al. 2017).

The chances of halting deforestation are clearly at risk. Environmental conservation is now far from being a public policy priority in Brazil. Nevertheless, because deforestation in the Amazon causes severe climatic and ecological disruptions that may ultimately trigger CCC, current trend urgently needs to be reversed. Brazil must take bold steps to eradicate both illegal and legal deforestation in the region over the next years. However, in an epoch marked by unprecedented human-produced developments, lessons of the past offer little guidance in preparing for a "no analogue future" (Crawford 2016; Sample and Topik 2014); consequently, an innovative and revolutionary approach is needed if Brazil and the world are to conserve the most important tropical biome on the planet.

## 4 A New Development Paradigm for the Amazon and Novel Pathways for the Brazilian Diplomacy

What would be a sustainable development paradigm for the Amazon rainforest? As the Brazilian Academy of Sciences asserts, attributing greater economic value to the standing forest would allow it to compete with other uses that are based on its deforestation or degradation, and transdisciplinary science, technology, and innovation are the tools that can show us how we can take advantage of the vast Amazonian potential without destroying the forest (Academia Brasileira de Ciências 2008). Hence, the effective conservation and exploration of this potential require opening the region to the high-tech revolutionary currents of the Fourth Industrial Revolution,[22] which would entail moving towards zero deforestation and putting an end to illegal activities such as biopiracy.

In the twenty-first century, knowledge has become the most important and strategic element of the world economy. Knowledge advancements in various

[22]On the Fourth Industrial Revolution, see, for example, https://www.weforum.org/agenda/2016/01/the-fourth-industrial-revolution-what-it-means-and-how-to-respond/, accessed 14 March 2017.

areas are increasingly influencing and modifying productive processes, the generation and diversification of products and the organization of enterprises and undertakings; as a result, the dominance of knowledge will determine the competitiveness of states and companies in the world market (De Mello 2015). The acceleration of emerging technological advances, encompassing broad fields such as artificial intelligence, robotics, nanotechnology, synthetic biology, DNA editing, biomimetics, advanced material science, energy storage, and quantum computing, will shape the future (Nobre et al. 2016). Hence, overcoming low economic development and poverty seems inseparable from the promotion of a green knowledge economy.

According to Nobre et al. (2016), if the Amazon's biodiversity is protected, the forest may hold the "biological keys" to boost the Fourth Industrial Revolution, since its extraordinary biomimetic resources and biodiversity can lead us to revolutionary developments. If combined with innovative technologies, the forest's fauna and flora could be the basis for significant scientific advances and high added-value products and services. Exploring the Amazon's potential through Fourth Industrial Revolution technology on a sustainable basis would imply prioritizing low-carbon developments, namely, green infrastructure, low-carbon forestry, international green tourism (strictly controlled to prevent ecosystem degradation), as well as information- and knowledge-intense production systems linked to forests, biodiversity, water, and climate (Viola and Franchini 2018). As La Rovere (2017) stresses, the current economic crisis in Brazil demonstrates that the country's previous cycle of economic growth has been exhausted, which means that a new economic growth cycle must be built through another basis. According to the author, national sectorial priorities include agriculture, forestry, and other land use, renewable energy, and energy efficiency. The Amazon has a very important role to play in these areas, since the region has, for example, significant potential in terms of biomass, wind energy, and solar energy.[23] The implementation of this new development paradigm could make Brazil the first fully developed tropical country in the world.[24]

In a world threatened by the global environmental crisis and in a country in which rural poverty is a reality, activating Amazonian wealth according to a paradigm based on the intelligent, selective, and environmentally safe use of the forest's natural resources—which would also integrate the vast knowledge of rural

[23]Interview with Professor Roberto Schaeffer (from the Federal University of Rio de Janeiro), 16 April 2017.

[24]It is worth noting, however, that technological development per se is an insufficient condition for protecting nature. Considering the severity of the global ecological crisis and the urgency of concrete answers, technology is a key part of solving the problem. Nevertheless, a paradigm shift in the way in which humanity relates to nature is absolutely essential; otherwise, instead of challenging the current development paradigm, the revolutionary technologies of the Fourth Industrial Revolution may convert into vehicles for reinforcing the prevailing consumption patterns and lifestyles. Discussing the new ethical and societal approach needed to protect the Amazon is beyond the scope of this chapter, whose focus relies only on the technological conditions for preserving the forest. On the ethical and societal conditions for conserving nature, see, for instance, Burke et al. (2016), Pereira (2017a, b), Pereira and Freitas (2017), and *New Earth Politics: Essays from the Anthropocene*, edited by Nicholson and Jinnah (2016).

populations and indigenous peoples regarding the Amazon and its resources—would be an important engine for new and inclusive opportunities for Brazilian society. Recognizing that the economic value of the conservation of the forest is greater than that of its destruction is thus fundamental. Nevertheless, as we have seen, there are currently no legal tools sufficiently strong to protect the Amazonian biome, which reinforces the need for the real, effective, and environmentally conscious presence of the state in the region (Penna Filho 2013). This presence is also essential for ensuring that nature protection is the absolute priority of future projects and activities as well as for attracting the necessary local and international investments to fuel a technological revolution in the forest. Since the country faces a severe economic recession, the Brazilian government should create favourable conditions for foreign direct investment, establishing clear competition rules and ensuring strong rule of law in the region. In fact, in addition to requiring the creation of new public universities and scientific and technological institutes, which would train, attract, and establish high quality human resources that would be inseparable from very expressive investments, deploying this new development paradigm would also necessitate a close connection with the business sector. Therefore, it is essential that Brazil promotes the necessary conditions to make the scientific revolution of the Amazon a reliable project. According to the Brazilian Academy of Sciences, these large investments would be fully offset by the emergence of new economic activities, accelerated innovate industrialization, and the expansion of employment opportunities and a highly qualified workforce (Academia Brasileira de Ciências 2008).

Conserving the Amazon and attracting financial resources to revolutionize the forest would entail not only the successful promotion of the necessary internal conditions already mentioned but also an avant-garde and activist attitude on the part of Brazilian diplomacy in the regional and international arenas of climate change discussion as soon as possible. Consequently, Brazil should follow a diplomatic path that upholds the fundamental ecological principles for creating a safe operating space for humanity. These principles should focus on recognizing that the new conditions of the Anthropocene are the greatest global challenge of the twenty-first century, that the Amazon is a global natural asset needed to avoid potential CCC, and that the conservation of the forest has economic value within the Fourth Industrial Revolution framework, which needs to build coalitions and partnerships to attract the best investments to manage, preserve, and develop a green knowledge economy in the region. As we have seen, for Amazonia, Brazil, and the world, it is essential to limit global warming at a level that ensures the survival of the forest. It is also important to include Amazon forest maintenance as a mitigation option on a level that adequately halts deforestation. Brazil could take advantage of its extremely rich natural patrimony and the fact that 70% of one of the forest tipping points of the Earth's climate system is in its national territory.

At the regional level, Brazilian diplomacy should become the "voice" of the Anthropocene's challenges in the arenas of South American environmental cooperation, especially with respect to the Amazon Cooperation Treaty Organization (ACTO), which seeks to gather the support of Brazil's Amazonian neighbours to form a collective international strategy, which would help them strengthen their presence in the world. Although ACTO is an organization that aims to protect the Amazonian biome and promote sustainable development, it has not established long-term policies that promote real changes in the region, remaining essentially a platform for the dissemination of information and thus being widely underestimated despite its potential (Tigre 2016). Strengthening ACTO is a fundamental condition for controlling illegal activities in the region because building common policies and unifying efforts among the different Amazonian countries would allow them to overcome some of the constraints they all face, such as the scarcity of financial resources (Becker 2016). As stated by Viola and Franchini (2018), since two of the major ACTO's members, Colombia and Peru, are cooperative in environmental matters and the organization's headquarters are located in Brasília, Brazil could, investing relatively small resources, build a sustainable leadership among the Amazonian countries by promoting forest conservation and the technological revolution of the region. Brazil and its neighbours could also use to good advantage the fact that the critical role of forests in fighting climate change was formally recognized in the Paris Climate Agreement (article 5). This declaration could help Brazil raise significant funds to help conserve the Amazon.

Internationally, Brazil should promote discussion on PBs, forest tipping points, and CCC in the agendas of the World Trade Organization, the International Monetary Fund, and the World Bank while also advocating for the creation of a global environmental organization whose powers would be equivalent to those of the World Trade Organization (Viola and Franchini 2018). Brazil should foster programmes such as the creation of a sustainable forestry coalition and a sustainable energy technology coalition as channels through which to scale up the investment in sustainable infrastructure and to encourage technological cooperation capable of triggering the deployment of forestry and sustainable energy technologies. These coalitions could potentially accelerate technological development and stimulate the deployment of renewable energy (wind energy, solar energy, second-generation biofuels, and renewable charcoal production) in Brazil and abroad while also playing an important role in helping Brazil find new short-term practices to control deforestation and conserve biodiversity that draw from the good practices of other forest countries. The experience that Brazil acquired during the second half of the 2000s in terms of forest management and conservation (for instance, the inventory of GHG emissions from land use change, early warning systems for forest clearance and fire detection via satellite imagery, and low-carbon agricultural practices) could also be shared with others (especially Southeast Asian and African countries) (La Rovere 2017).

Relevant and potentially cooperative partners for those coalitions would include countries and groups such as the European Union, Norway, Canada (whose boreal forests remain largely intact), China (because the Chinese government is interested in moving towards a green energy transition as soon as possible), the Independent Association of Latin America and the Caribbean (a reformist group within the UNFCCC), Mexico (an important green pole in the American continent given its low-carbon potential), and Indonesia (whose rainforests are home to some of the highest levels of biodiversity on the planet). Within these coalitions, Brazil and its partners should make efforts to challenge the conservative climate tendencies of decisive players in climate change issues, such as the USA and India, and deepen environmental global governance structures (Viola and Franchini 2018). Brazilian diplomacy should act with the double aim of ensuring that global warming does not reach a level that leads to the savannization of the Amazon and gathering enough international support to deploy a scientific revolution in the region, thus ensuring zero deforestation as well as reforestation of the deforested area, which would greatly enhance the role of the Amazon in fighting potential CCC.

This strategy would obviously first require a substantial improvement in environmental conservation in Brazil, so that the country could become a national example in environmental matters, legitimizing its arguments and position in regional and international cooperation spaces. Brazil is unfortunately and dramatically following the opposite path. The scientific community, the reformist forces in Brazilian society, national and international environmental NGOs, and the international community as a whole must act to put the country back onto the path it was following in the second half of the 2000s.

## 5 Conclusion

This chapter aimed to draw attention to the fact that CCC may be closer than most of humanity is able to perceive and to highlight the key role of the Amazon rainforest as both a potential trigger of a global climate catastrophe and a powerful tool to help limit catastrophic climate risk by conserving carbon stored in soils and vegetation, enhancing carbon uptake through reforestation of deforested areas, and potentially paving the way for disruptive sustainable energy technologies. Focusing on the Brazilian Amazon, this chapter explored the challenge of sustainable development in the forest, addressing the issues that have marked the federal politics and policies towards the region from 2005 to 2018 while framing them in the broader national environmental political context. This chapter thus sought to provide a comprehensive overview of the Brazilian governance of the Amazon as well as the obstacles facing the conservation of the forest. The future of this crucial biome is highly dependent on the development paradigm followed by Brazil as well as on the performance of Brazilian diplomacy in international environmental arenas. Considering the recent deterioration in the quality of the Brazilian governance of the forest, bringing the Amazon to the forefront of the academic, political, and public debate is consequently of vital importance. Within this context and aiming to begin a dialogue on how to manage the Amazon under the challenging conditions of the Anthropocene, conserving the forest was suggested as inseparable from an innovative development paradigm based on the revolutionary technologies of the Fourth Industrial Revolution. Based on this assumption, novel pathways for Brazilian diplomacy were suggested.

As we have seen, humanity might have already crossed core PBs such as climate change and biosphere integrity; the Earth is experiencing record-breaking high temperatures, and even a 2 °C global warming may trigger tipping points in the Earth system. The prevailing estimates and assumptions are based on the stability of threatened vital terrestrial biomes, such as the Amazon rainforest, and the IPPC's calculation of the remaining global carbon budget to stand a reasonable chance of staying below 2 °C of global warming does not consider carbon loss associated with forest dieback. In addition, the Paris Climate Agreement is clearly insufficient to prevent dangerous/catastrophic climate change. Given this overview, it would be unwise to overlook the prospect of transcending the IPCC's mid-range RCPs. The fact that environmental conservation is far from being a political priority on the agenda of the current Brazilian government and that the unsustainable exploitation of the Amazon in particular is being used as a means of reviving Brazil's recessive economy reinforces the urgency of an informed and resolute debate at the regional, national, and international levels on the prospect of the dieback of the forest.

Rising temperatures, droughts, and extensive fires; the possible construction of hydropower plants and highways; the expansion of biofuel production; governmental efforts to weaken environmental licensing, erode protected areas, and suspend the ratification of indigenous lands; the strength of the Brazilian agribusiness lobby; profound budget cuts in the fields of science and technology; and future prospects for the expansion of palm oil exploitation emerge as the main risks to the resilience of the Amazon. Regarding hydropower plants, projects may be blocked due to substantial societal resistance despite the Brazilian government's plans for the construction of new dams in the Amazonian region. In addition, the fact that Brazil's largest construction firms are struggling with serious corruption scandals may postpone large infrastructure projects. Concerning the government's intent to promote the expansion of biofuel production under the RenovaBio programme, ensuring the sustainable cultivation of sugarcane and soybeans as well as investing in second-generation biofuels to avoid undesirable trade-offs between emission mitigation and the conservation of the forest is essential because these fuels are an important part of reducing GHG energy emissions throughout the country. Diversifying the renewable energy matrix would be the best option to prevent energy security-related threats to the forest. Although biofuels were never the main driver of Amazonian deforestation even when their production was greatly encouraged in Brazil during the 2000s, the profound conservative environmental tendencies of the current Brazilian government may translate into unsustainable practices. Countering these conservative tendencies is key to overcoming all of the risks mentioned above. These successes can only be achieved if the Brazilian government recognizes the great economic value of the standing forest, which is where the Fourth Industrial Revolution enters the discussion. Once Brazil acknowledges the benefits of protecting the forest's biodiversity as well as the crucial role that the Amazon plays in limiting catastrophic climate risk, an ambitious national plan to eradicate deforestation and an avant-garde environmental diplomatic strategy should be developed and implemented.

Social science research on the relationship between forest dieback and CCC would have a positive impact on national climate policies in the Amazonian countries—especially in Brazil, Colombia, Peru, Bolivia, and Ecuador—as well as on the performance of ACTO. Social scientists have an important role to play in working with this regional organization to build constructive and reformist political proposals and practices that consider catastrophic climate risk as well as the urgent need of preserving one of the most important forest tipping points of the Earth's climate system.

**Funding**

This work was funded by Portuguese National Funds through the Fundação para a Ciência e a Tecnologia (FCT) in the framework of the project "UID/CPO/04627/2013".

## References

Abe, N., Aguirre, A., Andersson, M., Baum, S., Benedict, K., Conn, A., et al. (2017). *Global catastrophic risks 2017*. Stockholm.

Academia Brasileira de Ciências. (2008). *Amazônia: Desafio Brasileiro do Século XXI. A Necessidade de uma Revolução Científica e Tecnológica*. São Paulo: Academia Brasileira de Ciências e Fundação Conrado Wessel.

Allen, C. D., Breshears, D. D., & McDowell, N. G. (2015). On underestimation of global vulnerability to tree mortality and forest die-off from hotter drought in the Anthropocene. *Ecosphere, 6*(8), 1–55. https://doi.org/10.1890/ES15-00203.1

Aragão, L., Poulter, B., Barlow, J. B., Anderson, L. O., Malhi, Y., Saatchi, S., et al. (2014). Environmental change and the carbon balance of Amazonian forests. *Biological Reviews, 89*(4), 913–931. https://doi.org/10.1111/brv.12088

Baccini, A., Walker, W., Carvalho, L., Farina, M., Sulla-Menashe, D., & Houghton, R. A. (2017). Tropical forests are a net carbon source based on aboveground measurements of gain and loss. *Science, 358*, 230–234. https://doi.org/10.1126/science.aam5962

Baum, S. D., & Handoh, I. C. (2014). Integrating the planetary boundaries and global catastrophic risk paradigms. *Ecological Economics, 107*, 13–21. https://doi.org/10.1016/j.ecolecon.2014.07.024

Becker, B. K. (2016). Geopolitics of the Amazon. *Area Development and Policy, 1*(1), 15–29. https://doi.org/10.1080/23792949.2016.1149435

Benavidez, J., & Cadena, A. (2011). Políticas y capacidades de investigación y desarrollo e innovación (I&D+i) para el desarrollo de biocombustibles en América Latina y el Caribe. In A. Rodríguez (Ed.), *Investigación y desarrollo e innovación para el desarrollo de los biocombustibles en América Latina y el Caribe* (pp. 11–48). Santiago de Chile: CEPAL.

Bostrom, N. (2013). Existential risk prevention as global priority. *Global Policy, 4*(1), 15–31. https://doi.org/10.1111/1758-5899.12002

Bostrom, N., & Cirkovic, M. M. (2008). Introduction. In N. Bostom & M. M. Cirkovic (Eds.), *Global catastrophic risks* (pp. 1–29). Oxford: Oxford University Press.

Boucher, D., Roquemore, S., & Fitzhugh, E. (2013). Brazil's success in reducing deforestation. *Tropical Conservation Science, 6*(3), 426–445. https://doi.org/10.1177/194008291300600308

Brazil Ministry of Science, Technology, and Innovation. Secretariat of Policies and Programs of Research and Development. General Coordination of Global Climate Change. (2016). *Third national communication of Brazil to the United Nations framework convention on climate change – Volume III*. Retrieved from Brasília http://sirene.mcti.gov.br/documents/1686653/1706740/Volume+3.pdf/6bb2f28a-02ca-4072-980b-1b2514a87cab

Burke, A., Fishel, S., Mitchell, A., Dalby, S., & Levine, D. J. (2016). Planet politics: A manifesto from the end of IR. *Millennium-Journal of International Studies, 44*(3), 499–523. https://doi.org/10.1177/0305829816636674

Brienen, R. J. W., Phillips, O. L., Feldpausch, T. R., Gloor, E., Baker, T. R., Lloyd, J., et al. (2015). Long-term decline of the Amazon carbon sink. *Nature, 519*(7543), 344–348. https://doi.org/10.1038/nature14283

Carvalho, F. V. (2012). The Brazilian position on forests and climate change from 1997 to 2012: From veto to proposition. *Revista Brasileira de Política Internacional, 55*, 144–169.

Ceballos, G., Ehrlich, P. R., & Dirzo, R. (2017). Biological annihilation via the ongoing sixth mass extinction signaled by vertebrate population losses and declines. *Proceedings of the National Academy of Sciences, 114*, E6089–E6096. https://doi.org/10.1073/pnas.1704949114

Chazdon, R. L., Broadbent, E. N., Rozendaal, D. M. A., Bongers, F., Zambrano, A. M. A., Aide, T. M., et al. (2016). Carbon sequestration potential of second-growth forest regeneration in the Latin American tropics. *Science Advances, 2*(5), 10. https://doi.org/10.1126/sciadv.1501639

Cotton-Barratt, O., Farquhar, S., Halstead, J., Schubert, S., & Snyder-Beattie, A. (2016). *Global catastrophic risks 2016*. Stockholm.

Crawford, N. C. (2016). Studying world politics as a complex adaptive system. In K. Booth & T. Erskine (Eds.), *International relations theory today* (pp. 263–267). Cambridge: Polity Press.

Crouzeilles, R., Feltran-Barbieri, R., Ferreira, M. S., & Strassburg, B. B. N. (2017). Hard times for the Brazilian environment. *Nature. Ecology & Evolution, 1*, 1213. https://doi.org/10.1038/s41559-017-0303-7

Crutzen, P. J., & Stoermer, E. F. (2000). The Anthropocene. *Global Change Newsletter, 41*, 17–18.

De Mello, A. F. (2015). The dilemmas and challenges of Amazon's sustainable development: The example of Brazil. *Revista Crítica de Ciências Sociais*, (107), 91–108.

Federative Republic of Brazil. (2015). *Intended nationally determined contribution towards achieving the objective of the united nations framework convention on climate change*. Retrieved from http://www4.unfccc.int/submissions/INDC/Published%20Documents/Brazil/1/BRAZIL%20iNDC%20english%20FINAL.pdf

El Bizri, H. R., Macedo, J. C. B., Paglia, A. P., & Morcatty, T. Q. (2016). Mining undermining Brazil's environment. *Science, 353*(6296), 228–228. https://doi.org/10.1126/science.aag1111

EPE. (2007). *Plano Nacional de Energia 2030*. Rio de Janeiro: Ministério de Minas e Energia.

EPE. (2011). *Análise de Conjuntura dos Biocombuestíveis (Janeiro 2010–Dezembro 2010)*. Brasília: Ministério de Minas e Energia.

EPE. (2017). *Balanço Energético Nacional 2017: Relatório Síntese*. Retrieved from Rio de Janeiro.

FAO. (2012). *Forest management and climate change: A literature review*. Rome: FAO.

Fearnside, P. M. (2012). Brazil's Amazon forest in mitigating global warming: unresolved controversies. *Climate Policy, 12*(1), 70–81. https://doi.org/10.1080/14693062.2011.581571

Fearnside, P. M. (2013). The evolving context of Brazil's environmental policies in Amazonia. *Novos Cadernos NAEA, 16*(2), 9–25.

Fearnside, P. M. (2016a). Brazil's Amazonian forest carbon: The key to Southern Amazonia's significance for global climate. *Regional Environmental Change*, 1–15. https://doi.org/10.1007/s10113-016-1007-2

Fearnside, P. M. (2016b). Brazilian politics threaten environmental policies. *Science, 353*(6301), 746–748. https://doi.org/10.1126/science.aag0254

Feldpausch, T. R., Phillips, O. L., Brienen, R. J. W., Gloor, E., Lloyd, J., Lopez-Gonzalez, G., et al. (2016). Amazon forest response to repeated droughts. *Global Biogeochemical Cycles, 30*(7), 964–982. https://doi.org/10.1002/2015gb005133

Figueres, C., Schellnhuber, H. J., Whiteman, G., Rockström, J., Hobley, A., & Rahmstorf, S. (2017). Three years to safeguard our climate. *Nature, 546*(7660), 593–595. https://doi.org/10.1038/546593a

Fraser, B. (2016, August 16). Q&A: Amazon 'tipping point' may be closer than we think, Thomas Lovejoy says. *Science*. Retrieved from http://www.sciencemag.org/news/2016/08/qa-amazon-tipping-point-may-be-closer-we-think-thomas-lovejoy-says

Gaffney, O., & Steffen, W. (2017). The Anthropocene equation. *The Anthropocene Review, 4*(1), 53–61. https://doi.org/10.1177/2053019616688022

Goldemberg, J. (2008). The Brazilian biofuels industry. *Biotechnology for Biofuels, 1*(1), 6. https://doi.org/10.1186/1754-6834-1-6

Goldemberg, J., Coelho, S. T., & Guardabassi, P. M. (2008). The sustainability of ethanol production from sugarcane. *Energy Policy, 36*, 2086–2097.

Goldemberg, J., Schaeffer, R., Szklo, A., & Lucchesi, R. (2014). Oil and natural gas prospects in South America: Can the petroleum industry pave the way for renewables in Brazil? *Energy Policy, 64*, 58–70. https://doi.org/10.1016/j.enpol.2013.05.064

Golladay, S. W., Martin, K. L., Vose, J. M., Wear, D. N., Covich, A. P., Hobbs, R. J., et al. (2016). Achievable future conditions as a framework for guiding forest conservation and management. *Forest Ecology and Management, 360*, 80–96. https://doi.org/10.1016/j.foreco.2015.10.009

Gupta, J. (2012). Glocal forest and REDD+ governance: Win–win or lose–lose? *Current Opinion in Environmental Sustainability, 4*(6), 620–627. https://doi.org/10.1016/j.cosust.2012.09.014

Henley, B. J., & King, A. D. (2017). Trajectories toward the 1.5°C Paris target: Modulation by the Interdecadal Pacific Oscillation. *Geophysical Research Letters, 44*(9), 4256–4262. https://doi.org/10.1002/2017GL073480

IPCC. (2013). *Climate change 2013: The physical science basis. Contribution of Working Group I to the Fifth Assessment Report of the Intergovernmental Panel on Climate Change*. Cambridge and New York.

IPCC. (2014a). *Climate Change 2014: Impacts, Adaptation, and Vulnerability: Summary for Policymakers (Contribution of Working Group II to the Fifth Assessment Report of the Intergovernmental Panel on Climate Change*. Cambridge.

IPCC. (2014b). *Climate Change 2014: Synthesis Report (Contribution of Working Papers I, II, III to the Fifth Assessment Report of the Intergovernmental Panel on Climate Change)*. Geneva.

IPEA. (2010). *Biocombustíveis no Brasil: Etanol e Biodiesel*. Retrieved from http://www.ipea.gov.br/portal/images/stories/PDFs/comunicado/100526_comunicadoipea53.pdf

Jiménez-Muñoz, J. C., Mattar, C., Barichivich, J., Santamaría-Artigas, A., Takahashi, K., Malhi, Y., et al. (2016). Record-breaking warming and extreme drought in the Amazon rainforest during the course of El Niño 2015–2016. *Scientific Reports, 6*, 33130. https://doi.org/10.1038/srep33130

King, D., Schrag, D., Dadi, Z., Ye, Q., & Ghosh, A. (2015). *Climate change: A risk assessment*. Cambridge: Center for Science and Policy, University of Cambridge.

La Rovere, E. L. (2017). Low-carbon development pathways in Brazil and 'Climate Clubs'. *Wiley Interdisciplinary Reviews: Climate Change, 8*(1), e439. https://doi.org/10.1002/wcc.439

La Rovere, E. L., Burle Dubeux, C., Pereira, A. O., & Wills, W. (2013). Brazil beyond 2020: From deforestation to the energy challenge. *Climate Policy, 13*(sup01), 70–86. https://doi.org/10.1080/14693062.2012.702018

Levitt, T., & Araujo, H. (2017, 29 June). The Amazon's new danger: Brazil sets sights on palm oil. *The Guardian*. Retrieved from https://www.theguardian.com/sustainable-business/2017/jun/29/brazil-palm-oil-amazon-rainforest-deforestation-temer-farming-para-cerrado

Lucena, A. F. P., Clarke, L., Schaeffer, R., Szklo, A., Rochedo, P. R. R., Nogueira, L. P. P., et al. (2016). Climate policy scenarios in Brazil: A multi-model comparison for energy. *Energy Economics, 56*, 564–574. https://doi.org/10.1016/j.eneco.2015.02.005

Lucena, A. F. P., Schaeffer, R., Szklo, A., Soria, R., & Chávez-Rodriguez, M. (2013). *Energy security in the Amazon. A report for the Amazonia Security Agenda Project*. Retrieved from http://globalcanopy.org/publications/energy-security-amazon

Malhi, Y., Gardner, T. A., Goldsmith, G. R., Silman, M. R., & Zelazowski, P. (2014). Tropical forests in the Anthropocene. *Annual Review of Environment and Resources, 39(39)*, 125–159. https://doi.org/10.1146/annurev-environ-030713-155141

Marengo, J. A., Nobre, C. A., Sampaio, G., Salazar, L. F., & Borma, L. S. (2011). Climate change in the Amazon Basin: Tipping points, changes in extremes, and impacts on natural and human systems. In M. Bush, J. Flenley, & W. Gosling (Eds.), *Tropical rainforest responses to climatic change* (pp. 259–283). Berlin: Springer.

Monteiro, M. S. A., Seixas, S. R. D., & Vieira, S. A. (2014). The politics of Amazonian deforestation: environmental policy and climate change knowledge. *Wiley Interdisciplinary Reviews-Climate Change, 5*(5), 689–701. https://doi.org/10.1002/wcc.298

Moutinho, P., Guerra, R., & Azevedo-Ramos, C. (2016). Achieving zero deforestation in the Brazilian Amazon: What is missing? *Elementa. Science of the Anthropocene, 4*, 000125. https://doi.org/10.12952/journal.elementa.000125

Nepstad, D., Schwartzman, S., Bamberger, B., Santilli, M., Ray, D., Schlesinger, P., et al. (2006). Inhibition of Amazon deforestation and fire by parks and indigenous lands. *Conservation Biology, 20*(1), 65–73. https://doi.org/10.1111/j.1523-1739.2006.00351.x

Nicholson, S., & Jinnah, S. (2016). *New earth politics: Essays from the Anthropocene*. Cambridge: MIT.

NOAA. (2017). *Global climate report – June 2017*. Retrieved from https://www.ncdc.noaa.gov/sotc/global/201706

Nobre, C. A., Sampaio, G., Borma, L. S., Castilla-Rubio, J. C., Silva, J. S., & Cardoso, M. (2016). Land-use and climate change risks in the Amazon and the need of a novel sustainable development paradigm. *Proceedings of the National Academy of Sciences of the United States of America, 113*(39), 10759–10768. https://doi.org/10.1073/pnas.1605516113

Ometto, J. P., Aguiar, A. P. D., & Martinelli, L. A. (2011). Amazon deforestation in Brazil: Effects, drivers and challenges. *Carbon Management, 2*(5), 575–585. https://doi.org/10.4155/cmt.11.48

Overbeck, G. E., Bergallo, H. G., Grelle, C. E. V., Akama, A., Bravo, F., Colli, G. R., et al. (2017). Global biodiversity threatened by science budget cuts in Brazil. *BioScience, 68*, 11–12. https://doi.org/10.1093/biosci/bix130

Penna Filho, P. (2013). Reflexões sobre o Brasil e os desafios Pan-Amazônicos. [Reflections on Brazil and the Pan-Amazonian challenges]. *Revista Brasileira de Política Internacional, 56*(2), 94–111. https://doi.org/10.1590/s0034-73292013000200006

Pereira, J. C. (2013a). Brazil and global climate governance (2020–2024): Between conservatism and avant-garde reformism. *Carta Internacional, 8*(1), 91–110.

Pereira, J. C. (2013b). *Climate security and governance: Brazil in the International Arena*. PhD thesis, New University of Lisbon, Lisbon. Retrieved from http://hdl.handle.net/10362/11802

Pereira, J. C. (2017a). *Low carbon development in Latin America*. European Union Institute for Security Studies Issue Alert(15). https://doi.org/10.2815/05137

Pereira, J. C. (2017b). The limitations of IR theory regarding the environment: Lessons from the Anthropocene. *Revista Brasileira de Política Internacional, 60*(1), e018. https://doi.org/10.1590/0034-73292017001019

Pereira, J. C., & Freitas, M. R. (2017). Cities and water security in the Anthropocene: Challenges and opportunities for International Relations. *Contexto Internacional, 39*(3), 521–544. https://doi.org/10.1590/s0102-8529.2017390300004

Pinto, H. S., Assad, E. D., Zullo Jr., J., Martins, S. C., & Ávila, A. M. H. (2012). *Diretrizes para uma Economia Verde no Brasil: Avanços Tecnológicos para a Agricultura Familiar*. Rio de Janeiro: Fundação Brasileira para o Desenvolvimento Sustentável.

Pistorius, T. (2012). From RED to REDD+: The evolution of a forest-based mitigation approach for developing countries. *Current Opinion in Environmental Sustainability, 4*(6), 638–645. https://doi.org/10.1016/j.cosust.2012.07.002

Read, P. (2008). Biosphere carbon stock management: Addressing the threat of abrupt climate change in the next few decades: An editorial essay. *Climatic Change, 87*(3), 305–320. https://doi.org/10.1007/s10584-007-9356-y

Rees, M. (2013). Denial of catastrophic risks. *Science, 339*(6124), 1123–1123. https://doi.org/10.1126/science.1236756

Reyer, C. P. O., Rammig, A., Brouwers, N., & Langerwisch, F. (2015). Forest resilience, tipping points and global change processes. *Journal of Ecology, 103*(1), 1–4. https://doi.org/10.1111/1365-2745.12342

Rockström, J., Gaffney, O., Rogelj, J., Meinshausen, M., Nakicenovic, N., & Schellnhuber, H. J. (2017). A roadmap for rapid decarbonization. *Science, 355*(6331), 1269–1271. https://doi.org/10.1126/science.aah3443

Rockström, J., Schellnhuber, H. J., Hoskins, B., Ramanathan, V., Schlosser, P., Brasseur, G. P., et al. (2016). The world's biggest gamble. *Earths Future, 4*(10), 465–470. https://doi.org/10.1002/2016ef000392

Rockström, J., Steffen, W., Noone, K., Persson, A., Chapin, F. S., Lambin, E. F., et al. (2009). A safe operating space for humanity. *Nature, 461*(7263), 472–475. https://doi.org/10.1038/461472a

Sample, V. A., & Topik, C. (2014). Summary and synthesis. In V. A. Sample & R. P. Bixler (Eds.), *Forest conservation and management in the Anthropocene: Adaptation of science, policy, and practices* (pp. 471–494). Fort Collins, CO: US Department of Agriculture, Forest Service. Rocky Mountain Research Station.

Schaeffer, R., Lucena, A. F. P., Rathmann, R., Szklo, A., Soria, R., & Chavez-Rodriguez, M. (2015). *Who drives climate-relevant policies in Brazil?* Retrieved from http://www.ids.ac.uk/publication/who-drives-climate-relevant-policies-in-brazil

Schaeffer, R., Szklo, A., Lucena, A. F. P., Soria, R., & Chávez-Rodriguez, M. (2013, May/June). The impact of climate change on the untapped potential of hydropower systems. *IEE Power & Energy Magazine*.

Smith, P., Bustamante, M., Ahammad, H., Clark, H., Dong, H., Elsiddig, E. A., et al. (2014). Agriculture, forestry and other land use (AFOLU). In O. Edenhofer, R. Pichs-Madruga, Y. Sokona, E. Farahani, S. Kadner, K. Seyboth, A. Adler, I. Baum, S. Brunner, P. Eickemeier, B. Kriemann, J. Savolainen, S. Schlömer, C. von Stechow, T. Zwickel, & J. C. Minx (Eds.), *Climate change 2014: Mitigation of climate change. Contribution of working group III to the fifth assessment report of the intergovernmental panel on climate change*. Cambridge: Cambridge University Press.

Stanfurt, J. A., Kant, P., Lillesø, J. P. B., Mansourian, S., Kleine, M., Graudal, L., et al. (2015). *Forest landscape restoration as a key component of climate change mitigation and adaptation*. Vienna: International Union of Forest Research Organizations (IUFRO).

Steffen, W., Richardson, K., Rockström, J., Cornell, S. E., Fetzer, I., Bennett, E. M., et al. (2015). Planetary boundaries: Guiding human development on a changing planet. *Science, 347*(6223), 1259855. https://doi.org/10.1126/science.1259855

Tigre, M. A. (2016). Cooperation for climate mitigation in Amazonia: Brazil's emerging role as a regional leader. *Transnational Environmental Law, 5*(2), 401–425. https://doi.org/10.1017/s2047102516000297

Tundisi, J. G., Goldemberg, J., Matsumura-Tundisi, T., & Saraiva, A. C. F. (2014). How many more dams in the Amazon? *Energy Policy, 74*, 703–708. https://doi.org/10.1016/j.enpol.2014.07.013

Turnhout, E., Gupta, A., Weatherley-Singh, J., Vijge, M. J., de Koning, J., Visseren-Hamakers, I. J., et al. (2017). Envisioning REDD+ in a post-Paris era: Between evolving expectations and current practice. *Wiley Interdisciplinary Reviews: Climate Change, 8*(1), e425. https://doi.org/10.1002/wcc.425

UNEP. (2014). *Forests in a changing climate: A sourcebook for integrating REDD+ into academic programmes*. Nairobi: United Nations Environment Programme.

Vieira, M. A., & Dalgaard, K. G. (2013). The energy-security–climate-change nexus in Brazil. *Environmental Politics, 22*(4), 610–626. https://doi.org/10.1080/09644016.2013.806633

Viola, E. (2013). Transformations in Brazilian deforestation and climate policy since 2005. *Theoretical Inquiries in Law, 14*(1), 109–124.

Viola, E., & Basso, L. (2014, March). *Amazonian policy and politics, 2003–13: Deforestation, hydropower and biofuels*. Retrieved from https://noref.no/Publications/Themes/Emerging-powers/Amazonian-policy-and-politics-2003-13-deforestation-hydropower-and-biofuels

Viola, E., & Basso, L. (2015). Brazilian energy-climate policy and politics towards low carbon development. *Global Society, 29*(3), 427–446. https://doi.org/10.1080/13600826.2015.1028904

Viola, E., & Franchini, M. (2013). Brazil in the global governance of climate change, 2005–2012: The struggle between reformists and conservatives. *Contexto Internacional, 35*(1), 43–76. https://doi.org/10.1590/S0102-85292013000100002

Viola, E., & Franchini, M. (2014). Brazilian climate politics 2005–2012: Ambivalence and paradox. *Wiley Interdisciplinary Reviews: Climate Change, 5*(5), 677–688. https://doi.org/10.1002/wcc.289

Viola, E., & Franchini, M. (2018). *Brazil and climate change: Beyond the Amazon*. New York: Routledge.

Walker, W., Baccini, A., Schwartzman, S., Rios, S., Oliveira-Miranda, M. A., Augusto, C., et al. (2015). Forest carbon in Amazonia: The unrecognized contribution of indigenous territories and protected natural areas. *Carbon Management, 5*(5–6), 479–485.

Xu, Y., & Ramanathan, V. (2017). Well below 2 °C: Mitigation strategies for avoiding dangerous to catastrophic climate changes. *Proceedings of the National Academy of Sciences, 114*(39), 10315–10323. https://doi.org/10.1073/pnas.1618481114

Wiener, J. B. (2016). The tragedy of the uncommons: On the politics of Apocalypse. *Global Policy, 7*, 67–80. https://doi.org/10.1111/1758-5899.12319

Zalasiewicz, J., Williams, M., Steffen, W., & Crutzen, P. (2010). The new world of the Anthropocene. *Environmental Science & Technology, 44*, 2228–2231.

Zemp, D. C., Schleussner, C. F., Barbosa, H. M. J., & Rammig, A. (2017). Deforestation effects on Amazon forest resilience. *Geophysical Research Letters, 44*(12), 6182–6190. https://doi.org/10.1002/2017GL072955

# Climate Change Mitigation and Adaptation in Africa: Strategies, Synergies, and Constraints

**Linus M. Nyiwul**

## Contents

1 Introduction .......... 220
2 Climate Change Mitigation and Adaptation in Africa: Current Priorities .......... 220
2.1 Mitigation .......... 222
2.2 Adaptation .......... 225
3 Synergies in Climate Change Mitigation in Africa and Their Benefits to Sustainable Development .......... 227
4 Constraints to Climate Change Mitigation and Adaptation in Africa: Implications for Sustainable Development .......... 231
5 Conclusion .......... 236
Appendix .......... 238
References .......... 239

**Abstract** This chapter uses data on Intended Nationally Determined Contributions (INDCs) to examine the nature of climate change mitigation and adaptation actions being pursued in African countries and assesses the extent to which preferred mitigation and adaptation priorities advance the cause of sustainable development on the continent. The prospective synergies between mitigation and adaptation approaches and sustainable development are assessed. Also, the pathways through which resource constraints and institutional and policy environment affect Africa's ability to mitigate and adapt to climate change are examined, as well as the degree to which these constraints are being addressed. It is argued that Africa's ability to benefit from sustainable development synergies embedded in the mitigation and adaptation strategies in the INDCs will be greatly limited by institutional and policy environment that hinders funding, capacity building, and technological innovation systems development. The slow pace of efforts to address these impediments further erodes confidence that climate adaptation in Africa will be effective at sufficiently contributing to a reduction in climate change risks to the continent.

L. M. Nyiwul (✉)
Department of Economics, Africana Studies Program, Gettysburg College, Gettysburg, PA, USA
e-mail: lnyiwul@gettysburg.edu



T. Sequeira, L. Reis (eds.), *Climate Change and Global Development*, Contributions to Economics, https://doi.org/10.1007/978-3-030-02662-2_11

**Keywords** Africa · Climate change · Mitigation and adaption · Sustainable development · Climate policy

## 1 Introduction

Africa emits relatively a small proportion of aggregate greenhouse gases (GHG); hence its climate change problem is fundamentally a developmental one. That is, the approach to addressing climate change should have minimum impacts on development objectives. This is a significant challenge because economic development in Africa is driven by natural resources whose exploitation contribute to climate change and the external demand that underpins this trend. Hence, the region's approach to addressing climate change will necessarily require significant structural economic changes whose results may be uncertain. This chapter uses data on Intended Nationally Determined Contributions to examine how African countries are emphasizing these structural changes in their climate change mitigation and adaptation strategies and the extent to which these strategies suit sustainable development goals. Adaptation aims at helping communities deal with the climate change-related vulnerabilities and risks in the environmental and socioeconomic areas such as food security, health, species extinction, and water supply. Mitigation focuses on addressing causes of climate change, usually with heavy emphasis on greenhouse gas (GHG) emission reductions.

Existing data indicate that the energy sector dominates in mitigation actions and the agricultural sector is the main focus of adaptation measures. It is worth noting that the latter sector is a slightly larger source of greenhouse gases than the former. There are important synergies between various intended climate change mitigation and adaptation actions and sustainable development, at the sectoral and national levels, that if fully exploited can enhance the quality and increase the pace of development in Africa. However, there are significant obstacles to realizing these outcomes. These obstacles include the absence of suitable macro- and microlevel policy environment for desirable mitigation and adaptation actions, institutional and capacity deficiencies in both climate policy and scientific research, and, most importantly, the inability to bear the enormous cost of mitigation and adaptation.

## 2 Climate Change Mitigation and Adaptation in Africa: Current Priorities

In both relative and absolute terms, Africa contributes relatively a small share of the global greenhouse gas emissions; World Resources Institute (2017) data shows sub-Saharan Africa's contribution at about 7% but growing. Only a few countries (South Africa, Nigeria, Egypt, Algeria, and Morocco) on the continent are responsible for a large portion of this share of global emissions (Boden et al. 2011). At the same time, the continent is generally predicted to disproportionately bear the potential effects of climate change, with environmentally sensitive sectors such as water and agriculture among the most vulnerable. Sufficient water supplies in many countries of the continent heavily depend on adequate rainfall, raising not only issues of water security but the consequences of climate change-induced drought. The agricultural sector in many parts of the continent relies on rain-fed systems of agriculture, and climate change raises the risk of increased food insecurity. Overall, the cost of climate change is estimated to be up to an average of 7% of the region's gross domestic product (GDP) by 2100 (African Development Bank (AfDB) 2015).

The continent is exposed to a relatively large number of approaches to addressing this climate change problem and its risks. Some are feasible, while others are not, and some will be more beneficial than others. It is in this context that priorities of African countries on climate change-related actions should be understood; especially in relation to their pursuit of sustainable development objectives. In fact, there is evidence that African countries are emphasizing the links between their sustainable development planning and climate change mitigation and adaptation (Mbeva et al. 2015). This has been partially aided by the continent's active participation in initiatives at the national, regional, and global levels aimed at promoting reductions in greenhouse gas (GHG) and other emissions contributing to climate change. Under the UNFCCC's common but differentiated responsibility (CBDR) principles, a notion also maintained in Article 2 of the Paris Agreement (PA), there are no expectations of specific emission targets for Africa. However, they are not excluded from Intended Nationally Determined Contributions (INDCs)—post-2020 self-identified, voluntary national mitigation and adaptation targets, which become binding under the PA. With no legally binding obligations currently in place, Africa's mitigation and adaptation priorities can be understood in the context of their INDCs as well as their commitments in the Non-state Actor Zone for Climate Action (NAZCA)—a UNFCCC depository for climate change mitigation and adaptation initiatives undertaken by non-state entities (e.g., cities, investors, subnational regions, and companies) in their respective countries. As of June 2017, of the 165 countries that have submitted INDCs, nearly all African countries (53 in total) are among them. Note that most of these are still commitments rather than specific policy actions already taken and carefully designed to complement existing development objectives. Of the 517 (as of June 2017) climate actions already undertaken by various countries around the world and communicated to the UNFCCC, 77 of them are in Africa. These actions, according to the UNFCCC, include "good practices, approaches and technologies with significant mitigation potential, which could be tapped into up to 2020." In addition, the NAZCA portal shows that 66 cities and 5 subnational regions in Africa have committed to climate change actions of various types and scales.

Available data on INDCs show a clear pattern of climate change priorities for African countries. A casual look at the data on NAZCA also shows a similar pattern, but the analysis here will be limited to the data on INDCs simply because more data is available on them. In part because INDCs are voluntary, African countries have

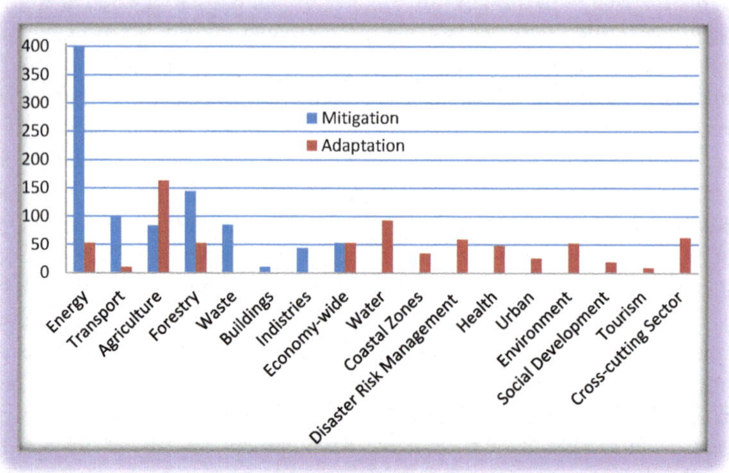


**Fig. 1** Sectoral balance in mitigation and adaptation actions in Africa. Source of Data: World Bank database on INDCs (as of September 9, 2017, http://spappssecext.worldbank.org/sites/indc/Pages/Adaptation.aspx). Note: each bar represents the number of mitigation or adaptation actions across 53 African countries. A country may have more than one action in the same sector

taken the initiative, and rightly so, to align mitigation and adaptation commitments with sustainable development goals. As can be seen in Fig. 1, top mitigation priorities across 53 African countries are in energy, forestry, transport, agriculture, waste, and to some degree buildings, economy-wide, and industries.[1] For adaptation, top priorities are concentrated in agriculture, water, energy, and forestry. Mbeva et al. (2015) conduct an analysis of INDCs for 43 African countries and identify a similar trend.

## *2.1 Mitigation*

In 2013, Africa's share of global energy-related $CO_2$ emissions stood at a mere 3% but projected to grow by 40% by 2030 (IEA 2015). According to recent data from the World Resources Institute database, energy consumption is the second largest source (after land use change and forestry) of GHG emissions in Africa, and this sector has witnessed the fastest growth in emissions between 1990 and 2013. This may partially explain why renewable energy is a leading mitigation approach in African countries, targeting emission reduction through increased consumption of

[1]Data based on dominant themes in INDCs. Extracting exact details from INDCs is complicated by the fact that there are no universal guidelines for designing and formatting them. Each country submits an INDC according to its understanding of the request from the UNFCCC.

energy from renewable sources, with more than 100 actions reported in this sector across the 53 African countries that have submitted INDCs (see Fig. 3 in the Appendix). The deep focus in Africa on renewable energy as a mitigation strategy is significant for obvious reasons. The primary one is that the energy needs in sub-Saharan Africa (SSA) alone are enormous. IEA (2014) estimates that only 32% of the population in SSA has access to electricity and about 80% of those without access reside in rural areas. Renewable energy is deemed to be a possible solution to this energy insecurity in rural areas since it is capable of providing electricity access both on the main grid and distributed forms. IEA (2015) projections through 2030 indicate that renewable energy will play an important role in improving access to energy in SSA.

Renewable energy use as a mitigation strategy is one area where Africa has gone beyond promises; SSA in particular has witnessed a rapid rise in the share of renewable energy generation and consumption beginning at the turn of the century. This growth in total renewable energy consumption, while driven mostly by hydroelectricity, can be observed in various renewable energy sources such as biomass and geothermal and less so in solar and wind. However, in levels and per capita basis, renewable energy consumption in SSA remains low relative to other developing regions, with total renewable energy generation in SSA just below 100 billion kilowatt hours and total installed capacity of 29 million kilowatts for the continent in 2012 (Nyiwul 2017).

The focus of mitigation on the energy sector emphasizes two aspects of energy use: substituting fossil fuels with energy from renewable sources and enhancing energy efficiency in general. While the sector is still dominated by hydropower, there are significant emphases on biofuels and solar and efficient cook stoves in the INDCs. Also, INDCs data show that energy efficiency tops the list of priorities in this sector. This in part is recognition of the current poor state of infrastructure which reduces energy efficiency in this sector. Some countries' energy supply is still largely based on colonial structures with major transmission and distribution losses, while the persistent practice of gas flaring in oil-producing countries represents lost opportunities to expand and improve energy supply chains.

Renewable energy target-setting in climate policy is common in African countries; with projected shares of renewable energy in total energy consumption ranging anywhere from 2% in Guinea-Bissau to 80% in Gabon (IRENA 2015). Per the INDCs, only a few African countries have set targets for carbon emissions. These include South Africa which proposes 34% emission reduction below business as usual by 2020 and 42% in 2025; Ethiopia which proposes a 65% emission reduction below business as usual by 2030; Morocco with a target of 17% reduction in GHG emissions below business as usual by 2030; and Gambia with a target of *44.4% reduction in* GHG emissions *below business-as-usual scenario by 2025.*[2]

[2]These percentages are extracted from the INDCs of the respective countries, as listed in the World Bank Database (as of July 6, 2017, http://spappssecext.worldbank.org/sites/indc/Pages/Adaptation.aspx).

Mitigation in Africa partly builds on prior conservation efforts which now take on additional significance in the face of climate change. This is especially true for forestry, with Africa being home to 17% of the world's forests (Pan et al. 2013). Land use change and forestry represent the largest source of GHG emissions in Africa, and research reports indicate that the forest sector in general is responsible for anywhere between 10 and 30% of the global total (Harris et al. 2012; van der Werf et al. 2009; IPCC 2007; FAO 2006), driven by high rates of deforestation, forest degradation, and other land use changes that serve to increase the concentration of GHG in the atmosphere. The forest functions as absorbers of carbon dioxide, a major GHG, and managing forests sustainably enhances this function and helps to mitigate climate change. This, perhaps, may explain why this sector is the second most important mitigation priority, as seen in Fig. 1. INDC data on this sector show reduced emissions from deforestation and forest degradation and the role of conservation, sustainable forest management and enhancement of forest carbon stocks (REDD+), afforestation, and reforestation as the dominant specific mitigation strategy in Africa (see Fig. 3 in the Appendix). There is also heavy emphasis on sustainable forest and land management. This is a lauded strategy because there is evidence showing higher aggregate mitigation potential in the forestry sector, at lower per ton costs of carbon (Sathaye and Ravindranath 1998). This approach suits sustainability if appropriately deployed. The risk however lies in the specific instruments adopted. For example, Hussein et al. (2013) show that forest carbon sequestration incentives can hurt the poor and vulnerable in developing countries, in part because of a rise in returns to land, a fall in agricultural output, and resulting first- and second-order effects of these changes. Furthermore, along with agriculture and fisheries, the forest sector employs a large proportion of the population, and thus mitigation and adaptation strategies have implications for equity.

The INDC data show the transportation sector as the third most important mitigation strategic focus in Africa. A detailed breakdown (see Fig. 3 in the Appendix) of the data on this sector indicates that specific commitments include the urban and public transport, fleet vehicle and freight vehicles. In addition to transportation planning and infrastructure (especially in road, transit, and rail), there are a considerable number of actions envisioned in transportation fuels.

Dealing with waste also figures prominently in the mitigation strategy in Africa. Specific actions on this are reported for solid waste, waste-to-energy, agricultural and water waste, recycling, and reuse (see Fig. 3 in the Appendix). Analysis in Nyiwul (2016) exposes problems of waste management in Africa whose health impacts would easily be exacerbated by climate change-related impacts.

Agriculture is a significant source of greenhouse emissions. The INDCs of African countries in this sector indicate that specific mitigation actions are dominated by climate-smart agriculture, agricultural waste, soil management, livestock, and to a less extent fisheries and aquaculture (see Fig. 3 in the Appendix)—worth noting that the limited focus on fisheries is disappointing because African countries have generally underperformed in the management of their fish stocks (Nyiwul 2016). Mitigation in the agricultural sector might require great sensitivity to local realities, particularly because this sector employs a large proportion of the population, especially in rural areas.

## 2.2 Adaptation

Figure 1 illustrates the importance of climate change adaptation to Africa. While the classification of all sectors and subsectors reported in the INDCs was not possible, considering only the most visible sectors reported gives a clear picture that African countries consider adaptation a top priority compared to mitigation in terms of sectors in which some form of action will be pursued. Given the different levels of vulnerabilities of various economic sectors, Fig. 1 illustrates that African countries are prioritizing adaptation in accordance with perceived climate change sectoral risk profiles. Adaptation actions, as expected, are observed in more sectors than mitigation actions, illustrating the degree of concerns about climate change risks in Africa. In all, agriculture, followed by water, consistently ranks as the most visible sector in which climate change consequences would be quite severe. This ranking is consistent with findings elsewhere on this topic; for example, see Mbeva et al. (2015) and Pauw and Pegels (2013). As can be seen in Fig. 4 in the Appendix, adaptation actions in this sector are oriented toward developing climate-smart agriculture; improving the agricultural sector in the areas of irrigation systems, land and soil management, crops, livestock, fisheries, and aquaculture; and ensuring food security. These areas are projected to be particularly affected by climate change, with sub-Saharan Africa expected to experience a 3.2% fall in average cereal yield by 2050 (Gerald et al. 2009; Ringler et al. 2010) and even greater decreases in sorghum, wheat, corn, and millet (Knox et al. 2012). These impacts would affect the region's ability to achieve its goal of sustainable development by securing food production and reducing vulnerability of livelihoods for low-income households. Thus, there is good reason to position the agriculture as top priority for adaptation: sustainable development in most of Africa depends on this sector. Per AfDB (2017), the share of agriculture in Africa ranges from 2.3% of gross domestic product (GDP) in South Africa to 80% in Liberia, with even higher numbers for employment in the sector.

Agriculture and water are intricately linked in Africa. The fraction of the population in sub-Saharan Africa that relies on rain-fed agriculture is at a high 96% (Madzwamuse 2010). A small decrease in rainfall can lead to a large reduction in river water volumes and hence endanger water supplies to a significant number of people, thus affecting food production and security. Data from the INDCs indicate that African countries are addressing these risks by taking adaptation actions that address water management issues, water supply and quality, increased water conservation and reuse, and water infrastructure and improving wastewater treatment capacity. These actions are expected to help reduce the risks associated with climate change, including extreme weather events such as drought and floods that would exacerbate food insecurity by reducing agricultural productivity and arable land, thereby increasing poverty and jeopardizing sustainable development. The Horn of

Africa is particularly prone to severe droughts, and many parts of the continent face frequent floods. Climate change-induced droughts can reduce water flows from lakes and rivers that feed hydropower dams, thus exacerbating energy insecurity. For example, the Nile region is estimated to experience a 75% decrease in river flow by 2100, along with depletion of biomass stock, with negative consequences for energy security and sustainable development (UNEP 2006). These risks of natural disasters may explain why disaster risk and management is another important focus of adaptation in Africa, with INDCs emphasizing adaptation actions on disaster preparedness, monitoring and evaluation, relief and recovery, and more importantly investments in early warning systems. These emphases point to the concerns for losses and damages resulting from climate change-induced natural disasters, particularly in countries with coastlines and those at risk of prolonged droughts.

Vulnerabilities in water systems, agriculture, and natural disasters also have implications for health in Africa. Poor water quality and insufficient supplies increase chances of waterborne diseases, and food insecurity due to poor agricultural performance induces malnutrition. Meanwhile floods and droughts from climate change impacts can result in vector-borne disease like malaria and cholera. African countries are planning adaptation to these risks by taking actions that emphasize disease surveillance and control, health services and assessments, and awareness and behavior change, as well as reducing malnutrition.

Other important priority areas of adaptation that are visible but less prominent in the INDCs include energy, environment, and forests. In energy, the adaptation focus is more generally on shifts to renewable sources of energy but also includes power systems planning. The latter is important because it aims to improve energy efficiency, which is one major problematic area of energy use in Africa. Upgrades in power systems as well as a revamp of regulatory regimes will improve climate adaptation measures in the region. The shift to renewable energy is fraud with risks because of the heavy reliance on hydropower as a source of renewable energy. Nyiwul (2018) presents a detailed critique of large hydropower as a renewable energy source, citing evidence of net negative environmental impact of large hydropower. With regard to the environment and forests, environment adaption actions in Africa are focused on watershed and river basin management, ecosystem, and biodiversity, while forest adaptation actions are dominated by sustainable forest and land management, along with reforestation, afforestation, and land degradation topping out the list. Biodiversity and forest management is especially important for sub-Saharan Africa because an unusually high proportion of its population relies on products from forest.

## 3 Synergies in Climate Change Mitigation in Africa and Their Benefits to Sustainable Development

Africa's efforts to address the climate change problem is rightly focused more on adaptation than mitigation since the continent's contribution to climate change-causing emissions is relatively small. The challenge is to align adaptation with sustainable development planning given the implied costs. The PA, like its predecessors, recognizes and emphasizes the implied links between climate change and sustainable development. However, such agreements do not establish systematic links between the two and how to identify and evaluate their effectiveness, trade-offs, and impacts of synergies. The existing broad spectrum of mitigation and adaptation measures, particularly those identified in the INDCs of African countries, has important synergies for sustainable development on the continent. Hence, to be realistic and practical, particular attention is devoted to synergies in mitigation and adaptation measures identified in the African INDCs. One such synergy crucial for sustainable development is in renewable energy development, which can be viewed from several perspectives. First, beyond the climate change mitigation role of renewable energy, there is the potential for stimulating local and national economies through microenterprise development (Foley 1995). Second, a major hurdle to the provision of electricity in rural areas is the cost concerns. Renewable energy resolves this problem through its ability to provide off-grid solutions. This is particularly important for areas where grid connection is too expensive—the AfDB (2010) estimates that areas with difficult access for electricity provision in SSA are home to 66% of the population. Off-grid renewable energy is considered to be among the easiest, fast, and cheap options for improving energy access in the developing world (REN21 2011). Also, empirical evidence in Chien and Hu (2007, 2008) suggests that increases in renewable energy consumption also increase technical efficiency. This result would seem to support the emphasis on renewable energy use as a legitimate approach to address the energy efficiency issue discussed in Sect. 2. Third, renewable energy, especially small hydropower, has the potential to contribute to poverty reduction because it possesses the many benefits of "distributed power" (El-Khattam and Salama 2004) associated with the other forms of renewable energy. Specifically, decentralized renewables are likely to lead to increased locally available resource use and employment (directly in projects and indirectly in ancillary services) and increase technical skills required for the efficient management of energy-generating plants. This is in addition to technical benefits such as energy savings, power quality, voltage control, market liberalization, and efficiency.[3] This is crucial for SSA because about 80% of those lacking electricity access live in the rural areas of the region (IEA 2014); many of whom are capable of adopting and benefiting from various forms of sustainable energy (Bensch et al. 2017). Hence, distributed generation is clearly among the greatest potential to achieve sustainable development. In

[3]See Pepermans et al. (2005), Jordehi (2016), and El-Khattam and Salama (2004) for extended definitions and analysis of the types of distributed generation and their associated benefits.

fact, available data show relatively high concentration of small hydropower potential in some countries of the continent, accounting for a majority of total renewable electricity installed capacity in some cases (Nyiwul 2018).

Furthermore, the Technology Mechanism (TM) of the PA, coupled with its financial mechanisms, could have implications for the renewable energy sector as well as economy-wide benefits in Africa. Renewable energy production in Africa, like most of the developing world, is still costly compared to fossil fuel sources. Both technology transfer and funding mechanisms embedded in the PA could help reduce costs and provide scale economies to local economies, thus enhancing sustainable development. Of the estimated $783 billion in climate-smart investments in Africa, close to 16% (or $123 billion) are in the clean energy (IFC 2016). These investments have the potential to not just expand and modernize the energy sector but also to enhance market development, particularly in the market for financial intermediary and banking technology services, a key feature grossly lacking in most of the African economies. In addition, the design of instruments in climate markets and programs can have economy-wide positive externalities and foster domestic endogenous technologies that greatly enhance sustainable development in Africa.

The PA maintained and enhanced previous articles on technology transfer to aid mitigation and adaptation in developing countries. Among the most contentious barriers to such transfer in general has been the issue of weak property rights protections in many of the receiving developing countries. Climate change mitigation technologies are among the fast-proliferating technologies, coming after leading sectors like information communication technology that has transformed banking in ways unimaginable a decade ago. Costs seem to be the impending barrier to them than the technological know-how. Many less developed countries are cultivating domestic energy technologies at rates that may serve as a springboard for other industrial subsectors. Population dynamics will continue to favor this trend in Africa and help to support the emergence of market mechanisms of the types envisioned in climate change mitigation and adaptation.

Among these market mechanisms is the Clean Development Mechanism, originally designed under the Kyoto Protocol as an offset program and updated in Article 6 of the PA. It is essentially a market mechanism for GHG abatement. Specifically, it is a policy instrument that allows emission reduction projects in developing countries to earn certified emission reduction credits which can be traded and/or used by developed countries to meet emission reduction targets under the Paris and preceding climate agreements. It gives developed economies flexibility in their emission reduction efforts while supporting sustainable development in less affluent countries. The CDM embodies the synergies inherent in climate change mitigation and adaptation. Beyond the emission reduction function of the CDM for developing countries, the inflow of financial resources and transfers of technology embedded in CDM projects have implications for local capacity building and the potential to transform domestic economies. A large base of potential CDM projects, evident in Africa, provides opportunities for technology spillover and acquisition through learning by doing. There are significant opportunities for Africa to realize these potential benefits because it currently hosts only a minor fraction of implemented

CDM projects, owing to poor institutional, risk, and regulatory environment (Burian and Arens 2014). It is worth noting that such benefits would be maximized with disembodied technology transfer (sharing the technical know-how) than under an embodied one.

Tripathi (2016) argues that CDM is developmental in three major ways:

First, socially by emphasizing equity and fairness, alleviating poverty, and improving quality of life. For example, clean cook stoves not only improves health and access to energy, they also serve to reduce inequity by saving wood collection time—an activity noted to be gender bias against children and women.

Second, CDM is developmental environmentally through its contribution to resource conservation and biodiversity protection.

Third, economically through technology, cost-effectiveness, and financial resources to local communities.

Another dimension of CDM with important synergies for sustainable development in Africa is the potential externalities that may emanate from the markets, funding flows, and mechanisms of the CDM projects. Trading in certified emission reductions (CERs) would be a new market in the African economy characterized by perpetually and comparably underdeveloped market institutions. Other instruments of the CDM that generate CERs include carbon capture and storage and small afforestation and reforestation projects—those that will remove GHG by sinks of less than 16 kt of $CO_2$ annually or implemented by low-income communities. CDM projects are often thought of as aiming to meet the demand for carbon credits in developed economies, and there is also great potential for carbon markets in developing countries. There is an ongoing effort in Africa to develop carbon markets to support both climate change and sustainable development. The experience from developing such a market can create important economy-wide scale benefits, especially in terms of market development and intermediary services. The introduction of INDCs as binding commitments in the PA means that in eliminating double counting, CERs would require capacity building to facilitate the development of the CER market, which itself promotes important economic transformation.

Furthermore, like renewable energy, CDM also provides a unique vehicle through which technology transfer can foster sustainable development in Africa. Article 10 of the PA strengthened the TM of preceding climate agreements aimed at promoting climate change technologies and the transfer of mitigation and adaptation technologies to developing countries. Some of the functions of the TM, through its Technology Executive Committee (TEC), have broader development implications. For example, the committee aims to actively engage developing countries in their climate technology needs assessment and domestic barriers to innovation, facilitate north-south as well as south-south dialogue and cooperation in climate change technology transfer issues, and, mostly importantly, finance not just climate change technologies but also research and development in general. Thus, the opportunity exists to learn from the experience of TEC for the benefit of the rest of the economy.

Another market mechanism from which sustainable development in Africa will greatly benefit is the REDD+, though in only general terms, this mechanism is recognized under Article 5 of the PA and provides an impetus for further development of the workable approaches to the mechanism. While REDD+, as part of INDCs, is largely envisioned as an initiative at the national level, the private sector can play a vital role in achieving its goals. In African countries, engaging the private sector in the REDD+ mechanism can be economically transformational and hence developmental in different aspects. For example, involving the private sector reduces the burden of capacity and resource needs of the bureaucracy while providing avenues for private sector innovation and development of new markets in the areas of carbon offset credit buyers, carbon market project developers, and carbon market intermediary by banks and other financial institutions. It is worth noting that no African country currently operates carbon trading systems at the local or national level; however Africa is 38% of the composition of the Forest Carbon Partnership Facility (FCPF)—a global partnership of governments, businesses, civil society, and indigenous peoples dedicated to REDD+ mechanisms. The fund provides payments for verified emission reductions from REDD+ programs. The experience garnered from participating in this global initiative will have positive impacts on the effort to develop and integrate local economies with global carbon markets. Also, REDD+, like other climate change mechanisms, is essentially a new form of public policy in Africa and thus provides an opportunity to learn valuable lessons for development planning. Another potential benefit of REDD+ in Africa is that it may finally be the driver of land tenure reforms in many countries and even increase the momentum for property rights reforms in general. Lastly, and more broadly, REDD+ will further support sustainable development in Africa through its contributions to national public policy objectives in the areas of biodiversity, poverty reduction, and limit rates of deforestation. This is especially so because it dovetails well with forest sector policy framework outlined in many African countries since the 1990s.

Successful climate change mitigation and adaptation depend to a large degree on public awareness and support. This likely requires devoted educational campaigns and programs by specialized, trained personnel integrated into climate change mitigation and adaptation. For example, community or social forestry as part of mitigation and adaptation planning will entail engaging local communities in both policy formulation and implementation, which itself will require its effort on practices, laws, and their enforcement. This reduces chances of conflicting goals between the community and government authority and minimizing property rights issues. This experience has positive societal and economy-wide effects.

## 4 Constraints to Climate Change Mitigation and Adaptation in Africa: Implications for Sustainable Development

Here, we consider the multitude of pathways through which weak institutions, policy environment, and resource constraints affect Africa's ability to mitigate and adapt to climate change and the extent to which these constraints are being addressed. The mitigation and adaptation efforts and objectives conveyed in the INDCs by African countries are clearly very optimistic given the full set of formidable obstacles to achieving them and thus limiting the ability to fully take advantage of the synergies discussed in Sect. 3. The INDCs themselves are as close a climate change policy as can be found in most African countries. Yet, while not the fault of the parties submitting them, they are also representative of the kind of general policy and institutional environment prevailing in many countries of the continent: void of details, incoherent, and ultimately ineffectual.

Just as important is the fact that given the current development challenges in Africa, effective mitigation and adaptation priorities and strategies should be indistinguishable from development planning. This is an exceedingly difficult task because of the current tendency to favor macroeconomic interests, for example, commercial agriculture taking precedence over subsistence farmers' needs in adaptation and vulnerability assessments, in the effort to attract foreign direct investment (Madzwamuse 2010). A full integration of adaptation and development planning will require a policy framework and governance structures that are contrasting versions of current ones. Very few countries on the continent have the necessary and coherent climate change policy framework capable of supporting adaptation for sustainable development. A review by Madzwamuse (2010) identifies a tendency to locate adaptation planning within the environmental sector, with little to no integration with other sectors of the economy. By itself, this demonstrates limited understanding of climate change, its impacts, and how to appropriately address it.

*Funding Climate Change Mitigation and Adaptation* One specific and obvious obstacle to operationalizing the steps outlined in the INDCs is funding. In their INDCs, African countries request financial and technical assistance to support their implementation. They have also indicated commitments to devoting domestic resources to achieve mitigation and adaptation goals. But given the level of resources needed, as seen in Fig. 2, it is quite evident that much of the mitigation and adaptation actions will be unsuccessful without substantially higher level of external support. It is worth noting that the INDCs were developed with solemn anticipation of such support in the form of technology transfer, capacity building, and finance.

There are simply no domestic savings in Africa to finance climate change. Even if such savings were available, the problem of information asymmetry, finance knowledge, and skills may prove costly and thus prevent the productive deployment of climate funding. Note that a major sticking point in climate change negotiations has been on bearing the costs of mitigation and adaptation, where developing countries

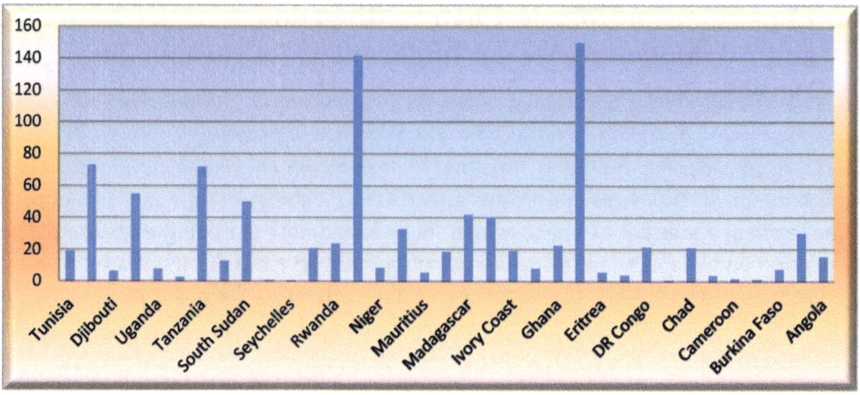


**Fig. 2** Mitigation and adaptation costs in Africa ($billion). Source of Data: World Bank Database (as of September 9, 2017, http://spappssecext.worldbank.org/sites/indc/Pages/Adaptation.aspx). (South Africa was excluded as an outlier, with $1688 billion in mitigation and adaptation costs)

insist on a compromise that does not impose undue burdens on their ability to address development challenges such as poverty reduction and health improvements given that their contributions to the climate change problem are small. Developing countries thus usually demand significant support from developed economies. This reliance on external support to achieve the goals outlined in the INDCs is a significant problem because there is little evidence that the kind of funding demands conveyed in the INDCs will be met. The region's negotiating position at climate talks is usually weak. This can be attributed to its limited capacity on climate change analysis to support negotiating positions and hence has to rely on external assistance in this regard. Equally important in climate negotiations is the issue of absorption capacity necessary to deal with large flows of adaptation funds. Developed countries which serve as the source of these funds would be focused on ensuring that mitigation and adaptation funding is effective. The inability to demonstrate the existence of absorptive capacity might be interpreted by donors as evidence to withhold financial assistance and thus undermining adaptation.

There is hope that the private sector will play a productive role in meeting climate change funding needs in Africa. There is some limited evidence that this indeed could materialize from external sources however. IFC (2016) estimates a $783 billion opportunity in Africa, but this is just about 3.4% of the total for emerging markets and may fail to be realized for a number of reasons. The inflow of foreign private climate investments will be curtailed by a public bureaucracy characterized by red tape, a banking sector whose infrastructure, technologies, and services are lagging behind other developing peer regions and with little financial intermediary of the type that is crucial in some elements of climate finance and other sectors in general. Bureaucratic bottlenecks are particularly important because of the stipulations of the financing mechanism: requiring developing countries receiving climate funds to demonstrate ability to effectively utilize them, though there is the question of how this ability will be evaluated and the potential impacts of noncompliance given the record of transparency in many African countries.

*Dearth of Expertise in Climate Change Science and Analysis* In addition to the climate change mitigation and adaptation goals being contingent on external funding, domestic constraints alone imply high likelihood of failure. In particular, many countries of the continent still face intractable problems in climate-related institutional capacity, quality, and effectiveness in numerous aspects. Reliable climate information and data analysis at the global, national, and local scales are crucial for understanding and planning for climate change, not just for the sake of minimizing negative impacts but also for maximizing potential development benefits from favorable climate conditions. In this respect, Africa is falling short, with many countries unable to base development planning on climate data because it simply does not exist or is inaccessible. While climate observation stations exist in many parts of the continent, the coverage is not only small but unevenly distributed, with limited availability in rural areas. Furthermore, even the data collected at existing observation stations is often of poor quality, with data collection systems mostly outdated, poorly maintained, and isolated with no capability of feeding data across networks in local or global systems (Dinku et al. 2016; UNECA 2011a, b). This problem is exacerbated by the dearth of technical expertise and resources at the agencies tasked with climate monitoring, thus making the funding and technology transfer mechanisms in the PA extremely important for Africa.

The consequence of limited climate technical expertise and funding in Africa is that scientific research on mitigation and adaptation is considerably constrained, thus building blind spots in national planning to reduce climate change risks. A recent joint initiative of the African Union (AU), the African Development Bank (AfDB), and the United Nations Economic Commission for Africa (UNECA), called the Climate for Development (ClimDev) Program in Africa, is a partial solution to this problem of climate change data generation and dissemination. It also aims at enhancing institutional capacity and policy support in integrating climate change data in development planning. Programs of this nature, themselves, require sustainable funding and technical support, especially in Africa to provide a base for national and domestic expertise to emerge. Such programs will also be essential in taking advantage of the mitigation, adaptation processes, and tools embedded in the PA.

Other recent efforts to help alleviate this problem include World Bank investments in hydrometeorological services on the continent and the Italian government-funded training program on climate change adaptation and disaster risk reduction in West African agriculture. These efforts are important because in the face of climate change sustainable development will require a systematic integration of climate information into crucial sectors such as agriculture and health. A related consequence of climate data unavailability in Africa is that climate change analyses of climate change impacts based on limited data from Africa and used in global agreements might be disadvantaging the continent both in terms of negotiating position and climate change risks. For example, the focus on climate change has placed a lot of emphasis limiting carbon emissions. Should this be Africa's priority at all or should it focus largely on adaptation?

Insufficient climate information and specifically environmental data in many African countries may partially explain the absence of legislation or other significant legally binding courses of action on climate change in Africa, even taking international climate change political economy considerations into account. This may also simply be an issue of general incompetence given the experience with environmental regulation in the some countries.[4] Climate change data may be available for some countries, mainly from outside establishments, but it does not advance the course of local human resources capable of analyzing, interpreting, and disseminating it to the communities that need it. There is little evidence that a domestic investment in scientific research on climate change is a priority.

*Bureaucratic Inertia and Policy Environment* Another challenge to effective climate change mitigation and adaptation in Africa lies in the policy environment necessary for climate-resilient development. Not only do the bureaucratic and legal frameworks require reforms to enhance efficiency, they also will certainly need to be updated to account for environmental and more generally climate change realities; for example, some mitigation and adaptation strategies will affect issues of existing property rights. Many countries of the continent are likely to lose out on the funding mechanisms outlined in the PA on account of their weak legal frameworks. Such low-performing legal systems are partly the reason for the small share of green investments in Africa, which so far is home to a mere 3% of registered CDM projects (OECD 2012). Existing intellectual property rights protections in Africa also means technology transfer envisioned under that PA will be unlikely to benefit Africa to the extent desired. External support is also likely to deteriorate in such a policy environment since it projects weak prospects of program implementation and success. Reforms serve to create a risk reduction environment in micro and macro terms, thereby creating attractive avenues for private initiative and confidence in the provision of assistance by external stakeholders.

The state of environmental regulation (lack of specificity in policy documentation, lax enforcement, under multiple government agencies with sometime conflicting visions and goals, etc.) may be replicated in climate change mitigation and adaption if changes are not undertaken. Some adaptation efforts are decentralized into community projects that are not fed into national planning. As in the case of environmental regulation, there is a risk that adaptation policies and projects will overlap and conflict with each other, leading to duplication and hence putting unnecessary pressures on limited institutional capacity.

Forest-based industries on the continent should require significant reforms as part of mitigation and adaptation strategies. The practice of low royalties in exchange for large concessions through timber extraction from forests is still prevalent in countries endowed with vast forests. Reforms may involve providing incentives to

[4]See Nyiwul (2016) for a detailed critique of environmental regulation in African countries.

operators of forestry firms that source materials from non-forests such as farmland or community forestry or encourage private forestry.

While the mitigation actions in the energy sector appear to already be yielding results, such success will be limited unless extensive effort is made to lift various cost-effective barriers in the sector in general. These barriers include state interference in the form of ineffective regulation, subsidization of fuels and electricity, import tariffs, and insufficient information dissemination on renewable energy technologies. These barriers raise the risk of expansion in the carbon-intensive energy consumption; for example, ineffective regulation and enabling of monopolies in the sector lead to higher energy prices which in turn force low-income consumers into cheap, high-carbon-content, and low-grade fuels.

Although CDM projects have the potential to expand at multiples of current numbers with increased foreign investments and assistance, the resulting CERs will likely continue to be used to fill the demand for them elsewhere. The development of carbon markets on the continent will take longer. This is in part due to constraints on the technical, institutional, and informational requirements. While the ease of current global fast information flows on carbon markets can reduce the burden on information acquisition, the private sector will continue to be faced with limited financial capacity, technical know-how, and bureaucratic bottlenecks that limit exposure to external access to solutions, especially in accessing international financing in this market. Continental bodies specifically targeting these issues would go a long way to address these problems. For example, the Africa Carbon Forum, which at the moment is largely focused on sharing experiences, can be given a larger role in developing carbon markets on the continent. A more general problem that may slow down CDM projects geared toward generating CERs in Africa is the persistent low pricing that may erode investor confidence and result in limiting flows of financing to the region.

It is worth recognizing the uphill task of TEC in successfully carrying out its functions in Africa given the long-standing obstacles in developing a requisite environment for innovation and technology transfer in the region. For example, it is not clear to what extent supporting infrastructure for climate change technology will be a priority since resource constraints are a factor. While TEC is also tasked with improving the capacity of developing countries to effectively adopt climate change technologies, Africa's capacity needs in this regard are not just in climate change but economy-wide.

The TM will take on additional significance with the PA, since it will be instrumental in helping African countries achieve their INDC goals. But it is easy to see here that the TM is less likely, given past experience outside climate change debate, to achieve desired results. To see why, we can consider what TM envisions and the potential sources of obstacles to the vision. The IPCC Special Report (IPCC 2000), from which the TM eventually emerged, defined technology transfer as "a broad set of processes covering the flows of know-how, experience and equipment for mitigating and adapting to climate change amongst different stakeholders such as governments, private sector entities, financial institutions, NGOs and research/education institutions. It comprises the process of learning to understand, utilize, and replicate the technology, including the capacity to choose it, adapt it to local conditions, and integrate it with indigenous technologies." Each of these stakeholders is imbued with serious limitations that would hinder TM's ability to carry out its mission. First among these limitations is an inefficient bureaucracy that inhibits private sector initiative. The private sector is already facing a shortage of needed skills, and this makes the capacity building mission of TM (specifically its implementation arm, the Climate Technology Center and Network (CTCN)) quite difficult. The TM will also be faced with NGOs, educational and research institutions that are underfunded, and infrastructure needs unmet or neglected, a financial sector with weak and unfledged markets. This is hardly a conducive environment for innovation systems to emerge. In these circumstances, the TM is unlikely to achieve desirable outcomes in its efforts to help African countries realize their INDC goals. Moreover, even the definition of technology transfer above appears to have different meanings to different parties to the climate change agreement, with developed countries seeming to adopt a much narrower interpretation of the concept than developing countries (de Coninck and Sagar 2015). This will likely hamper both the policy and implementation arms of the TM. And given the typically weak climate change negotiating position of African countries, the expected TM contributions of developed economies embedded in the INDCs may be currently vastly overestimated.

It is worth noting that due to the heavy reliance on agriculture and forestry and given the intertwined nature of the two sectors, mitigation and adaptation strategies may overlap. Avoiding such overlap would require extensive coordination at various levels to avoid duplication and waste. This is a challenge given the existing institutional fragmentation, with government bodies often operating independently of each other and jurisdictional overlaps that lead to power struggles within the bureaucracy that paralyze policy [see Nyiwul (2016) for a detailed description of this dynamic].

Incorporating REDD+ into sustainable development in Africa is a challenge. While general frameworks are prevalent, implementation has mostly been in the form of patchwork of projects and practices at local levels, with little national focus and coordination. Nevertheless, a carbon-centered REDD+ has sustainable development potential if it becomes the impetus for instituting well-regulated mechanism that results in proper land tenure reforms and clear rules that guarantee fairness in resource distribution and governance. Other policy and bureaucratic barriers to achieving the INDC goals in the forest sector include absence of institutions to promote participation of local communities and private sector stakeholders.

## 5 Conclusion

Africa contributes only a small share of the global greenhouse gas emissions responsible for climate change but stands to suffer disproportionately from the negative impacts of climate change. These negative impacts include shape falls in agricultural output and productivity, deterioration in health outcomes and water supplies that affect food and energy security, and increased frequencies of natural disasters. Policy-makers on the continent recognize these threats and are making efforts to address them. Across 53 countries in Africa, climate change mitigation has focused on energy, forestry, transport, agriculture, waste, and to some degree buildings, economy-wide, and industries, while adaptation priorities are concentrated in agriculture, water, energy, and forestry. In general, the balance of these mitigation and adaptation actions generally favors adaptation in Africa—this recognizes the reality that the continent faces negative impacts of climate change that have little to do with its contribution to the phenomenon. In some cases, mitigation and adaptation are not mutually exclusive. For example, using renewable energy technologies such as improved biomass stoves does have significant implications for deforestation. Similarly, mitigation and adaptation in agriculture has implications for forestry, where a disproportionately large share of deforestation is attributed to agriculture. This merge and integration of mitigation and adaptation priorities offer important synergies for sustainable development in African countries. These synergies are embedded in the nature and forms of technology transfer and funding mechanisms of the Paris Agreement and the potential externalities of such market mechanisms as CDM and REDD+. However, achieving these synergies will require a successful implementation of the mitigation and adaptation priorities. This successful implementation in Africa will require fundamental transformations in the institutional and policy environment. The difficult question is whether this requirement will be met.

# Appendix

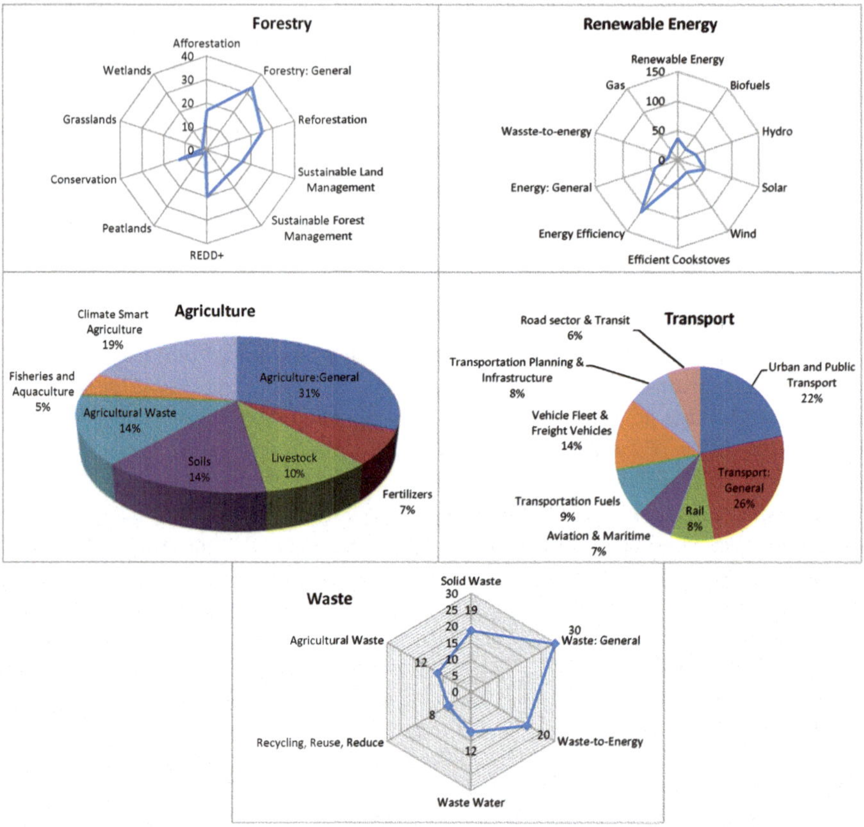


**Fig. 3** Mitigation priorities in Africa

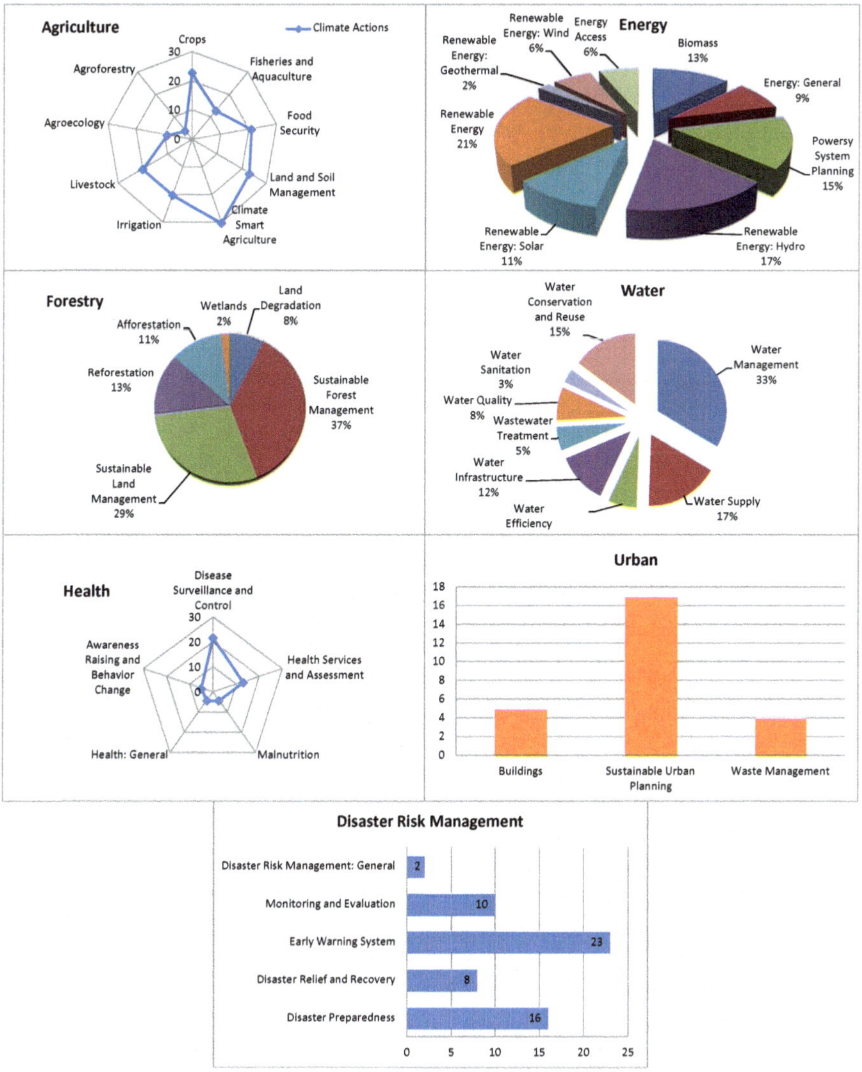


**Fig. 4** Adaptation priorities

# References


AfDB. (2010, October). *Financing of sustainable energy solutions* (Committee of Ten Policy Brief). Washington, DC

African Development Bank. (2015). *The African Development Bank at the UNFCCC COP21 meeting – Africa's climate opportunity: Adapting and thriving*. Abidjan: African Development Bank Group.

African Development Bank. (2017). *Statistical yearbook 2017*. Abidjan: African Development Bank Group.

Bensch, G., Peters, J., & Sievert, M. (2017). The lighting transition in rural Africa—From kerosene to battery-powered LED and the emerging disposal problem. *Energy for Sustainable Development, 39*, 13–20.

Boden, T. A., Marland, G., & Andres, R. J. (2011). *Global, regional, and national fossil-fuel $CO_2$ emissions*. Oak Ridge, TN: Carbon Dioxide Information Analysis Center, Oak Ridge National Laboratory, U.S. Department of Energy.

Burian, M., & Arens, C. (2014). The clean development mechanism: A tool financing low carbon development in Africa? *International Journal of Climate Change Strategies and Management, 6*(2), 166–191. https://doi.org/10.1108/IJCCSM-03-2013-0033

Chien, T., & Hu, J.-L. (2007). Renewable energy and macroeconomic efficiency of OECD and non-OECD economies. *Energy Policy, 35*(7), 3606–3615. https://doi.org/10.1016/j.enpol.2006.12.033

Chien, T., & Hu, J.-L. (2008). Renewable energy: An efficient mechanism to improve GDP. *Energy Policy, 36*(8), 3045–3052. https://doi.org/10.1016/j.enpol.2008.04.012

de Coninck, H., & Sagar, A. (2015). *Technology in the 2015 Paris Climate Agreement and beyond* (Issue Paper No. 42). Geneva: International Centre for Trade and Sustainable Development.

Dinku, T., Cousin, R., del Corral, J., Ceccato, P., Thomson, M., Faniriantsoa, R., et al. (2016, March). *The enacts approach: Transforming climate services in Africa one country at a time* (World Policy Paper). World Policy Institute.

El-Khattam, W., & Salama, M. M. A. (2004). Distributed generation technologies, definitions and benefits. *Electric Power Systems Research, 71*(2), 119–128.

Foley, G. (1995). *Photovoltaic applications in rural areas of the developing world* (World Bank Technical Paper No. 304). Washington DC: World Bank.

Food and Agricultural Organization (FAO). (2006, September 4). *Deforestation causes global warming: Key role for developing countries in fighting greenhouse gas emissions*.

Gerald, C. N., Rosegrant, M. W., Koo, J., Robertson, R., Sulser, T., Zhu, T., et al. (2009). *Climate change: Impact on agriculture and costs of adaptation*. Washington, DC: International Food Policy Research Institute.

Harris, N., Brown, S., Hagen, S. C., Baccini, A., & Houghton, R. (2012). Progress toward a consensus on carbon emissions from tropical deforestation. *Policy Brief*

Hussein, Z., Hertel, T., & Golub, A. (2013). Climate change mitigation policies and poverty in developing countries. *Environmental Research Letters, 8*(3). https://doi.org/10.1088/1748-9326/8/3/035009

International Energy Agency (IEA). (2014). *Africa energy outlook: A focus on energy prospects in Sub-Saharan Africa*. Paris: OECD/IEA.

International Energy Agency (IEA). (2015). *Energy and climate change* (World Energy Outlook, Special Report). Paris: OECD/IEA.

International Finance Corporation. (2016). *Climate investment opportunities in emerging markets: An IFC analysis*. Washington DC: World Bank Group.

IPCC. (2000). Methodological and technological issues in technology transfer. In B. Metz, O. Davidson, J.-W. Martens, S. Van Rooijen, & L. Van Wie Mcgrory (Eds.). Cambridge: Cambridge University Press. 432pp.

IPCC. (2007). Technical summary. In *Climate change 2007: Mitigation. Contribution of Working Group III to the Fourth Assessment Report of the Intergovernmental Panel on Climate Change*. Cambridge: Cambridge University Press.

IRENA. (2015). *Africa 2030: Roadmap for a renewable energy future*. Abu Dhabi: IRENA.

Jordehi, A. R. (2016). Allocation of distributed generation units in electric power systems: A review. *Renewable and Sustainable Energy Reviews, 56*, 893–905.

Knox, J., Hess, T., Daccache, A., & Wheeler, T. (2012). Climate change impacts on crop productivity in Africa and South Asia. *Environmental Research Letters, 7*(3), 034032.

Madzwamuse, M. (2010). *Climate governance in Africa: Adaptation strategies and institutions*. Nairobi: Heinrich Böll Stiftung.

Mbeva, K., Ochieng, C., Atela, J., Khaemba, W., & Tonui, C. (2015, November). *Intended nationally determined contributions as a means to strengthening Africa's engagement in international climate negotiations* (Working Paper 001/2015). African Center for Technology Studies, Climate Resilient Economics Programme.

Nyiwul, L. (2016). Environmental policy and performance in Africa: Implications of structural economic transformation for sustainable development. *Interdisciplinary Environmental Review, 17*(3–4). https://doi.org/10.1504/IER.2016.080229

Nyiwul, L. (2017). Economic performance, environmental concerns, and renewable energy consumption: Drivers of renewable energy development in Sub-Sahara Africa. *Clean Technologies and Environmental Policy, 19*(2), 437–450.

Nyiwul, L. (2018). Income, environmental considerations, and sustainable energy consumption in Africa. *International Journal of Green Energy, 15*(4), 264–276. https://doi.org/10.1080/15435075.2018.1439037

OECD. (2012). Carbon Finance in Africa. *OECD Journal: General Papers, 2010*(4), 143–162.

Pan, Y., Birdsey, R. A., Phillips, O. L., & Jackson, R. B. (2013). The structure, distribution, and biomass of the world's forests. *Annual Review of Ecology, Evolution, and Systematics, 44*, 593–622. https://doi.org/10.1146/annurev-ecolsys-110512-135914

Pauw, P., & Pegels, A. (2013). Private sector engagement in climate change adaptation in least developed countries: An exploration. *Climate and Development, 5*(4), 257–267. https://doi.org/10.1080/17565529.2013.826130

Pepermans, G., Driesen, J., Haeseldonckx, D., Belmans, R., & D'haeseleer, W. (2005). Distributed generation: Definition, benefits and issues. *Energy Policy, 33*(6), 787–798.

REN21. (2011). *Renewables 2011 global status report*. Paris: REN21 Secretariat.

Ringler, C., Zhu, T., Cai, X., Koo, J., & Wang, D. (2010). *Climate change impacts on food security in Sub-Saharan Africa* (IFRPI Discussion Paper 01042). Washington, DC: International Food Policy Research Institute.

Sathaye, J., & Ravindranath, N. H. (1998). Climate change mitigation in the energy and forestry sectors of developing countries. *Annual Review of Energy and the Environment, 23*, 387–437. https://doi.org/10.1146/annurev.energy.23.1.387

Tripathi, A. (2016, September 10). *Future of the clean development mechanism after Paris Convention*. LinkedIn Pulse.

UNECA. (2011a). *Assessment of Africa's climatic records and recording networks including strategic for rescuing of climatic data* (Working Paper 3). Africa Climate Policy Center.

UNECA. (2011b). *Integrating renewable energy and climate change policies exploring policy options for Africa* (Working Paper 10). Africa Climate Policy Center

United Nations Environment Programme (UNEP). (2006). *Report on climate change. African regional implementation review for the 14th session of the commission on sustainable development (CSD-14).*

van der Werf, G. R., Morton, D. C., DeFries, R. S., Olivier, J. G. J., Kasibhatla, P. S., Jackson, R. B., et al. (2009). $CO_2$ emissions from forest loss. *Nature Geoscience, 2*, 737–738. https://doi.org/10.1038/ngeo671

World Resources Institute. (2017). *CAIT climate data explorer*. Washington DC: World Resources Institute. Available online at: http://cait.wri.org

# Part III
# Climate Change Mitigation: Micro Evidence

# Climate Change Mitigation Through the Application of LCA Methodology on the Environmental Performance of Two Vehicles with Distinct Engines

**Bruno Pereira, Radu Godina, João C. O. Matias, and Susana Garrido Azevedo**

## Contents

1 Introduction .......... 246
2 Life-Cycle Assessment .......... 249
3 Study of the LCA Application .......... 250
  3.1 Definition of the Objective and Scope .......... 251
  3.2 Design of the Process Tree .......... 251
  3.3 Life-Cycle Inventory .......... 252
  3.4 Evaluation of LCA Impact .......... 253
  3.5 Interpretation of Results .......... 254
4 Conclusion .......... 257
Bibliography .......... 258

**Abstract** The purpose of this paper is to compare the environmental performance of two internal combustion engine vehicles (ICEVs) with distinct engines, one with the conventional diesel engine and the other with an electric motor supply (hybrid

B. Pereira (✉)
Department of Electromechanical Engineering, University of Beira Interior, Edifício I Das Engenharias, Covilhã, Portugal

R. Godina
C-MAST—Department of Electromechanical Engineering, University of Beira Interior, Edifício I Das Engenharias, Covilhã, Portugal
e-mail: rd@ubi.pt

J. C. O. Matias
Economics, Management, Industrial Engineering and Tourism Department, GOVCOPP – The Research Unit on Governance, Competitiveness and Public Policies, C-MAST—UBI, Universidade de Aveiro, Campus Universitário de Santiago, Aveiro, Portugal
e-mail: jmatias@ua.pt

S. G. Azevedo
CEFAGE-UBI Unit Research, Management and Economics Department, University of Beira Interior, Covilhã, Portugal
e-mail: sazevedo@ubi.pt

 
T. Sequeira, L. Reis (eds.), *Climate Change and Global Development*, Contributions to Economics, https://doi.org/10.1007/978-3-030-02662-2_12

diesel) and a diesel engine, and then to assess which one has less environmental impact. The Portuguese automobile park was chosen for this study. The purpose of the comparison of the two researched ICEVs is that they show many similarities in terms of features by having the same type of body and varying only in the powertrain system. The life-cycle assessment (LCA) methodology is performed by utilizing one ICEV with 170 kW. In both ICEVs the direct environmental impact is evaluated through a quantitative analysis, and a comparison is made with the purpose to assess the ICEV with the best environmental performance. The inputs and outputs data were collected from different sources and treated by the software OpenLCA for each life-cycle stage. The results have shown that the hybrid ICEV has revealed to have generally a better environmental performance compared to the diesel ICEV.

**Keywords** Life-cycle assessment · Environmental performance · Climate change · Vehicle emissions · Greenhouse gases

## 1 Introduction

The average annual growth in passenger transportation in the period 1995–2013 in the EU was 1.0%, and the transport in the European Union is steadily continuing its growth. However, with this growth more greenhouse gases (GHGs) will be released in the atmosphere and thus add a negative impact to the climate change. Yet the climate change consequences of transportation can be diminished by decreasing the quantity of combusted fuel, which can be achieved by improving the efficiency of the sector, reducing the travelled distances, increasing vehicle fuel efficiency, and replacing fuels with less environmental harmful ones. Both policy makers and researchers are currently putting climate change as an overall target into action (García-Olivares et al. 2018).

Currently consumers are more demanding, both in the conservation of natural resources and in the protection of the environment, as well as in the quality of the products and services they receive. For this reason, the industry faces the challenge of producing with high quality while at the same time satisfying the expectations of consumers and other interested parties in the field of environmental protection (Meng and Sager 2017).

Climate change has been deemed as a global environmental threat caused by human activities. It is a critical problem to the world and the European Commission which considers the climate changes as the second grimmest problem faced by the planet (Nanaki and Koroneos 2016). The human activities have contributed to the augmentation of the greenhouse effect and for increasing the temperature of the planet. The Intergovernmental Panel on Climate Change (IPCC) has predicted a global escalation of temperature between 1.1 °C and 6.4 °C, and the most recent projections from the IPCC report consider a scenario of very high emissions and predict a global rise of 52–98 cm by the end of this century, which will endanger the viability of many coastal cities (Carrasco et al. 2016).

The top emitted GHGs into the atmosphere due to anthropological action are carbon dioxide ($CO_2$), methane ($CH_4$), nitrous oxide ($N_2O$), and fluorinated gases (Remuzgo et al. 2016). Among all these GHGs, the $CO_2$ is considered the main responsible for the global warming (Nel and Nienaber 2012).

The GHG emissions from transport sector, particularly the ICEVs, contribute significantly to the overall high levels of GHGs in the atmosphere. This is the second-largest sector and is responsible for as much as 23% of $CO_2$ emissions globally (Lullo et al. 2016). Meanwhile, in Europe the transport system is responsible for circa 30% of the European emission of GHGs and has still been steadily growing in recent years (Nocera and Cavallaro 2016). Due to the fast increase of ICEVs and due to a progress at slow pace of emission control technologies, the transport sector is revealed to be the largest source of urban air pollution, which is a worrying public health obstacle in many cities around the world, essentially the ones from the developing world. The World Health Organization has projected that circa 2.4 million people pass away every year due to the pollutants generated by ICEVs (WHO 2009).

Air pollution and its impact on environment have been the subject of ongoing research in several areas, such as transport industry. The influence of air quality on public health has also been heavily investigated (Coria et al. 2015). Researchers also concluded that air quality is highly associated with mortality due to respiratory and cardiovascular diseases. The most significant gases with more environmental impact formed during a conventional ICEV life cycle are carbon dioxide ($CO_2$), ozone ($O_3$), carbon monoxide (CO), sulfur dioxide ($SO_2$), nitrogen oxides ($NO_x$), lead (Pb), and many others (Dalianis et al. 2016).

By the reason of the increasing concern of global pollution, the automotive industry has been incentivized to decrease the environmental impact that results from the production of ICEVs, as these are unsustainable in the medium and long term. The regulators have also forced the automotive industry for more sustainable solutions in an attempt to entice consumer buying patterns in order to stimulate the demand for more environmentally friendly solutions. Thus, the professionals from the automotive industry have been working together with academics to develop a clear and reliable way of measuring the environmental performance of ICEVs and analyze the progress and impact of alternative solutions (Bauer et al. 2015).

Diesel engine has a wide-ranging utilization due to higher energy efficiency, higher durability, low operating costs, and reliability (Liu et al. 2016). Even though the diesel engine has many advantages, it also has a substantial influence on air pollution and its impact on environment. In particular, diesel exhaust gas leads to higher levels of NOx emissions that are responsible of severe environmental and health problems (Cho et al. 2017). Researchers in health industry concluded that pollutants emitted by diesel engines of ICEVs have a negative impact on human health. Studies have also shown that a prolonged exposure to diesel exhaust gas produced by ICEVs gives origin to respiratory diseases, even lung cancer (Reşitoğlu et al. 2015; Kałuża 2016).

In this context it is important to analyze the impact that the production and use of such products have on the environment. The LCA is one of the methodologies used to analyze the environmental impact of a product, a set of processes, or to make a comparison between products or processes. The LCA allows detailed evaluation and measurement of the impact caused by a product or process in a "cradle to cradle" cycle (Branker et al. 2011). The product designers usually use this technique to take advantages of their competitors, to improve the characteristics of their products, to trace a strategic plan for the company, or even to develop their product to have a better environmental performance. This opens the way for the automotive industry to lunch ICEVs and several innovations to the market with environmental concerns by having better performances in terms of fuel consumption or with alternative energy sources. Examples of such types of innovations are ICEVs' internal combustion engines (diesel and gasoline) with better performance, electric ICEVs which have almost zero emissions during their use, hybrid gasoline, plug-in hybrids, and, more recently, hybrid diesel and fuel cell (Karakaya et al. 2014).

This study is intended to study the environmental performance using the LCA methodology of a diesel and a hybrid-diesel ICEV during their life cycle and compare them. The Portuguese automobile park was chosen for this study. The data utilized in this study was obtained from the manufacturers of both ICEVs with their permission. It is analyzed if the hybrid-diesel ICEV contributes for additional environmental problems during its lifetime. The main objective of this research is to analyze the environmental performance of two different types of ICEV engines using the LCA methodology. The two researched ICEVs have similarities in terms of features, offering the same body, varying only in the power-train system. Both have also the same final power level, but one is diesel, and the other is a hybrid with diesel and electric engines. The inputs and outputs data were collected from different sources and treated by the software OpenLCA for each life-cycle stage. The databases used to collect data to perform this study were the ELCD (European Reference Life Cycle Database), BioenergyDat, NEEDS, and also the CPM database. Whenever data was not available, the worst scenario was assumed. For example, in the case of thermoplastics and the glasses, a percentage in the total weight of the ICEV was considered.

The rest of the paper is organized as follows. The LCA is briefly addressed in chapter "On the Edge of Climate Change: In a Search of an Adequate Agent-Based Methodology to Model Environmental Dynamics." In chapter "Greenhouse Gas, Carbon Dioxide Emissions and Economic Growth: Empirical Evidence from Threshold Effect," the overall LCA application in this study is explained. The evaluation of the impact of the life cycle on the studied ICEVs is shown in chapter "Effects of Climate Change to Industrial Outputs and Employment in Asian Emerging Economies." Finally, chapter "The Role of Carbon Markets in the Paris Agreement: Mitigation and Development" concludes the study.

## 2 Life-Cycle Assessment

The environmental impact of a product begins with the extraction of raw materials and ends when the useful life of the product ends, becoming a waste that has to be managed properly. During manufacturing, companies must assess the environmental impact of their process, and they also have responsibility for the impact caused by the parties involved in the process until the product reaches the consumer customer (e.g., suppliers, distributors, and consumers) (Bascoul et al. 2014).

The emergence of environmental concerns, including the overuse of natural resources (raw materials), in the 1960s created the necessity of assessing the consumption of raw materials and energy. This contributed for promoting the methodology of the LCA. The product-related life-cycle assessment with an emphasis on energy, resources, and waste started around 1970 (Levasseur et al. 2016). The Limits to Growth, a report by the Club of Rome (Ju et al. 2015), and the first oil crisis were responsible by the shortage of oil and the vulnerability of the global economic system. Twenty years later, the LCA methodology was developed and harmonized by the Society of Environmental Toxicology and Chemistry (SETAC) (Curran 2012) and later on standardized by the International Organization for Standardization (ISO 2006). The term life-cycle assessment was only known in the 1990s, because since 1970 the designation for such studies was REPA—Resource and Environmental Profile Analysis (Vieira et al. 2016). With this incentive and the increasing number of studies conducted between the years 1970 and 1975, in which there was not always an agreement on the obtained results, the methodology used needed to be standardized. This was accomplished in the 1990s by the International Organization for Standardization (ISO), which created 4 more standards to the 14,000 family (Castka and Balzarova 2008): (1) ISO 14040: 1997 Environmental management—Principles and framework; (2) ISO 14041: 1998 Environmental management—Goal and scope definition and inventory analysis; (3) ISO 14042: 2000 Environmental management—Life cycle impact assessment; and (4) ISO 14043: 2000 Environmental management—Life cycle interpretation (ISO 2006).

The LCA presents advantages when it comes to the evaluation of the environmental impact of a product or process, of which the following stand out (Klöpffer and Grahl 2014):

- It allows the analysis of flows of materials and energy that cannot be verified by other methodologies.
- It allows to identify and compare products and processes among themselves, through a systematic evaluation of the environmental impacts.
- It allows to identify opportunities to improve the environmental performance of products or processes at different stages of their life cycle.
- It allows a better perception, from the beginning, of the materials or activities that cause the greatest environmental impacts.

By having a notion of the importance of potential environmental impacts and where they occur, the decision-maker has the capability of being able to solve them from the beginning, even during the design or development phase. It is possible to reach a decision based on the results of the LCA of a system (product or process).

The LCA methodology has been used by the automotive industry to compare the environmental performance of its products with different types of engines. Some studies have compared these different types of engines, for example, comparing Li-ion batteries of personal electric vehicles (PEVs) with fuel cell electric vehicles (FCEV), showing that pure PEVs should hardly compete with FCEV in the presence of medium-to-long driving range requirements, but the FCEV with a long-range battery and plug-in technology should be a great solution with the best of two clean technologies (MacLean et al. 2000). Other studies compared other types of cars, for example, ICEV and liquid petrol gas (LPG) with the BEVs or with the plug-in hybrids. There is a consensus in most of the studies that the PEV and the plug-in hybrid have the best environmental performance, with considerable reductions in the emissions of pollutants only when the battery production is not considered in the study (Danilecki et al. 2017; Groenewald et al. 2017; Tagliaferri et al. 2016).

Several damaging gases are generated during the combustion of ICEV's diesel engines. The most noteworthy hurtful ones are CO, hydrocarbons (HC), $NO_x$, and particulate matter (PM). However, the pollutant emissions only share less than 1% of the diesel exhaust gas. Of those, $NO_x$ represents the highest quantity with 50% in case of diesel. PM is the second in line of pollutant emissions due to the reason of diesel engines being lean combustion engines. Finally, both CO and HC display a minimal concentration and can be neglected. Also, one of the generated damaging gases could comprise a modest $SO_2$ concentration depending on the quality of fuel and its specifications. However, since it is dangerous for health reasons, most of oil distributors and customers prefer the alternatively ultralow sulfur diesel (ULSD) in order to avoid the damaging effects of $SO_2$ (Farahani et al. 2011; Reşitoğlu et al. 2015).

In consequence of several negative effects of diesel emissions on environment and health, policy makers have been called to act and thus have implemented several requirements for permissible exhaust emission standards. Since the limitations have been stronger and stronger, the automotive industry was incited to put an increasing amount of effort in reducing pollutant emission from ICEVs. Yet, the implemented actions have been unsuccessful to achieve the reduction of emissions established by such standards (Crippa et al. 2016).

## 3 Study of the LCA Application

This study uses the LCA methodology and follows the standard patterns of the ISO 14040: 2006. This methodology follows five steps: (1) the definition of the objective and scope, (2) the design of the process tree, (3) the life-cycle inventory, (4) the evaluation of the life-cycle impact, and (5) the interpretation of results (ISO 2006). The inputs and outputs data were collected from different sources and treated by the software OpenLCA for each life-cycle stage.

Regarding the LCA inventory, it emerges as the nucleus of this study. It is composed of several sub-steps that are necessary to go through several times and which discriminate the collection and processing of the environmental impacts originated during the life cycle of the product or service under study. The sub-steps correspond to the definition of system boundaries, schematic representation of the system block diagram, information gathering, data processing, analysis of results, and eventual redefinition of system boundaries. The inventory focuses on the determination and quantification of inputs and outputs associated with the life cycle of the product or service being analyzed. This results in a significant amount of data which may, however, not be sufficient to correctly define the inventory of the process under study (Lu et al. 2017).

## 3.1 *Definition of the Objective and Scope*

The definition of the objective and scope of the study is divided into several parts. It starts by identifying the objects under study. The object chosen to perform this study is two ICEVs with similarities in terms of features, offering the same body, varying only in the power-train system, and both having the same final power level. One of the studied ICEVs is a diesel, and the other is a hybrid with diesel and electric engines.

The main objective of this paper is to analyze the environmental performance of two different types of ICEV engines using the LCA methodology to find out which one has a better environmental performance throughout its life cycle. Also, it intends to improve the knowledge on the negative impact of this new power solution (hybrid diesel) on the environment.

For the present study, an ICEV with a power capacity of 170 kW was defined as a functional unit. Two different ICEVs were selected. One is a diesel ICEV with a power capacity of 170 kW and the other an ICEV with a diesel engine with the aid of an electric motor (hybrid) with 150 kW to 20 kW, respectively, for a total of 170 kW. Thus, it is necessary to define the boundaries of the process, i.e., the phases of the life cycle of the ICEV engine that are focused. These phases are the production process, fuel production process, and ICEV use.

## 3.2 *Design of the Process Tree*

The process tree is one of the most important steps in a LCA study. It shows the processes involved in the life cycle of a product, and it is the best method to verify

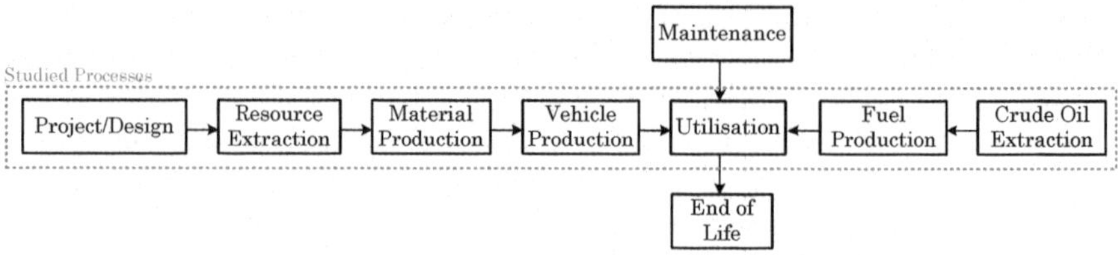


**Fig. 1** Studied processes

**Table 1** Main resources required in the production of an ICEV

| Resource | Units | Diesel | Hybrid |
|---|---|---|---|
| Bauxite | kg | 718 | 727 |
| Dolomite | kg | 110 | 128 |
| Iron ore | kg | 977 | 995 |
| Mixed ores (Cu, Pb, Zn) | kg | 120 | 146 |
| Precious metal ores | kg | 0.3 | 0.7 |
| Plastic (ABS) | kg | 217 | 218 |
| Other plastics | kg | 39 | 40 |
| Glass | kg | 58 | 58 |

the inputs and outputs necessary to develop the inventory. The process tree of the researched ICEVs engines is illustrated in Fig. 1.

*Resources Extraction Phase* The extraction of resources is included in the results of the material and automotive production phases. In this processes the transportation was not included since the related data was unavailable. The data used in this study was obtained from the manufacturers of both ICEVs with their permission (Table 1).

*Use Phase* The average consumption of the researched ICEVs was the maximum given by the brand. Accordingly, it was assumed that the average consumption of fuel for the diesel ICEV is 6.1 l per 100 km and for the hybrid-diesel ICEV is 4.3 l per 100 km.

*Fuel Production* Fuel production is a phase that depends on the use phase. Thus, during the lifetime it was considered a consumption of 15,250 l for the diesel ICEV and 10,750 l by the hybrid.

## 3.3 Life-Cycle Inventory

The Portuguese automobile park was chosen based on a convenience criterion since the research team is from Portugal. It was considered that a diesel ICEV drives an average of 25,000 km by year since the Portuguese automobile park has an average of 10 years (INE, I.P. 2012).

## 3.4 Evaluation of LCA Impact

After the LCA inventory, the evaluation of the impacts should be performed. This phase considers the selection of the categories (calculation method), the classification, and the characterization of the environmental impacts and could have optional steps such as normalizing, grouping, weighting, evaluating, and reporting LCIA (life-cycle inventory assessment) results.

Considering the impact categories, this study follows the environmental impact assessment of CML 2001 baseline. This involves the study of ten environmental impact categories: abiotic depletion potential (ADP), global warming potential ($GWP_{100}$), photochemical oxidation (POPC), acidification potential (AP), eutrophication potential (EP), ozone depletion potential (ODP), human toxicity potential (HTP), freshwater aquatic eco-toxicity potential (FAETP), marine aquatic eco-toxicity potential (MAETP), and terrestrial eco-toxicity potential (TETP).

In this phase the elements of ICEV as the constituents of the product life cycle are identified and grouped into different impact categories. Through the use of a LCA software, the process was performed automatically by combining the results of the ICEV with the environmental impact categories and displaying the results according to the selected methodology.

At this stage the environmental repercussions are calculated by category impacts and by multiplying the factors of environmental load obtained during the classification phase by their respective factor. This makes possible to compute the contribution of each element in each category for the environmental impact. Table 2 presents the results of each environmental impact category by using the LCA software.

According to Table 2, the diesel ICEV presents higher environmental impacts in all categories.

The ODP environmental impact category shows a value almost near zero in both cases, and for that reason, it was discarded. Also, the eco-toxicity categories that showed inconclusive results were not counted in the study. Beyond those impact categories, it is possible to identify pollutant emissions that contribute for this indicator, such as the air and water emissions. About the air emissions, the hybrid ICEV has less air emissions than the diesel one. However it can be seen that the diesel ICEV presents less emissions of $NO_2$[kg] (Table 3).

As for the water emissions, as illustrated in Table 4, the hybrid ICEV has more water emissions than the diesel, and also the last one has more $SO_4^{2-}$ [kg] and hydrocarbons [g].

**Table 2** Environmental impact categories by ICEV

| | ADP | GWP | POPC | AP | EP |
|---|---|---|---|---|---|
| ICEV | GJ | t $CO_2$ | kg $C_2H_4$ | kg $SO_2$ | kg $PO_4^{3-}$ |
| Diesel | 675.1 | 60.1 | 7.89 | 126 | 20.4 |
| Hybrid | 499.9 | 44.5 | 5.96 | 91.9 | 14.7 |

**Table 3** Results of air emissions

| | Diesel | Hybrid |
|---|---|---|
| Carbon dioxide $CO_2$ [t] | 58.2 | 42.9 |
| Carbon monoxide CO [kg] | 181 | 129 |
| Non-methane volatile organic compounds NMVOC [kg] | 0.73 | 0.702 |
| Methane $CH_4$ [kg] | 73.8 | 38.9 |
| Oxides of nitrogen $NO_x$ [kg] | 26.12 | 38.8 |
| Sulfur dioxide $SO_2$ [kg] | 22.62 | 18.21 |
| Ammonia $NH_3$ [g] | 250 | 243 |
| Nitrogen dioxide $NO_2$[kg] | 4.83 | 4.95 |
| Total | 617.3 | 516.462 |

**Table 4** Results of water emissions

| | Diesel | Hybrid |
|---|---|---|
| Nitrate $NO^{3-}$ [g] | 294 | 296 |
| Phosphate $PO_4^{3-}$ [g] | 314 | 330 |
| Sulfate $SO_4^{2-}$ [kg] | 9.47 | 9.27 |
| Hydrocarbons [g] | 126.7 | 89.4 |

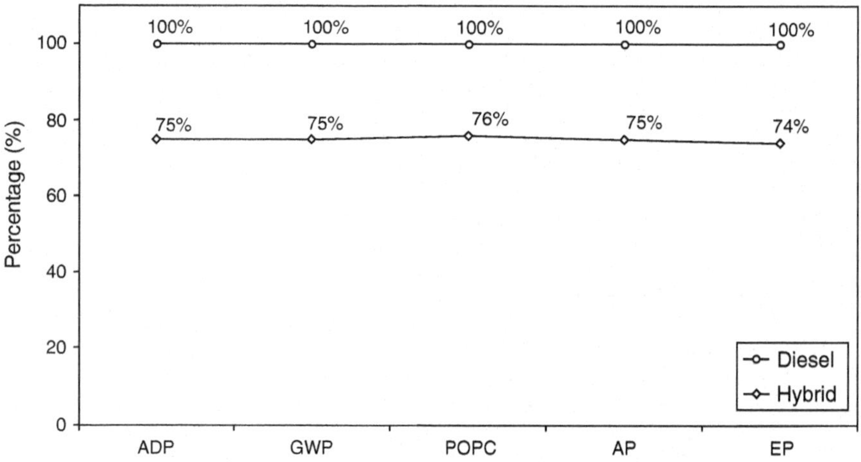


**Fig. 2** Categories of environmental impact by diesel and hybrid ICEV

## 3.5 *Interpretation of Results*

The main objective of the results interpretation stage is to verify which of the ICEVs in this study has a better environmental performance. Through Fig. 2, it is possible to observe that the values of the categories of environmental impact differ significantly between diesel and hybrid ICEV. Such differences happen due to the different

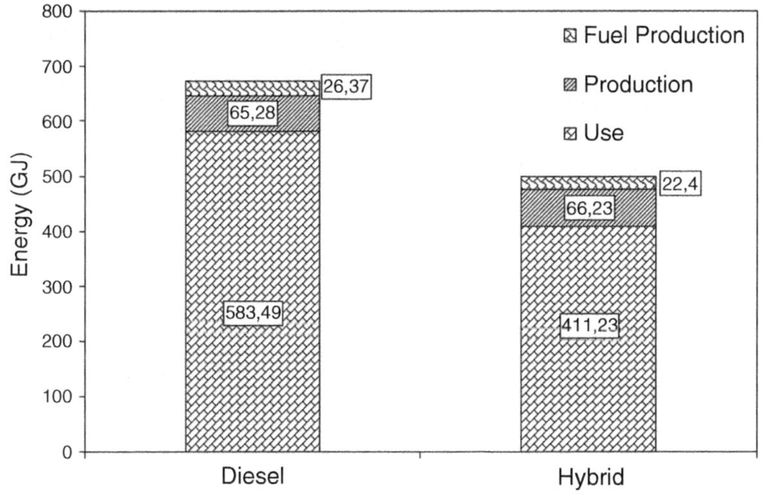


**Fig. 3** Differences between the two ICEVs by analyzing the ADP

components used in the production of automobiles and the consumption of fuel during their lifetimes.

By analyzing the ADP, particularly fossil fuels, between the two ICEVs, expressed in GJ, and using absolute values, it can be observed that the most significant differences between the two ICEVs are associated with their use (Fig. 3). This is due to the difference in intake observed between the two ICEVs. Consequently, the diesel ICEV has a greater contribution for this indicator in the production of fuel. During production, there is a contribution, however slight, in the case of the hybrid ICEV compared to the diesel one due to its superior requirement of materials and energy resources.

In Fig. 4 is shown the global warming potential between the two ICEVs, expressed in tons of carbon dioxide. It can be seen that the diesel ICEV has a larger contribution to global warming than the hybrid ICEV. This results from the difference between the emission of the main dangerous gases to the environment ($CO_2$ and $CH_4$) by showing a higher abundance in the life cycle of the diesel ICEV.

Figure 5 illustrates the POPC between the two ICEVs in kilograms of ethylene. By observing the figure, the diesel ICEV has a greater contribution for this environmental impact category than the hybrid one. This is explained by the gases that contribute to this indicator, which are $CH_4$, $NO_x$, and CO, which show different behaviors depending on the type of ICEV.

As can be observed in Fig. 6, the diesel ICEV has a larger contribution for the AP indicator. The main substances that contribute to this indicator are $SO_2$ and $NH_3$. The $SO_2$ causes the greatest impact since this indicator is in kilograms of $SO_2$, and as such it has a ratio of 1:1 contribution, and $SO_2$ emissions are produced when diesel is used as an energy source.

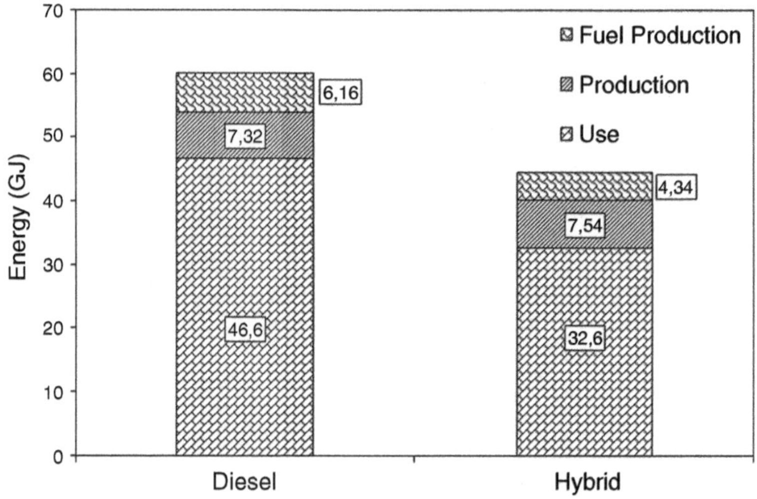


**Fig. 4** Potential of global warming between the two ICEVs in tons of carbon dioxide

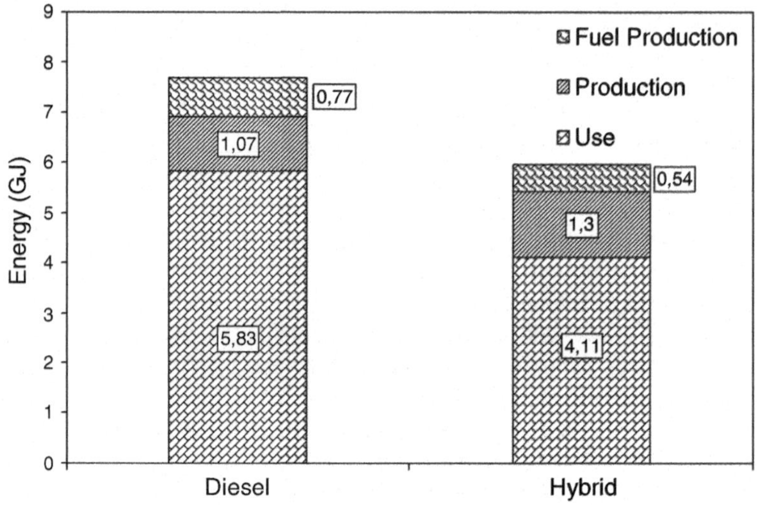


**Fig. 5** POPC between the two ICEVs in kilograms of ethylene

About the EP, it measures the excessive nutrient enrichment of water bodies, particularly with phosphate. It leads to the degradation of aquatic systems through increasing aquatic vegetation. In addition, it contributes for ammonia emission, nitrates, nitric acid, phosphate, and phosphorous, among others.

In this study the diesel ICEV presents a larger EP mainly during the utilization phase as can be seen in Fig. 7.

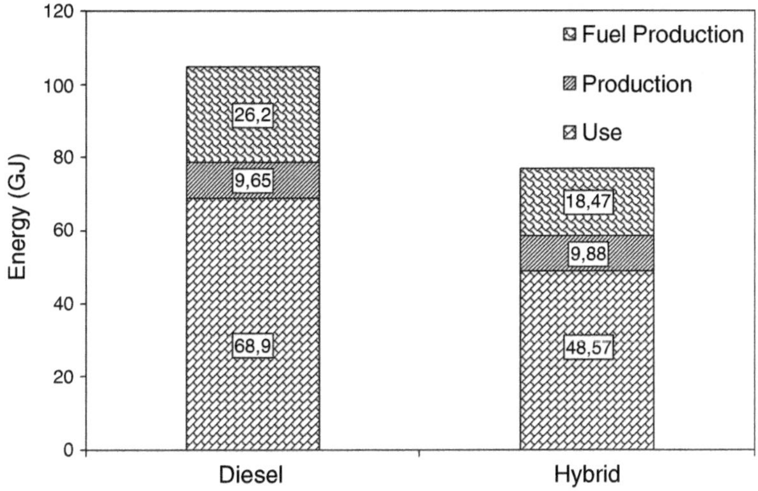


**Fig. 6** Contribution of the diesel ICEV for the AP indicator

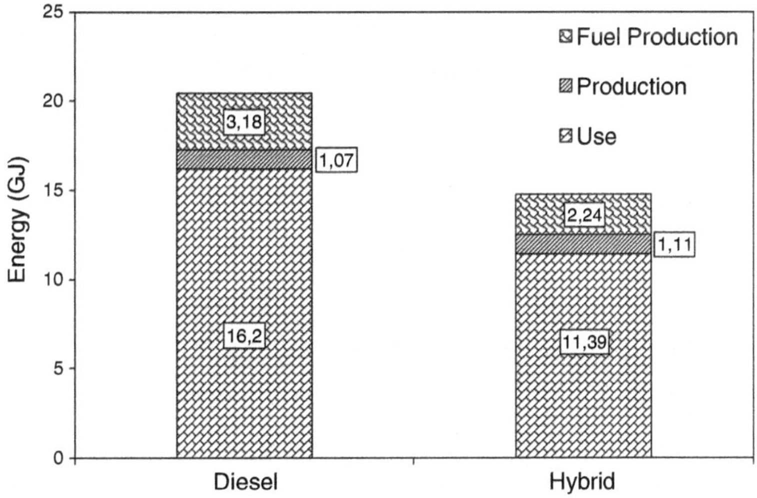


**Fig. 7** EP during the utilization phase by diesel ICEV

# 4 Conclusion

In this study the environmental performance of two ICEVs with different engines was analyzed: one with a diesel engine and the other with a diesel engine with an electric motor supply—a hybrid-diesel ICEV. The environmental impact of both ICEVs during their lifetime was assessed using the LCA methodology by employing the software OpenLCA. The Portuguese automobile park was chosen for this study.

The data utilized in this study was obtained from the manufacturers of both ICEVs with their permission. Then, by performing a quantitative analysis, a comparison was made with the purpose to assess the ICEV with the best environmental performance.

According to the results, the environmental impact differs significantly between diesel and hybrid ICEVs. This is due to the different components used during the production phase and the consumption of fuel during their lifetimes. Considering the ADP during production, there is a greater contribution of the hybrid ICEV compared to the diesel due to its superior requirements of materials and energy resources. However, the most significant differences between the two ICEVs, considering this indicator, were related to their use. Also, it can be stated that the diesel ICEV has a larger contribution to global warming than the hybrid ICEV. The potential impact of POPC of both ICEVs was also analyzed, and it was concluded that the diesel ICEV has a greater contribution for this environmental impact category than the hybrid one.

About the negative environmental impact of EP, the diesel ICEV presents a larger potential of eutrophication mainly during the utilization phase. The results indicated that the hybrid ICEV has shown generally a better environmental performance compared to the diesel ICEV. Results also indicated that the phase with the greatest impact on environment in both ICEVs is the use phase. The reduction in consumption by the hybrid ICEV due to the use of electrical energy generated by the power of the electric motor is a technological breakthrough of significant importance.

## Bibliography


Bascoul, G., Schmitt, J., & Rasolofoarison, D. (2014). Consumers' biased perceptions of the environmental impact of products, and their influence on advertising efficiency. In *The sustainable global marketplace* (pp. 400–404). Cham: Springer.

Bauer, C., Hofer, J., Althaus, H.-J., Duce, A. D., & Simons, A. (2015). The environmental performance of current and future passenger vehicles: Life cycle assessment based on a novel scenario analysis framework. *Applied Energy, 157*, 871–883.

Branker, K., Jeswiet, J., & Kim, I. (2011). Greenhouse gases emitted in manufacturing a product—A new economic model. *CIRP Annals – Manufacturing Technology, 60*(1), 53–56.

Carrasco, A., Ferreira, Ó., & Roelvink, D. (2016). Coastal lagoons and rising sea level: A review. *Earth-Science Reviews, 154*, 356–368.

Castka, P., & Balzarova, M. A. (2008). The impact of ISO 9000 and ISO 14000 on standardisation of social responsibility—An inside perspective. *International Journal of Production Economics, 113*(1), 74–87.

Cho, C. P., Pyo, Y. D., Jang, J. Y., Kim, G. C., & Shin, Y. J. (2017). NOx reduction and N2O emissions in a diesel engine exhaust using Fe-zeolite and vanadium based SCR catalysts. *Applied Thermal Engineering, 110*, 18–24.

Coria, J., Bonilla, J., Grundström, M., & Pleijel, H. (2015). Air pollution dynamics and the need for temporally differentiated road pricing. *Transportation Research Part A: Policy and Practice, 75*, 178–195.

Crippa, M., Janssens-Maenhout, G., Guizzardi, D., & Galmarini, S. (2016). EU effect: Exporting emission standards for vehicles through the global market economy. *Journal of Environmental Management, 183*(3), 959–971.

Curran, M. A. (2012). *Life cycle assessment handbook: A guide for environmentally sustainable products*. Beverly, MA: Scrivener Publishing.

Dalianis, G., Nanaki, E., Xydis, G., & Zervas, E. (2016). New aspects to greenhouse gas mitigation policies for low carbon cities. *Energies, 9*(3), 128.

Danilecki, K., Mrozik, M., & Smurawski, P. (2017). Changes in the environmental profile of a popular passenger car over the last 30 years – Results of a simplified LCA study. *Journal of Cleaner Production, 141*, 208–218.

Farahani, M., Pagé, D., & Turingia, M. (2011). Sedimentation in biodiesel and ultra low sulfur diesel fuel blends. *Fuel, 90*, 951–957.

García-Olivares, A., Solé, J., & Osychenko, O. (2018). Transportation in a 100% renewable energy system. *Energy Conversion and Management, 158*, 266–285.

Groenewald, J., Grandjean, T., & Marco, J. (2017). Accelerated energy capacity measurement of lithium-ion cells to support future circular economy strategies for electric vehicles. *Renewable and Sustainable Energy Reviews, 69*, 98–111.

INE, I.P. (2012). *Estatísticas dos Transportes – 2011*. Lisboa: Instituto Nacional de Estatística, I.P.

ISO. (2006). *ISO 14040:2006 – Environmental management – Life cycle assessment – Principles and framework*. Geneva: ISO Copyright Office.

Ju, K., Su, B., Zhou, D., Zhou, P., & Zhang, Y. (2015). Oil price crisis response: Capability assessment and key indicator identification. *Energy, 93*(2), 1353–1360.

Kałuża, A. (2016). General statistics of diesel engines' idle time: Shunting locomotives in industrial sidings in Poland 2009...2013. *Transportation Research Part D: Transport and Environment, 49*, 82–93.

Karakaya, E., Hidalgo, A., & Nuur, C. (2014). Diffusion of eco-innovations: A review. *Renewable and Sustainable Energy Reviews, 33*, 392–399.

Klöpffer, W., & Grahl, B. (2014). *Life cycle assessment (LCA): A guide to best practice*. Weinheim: Wiley.

Levasseur, A., et al. (2016). Enhancing life cycle impact assessment from climate science: Review of recent findings and recommendations for application to LCA. *Ecological Indicators, 71*, 163–174.

Liu, B., Yan, F., Hu, J., Turkson, R., & Lin, F. (2016). Modeling and multi-objective optimization of NOx conversion efficiency and NH3 slip for a diesel engine. *Sustainability, 8*, 478.

Lu, Y.-T., Lee, Y.-M., & Hong, C.-Y. (2017). Inventory analysis and social life cycle assessment of greenhouse gas emissions from waste-to-energy incineration in Taiwan. *Sustainability, 9*(11), 1959.

Lullo, G. D., Zhang, H., & Kumar, A. (2016). Evaluation of uncertainty in the well-to-tank and combustion greenhouse gas emissions of various transportation fuels. *Applied Energy, 184*, 413–426.

MacLean, H. L., Lave, L. B., Lankey, R., & Joshi, S. (2000). A life-cycle comparison of alternative automobile fuels. *Journal of the Air & Waste Management Association, 50*(10), 1769–1779.

Meng, L., & Sager, J. (2017). Energy consumption and energy-related CO2 emissions from China's petrochemical industry based on an environmental input-output life cycle assessment. *Energies, 10*(10), 1585.

Nanaki, E. A., & Koroneos, C. J. (2016). Climate change mitigation and deployment of electric vehicles in urban areas. *Renewable Energy, 99*, 1153–1160.

Nel, R., & Nienaber, G. (2012). Tax design to reduce passenger vehicle CO2 emissions. *Meditari Accountancy Research, 20*(1), 39–51.

Nocera, S., & Cavallaro, F. (2016). The competitiveness of alternative transport fuels for CO2 emissions. *Transport Policy, 50*, 1–14.

Remuzgo, L., Trueba, C., & Sarabia, J. M. (2016). Evolution of the global inequality in greenhouse gases emissions using multidimensional generalized entropy measures. *Physica A: Statistical Mechanics and its Applications, 444*, 146–157.

Reşitoğlu, İ. A., Altinişik, K., & Keskin, A. (2015). The pollutant emissions from diesel-engine vehicles and exhaust aftertreatment systems. *Clean Technologies and Environmental Policy, 17* (1), 15–27.

Tagliaferri, C., et al. (2016). Life cycle assessment of future electric and hybrid vehicles: A cradle-to-grave systems engineering approach. *Chemical Engineering Research and Design, 112*, 298–309.

Vieira, D. R., Calmon, J. L., & Coelho, F. Z. (2016). Life cycle assessment (LCA) applied to the manufacturing of common and ecological concrete: A review. *Construction and Building Materials, 124*, 656–666.

WHO. (2009). *Global health risks – Estimated deaths & Dalys attributable to selected environmental risk factors.* Geneva: World Health Organization.

# The Role of Marketing in Reducing Climate Change: An Approach to the Sustainable Marketing Orientation

**Cristina Calvo-Porral**

## Contents

1 Introduction .... 262
2 The Sustainable Marketing Orientation .... 263
2.1 Environmental-Conscious Marketing Orientations .... 263
2.1.1 Green Marketing Orientation .... 263
2.1.2 Social Marketing Orientation .... 263
2.1.3 Critical Marketing Orientation .... 264
2.1.4 Sustainable Marketing Orientation .... 264
2.2 New Ways of Consumption .... 266
2.2.1 Sustainable Consumption .... 266
2.2.2 Responsible Consumption .... 267
2.2.3 Anti-consumption .... 267
2.2.4 Mindful Consumption .... 268
3 How the Sustainable Marketing Orientation Helps Reducing Climate Change .... 268
3.1 Slow Food and Slow Fashion Movements .... 269
3.1.1 Slow Food .... 269
3.1.2 Slow Fashion .... 270
4 Methodology .... 271
5 Companies Developing Sustainable Marketing Actions .... 271
5.1 Panera: Reduction of Food Waste Through Donations .... 271
5.2 Patagonia: "Everything We Make Costs the Planet More Than It Gives Back" .... 272
5.3 Levi Strauss & Co.: Reducing Water Consumption .... 273
5.4 Ecoalf: "Because There Is No Planet B" .... 274
5.5 Tide: Laundry Washing in Cold Water .... 276
5.6 Carrefour: Removal of Disposable Plastic Bags .... 278
5.7 Bulk Purchase: The Removal of Food Packages in Grocery Shopping .... 280
6 Conclusion .... 281
References .... 282

**Abstract** "Don't buy this Jacket" is the advertising campaign of the clothing brand Patagonia asking customers not to buy its products. In this marketing campaign, the company requested customers to wrest the full life out of every Patagonia product by

C. Calvo-Porral (✉)
University of A Coruña, A Coruña, Spain
e-mail: ccalvo@udc.es



T. Sequeira, L. Reis (eds.), *Climate Change and Global Development*, Contributions to Economics, https://doi.org/10.1007/978-3-030-02662-2_13

buying used when possible and asking customers to reduce unnecessary consumption. "Baked before sunrise; donated after sunset" is the marketing campaign developed by the bakery company Panera. This company donates all the leftover bread after stores have shut down, and the unsold bread goes toward a great cause, reducing food waste, which is the cause of huge wastage of natural resources.

Nowadays, numerous companies are conscious that the consequences of marketing actions clearly extend beyond the company to society as a whole. Consequently, companies following the *sustainable marketing orientation* consider the environmental aspects in the exchange relationship between the company and the market. This orientation is grounded on the recognition that the best marketing strategy is being coherent with the long-term survival of the company and the natural environment. In this context, the main purpose of this study is the examination of some interesting examples of companies that are developing real marketing actions to reduce the climate change.

**Keywords** Marketing · Climate change · Sustainable Marketing · Anti-consumption

## 1 Introduction

We live on a finite planet, and this fact entails major challenges such as the depletion of natural resources, the climate change, or the ecological footprint. Likewise, the unsustainable nature of existing patterns of production, consumption, and development is increasing the actual concern about the environmental effects of companies' practices. The economic data report that the majority of economic trends continue to move away from sustainability and that consumption patterns and levels are not environmentally sustainable (Peattie and Peattie 2009). Therefore, consumption patterns cannot continue at their current rate; but such practices persist and are increasing due to the continued growth of the global economy (Lim 2017).

This is the main paradox of marketing. On one hand, marketing focuses on an increased consumption and economic development, being the marketing goal to encourage and increase consumption (Gordon et al. 2011). So, marketing is not being inherently focused to deliver sustainability. However, and on the other hand, marketing plays a key role in environmental issues, since marketing can influence and change consumer behavior. Moreover, marketing can encourage individuals to reuse, recycle, or save energy, provide ethical products, as well as address the goal of sustainability, serving consumer and society needs (Gordon et al. 2011). Finally, marketing can provide strategies for radical changes to the unsustainable consumption.

In this context, the present chapter considers and examines the potential of the discipline of marketing to contribute to climate change reduction, through a more sustainable consumption and through changes in consumer behavior.

# 2 The Sustainable Marketing Orientation

Sustainability embraces both production and consumption; and as a consequence, both sides will be examined: the companies' sustainable orientations and the new ways of consumption.

## *2.1 Environmental-Conscious Marketing Orientations*

### 2.1.1 Green Marketing Orientation

Peattie (1995) conceptualized the *green orientation* as the management orientation responsible for identifying, anticipating, and satisfying the requirements of consumers and society in a profitable and sustainable way, developing sustainable practices from production to post-purchasing services. Likewise, the *green marketing* orientation facilitates the development and commercialization of more sustainable products and services (Gordon et al. 2011). That is, *green marketing* addresses the gap between the current marketing practices and the ecological and social realities. Later, Papadas et al. (2017) conceptualized the *green marketing* orientation as the extent to which an organization engages in strategic, tactical, and internal processes and activities which aim to create, communicate, and deliver products and services with the minimal environmental impact.

However, keeping a conventional marketing orientation, while encouraging product substitutions in favor of *green products*, has neglected other important environmental issues, such as energy consumption, resource depletion, species extinction, or ecosystem destruction. In this vein, some authors pointed out that the *green marketing* orientation was entirely geared toward trying to persuade consumers to buy more green products or to dispose them more responsibly (Peattie and Peattie 2009), thus not making a substantive contribution toward sustainability.

Therefore, the *green marketing* orientation should imply a much broader adoption of sustainability behavior including product-related decisions to reduce the environmental footprint, tools in order to reduce the negative environmental impact of the company, environmentally responsible packaging, recyclable or reusable content, the re-examination of the product life cycle, and the reverse supply chain approach to recover the product's maximum possible value (Cronin et al. 2011).

### 2.1.2 Social Marketing Orientation

Kotler and Zaltman (1971) introduced the term *social marketing* as the use of marketing principles and techniques to advance a social idea, cause, or behavior. Later, Lazer and Kelley (1973) noted that the social marketing is concerned with the application of marketing knowledge, concepts, and techniques to achieve social and economic goals and with the analysis of the social consequences of marketing policies, decisions, and activities. Similarly, the social marketing orientation holds that marketing activities should take into consideration the welfare of society and the interests of consumers and business shareholders (Prothero 1990). Finally, Kotler et al. (2002) defined social marketing as "the use of marketing principles and techniques to influence a target audience to voluntarily accept, reject, modify or abandon a behavior for the benefit or society as a whole."

So, the *social marketing orientation* involves using the power of marketing in order to encourage sustainable behavior among consumers and to evaluate the impact of current marketing practices on sustainability (Gordon et al. 2011). That is, social marketing seeks to go beyond changing attitudes to changing human behavior, as a way to encourage sustainable marketing solutions for the benefit of the society as a whole. Further, the principles and practices of the *social marketing* orientation provide solutions to environmental problems and to tackle climate change, such as recycling, composting, energy efficiency and sustainable transport use, and the promotion of sustainable lifestyles.

### 2.1.3 Critical Marketing Orientation

The *critical marketing* orientation entails a critical appraisal of marketing theory and encourages a marketing system in which sustainability is the key goal (Gordon et al. 2011). Similarly, this marketing orientation focuses less on encouraging unnecessary consumption and more toward encouraging sustainability, stimulating sustainable marketing practices, and supporting customers to make informed choices that do not harm societal well-being. That is, critical marketing encourages sustainability, rather than unnecessary consumption.

According to Peattie (2007), the role of *critical marketing* is not only the application of a critical viewpoint to the marketing discipline but also the examination of some of the fundamental principles and concepts of marketing. Thus, critical marketing is involved with the evaluation of the impact of conventional marketing on society (Hastings 2009). Finally, *critical marketing orientation* could be used to challenge the dominant marketing theory and practice, shifting the focus from encouraging consumption to other marketing goals such as satisfaction, sustainability, and quality of life (Peattie and Peattie 2009). So, according to this marketing orientation, marketing should focus on sustainability, quality of life, quality of services and products, and consumer satisfaction (Peattie 2007).

### 2.1.4 Sustainable Marketing Orientation

Marketing practices today focus on an increased consumption and economic development, being the marketing goal to encourage consumption (Gordon et al. 2011), and in this context, the sustainability marketing orientation could be considered an alternative approach to marketing theory and practice. And it is evident that extensive environmental damage has been caused by continuous manufacturing, processing, consumption, and discarding (Saha and Darnton 2005).

According to the OECD (2002), the sustainability concept could be defined as "the consumption of goods and services that meet basic needs and quality of life, without jeopardizing the needs of future generations," thus considering both on the needs of current and future generations. Therefore, sustainability could become a key component of marketing theory and practice (Peattie and Peattie 2009).

The concept of *sustainable marketing orientation* recognizes that current consumption patterns are unsustainable and the requirement of changes in consumer behavior (UNEP, 2005). So, sustainable marketing offers products and services marketed in a responsible way that does not adversely impact upon environment sustainability. Further, the sustainable marketing orientation develops strategies and actions so that only items that are considered as "needs" are commercialized and marketed, while those considered as "wants" are not (Schor, 1998), encouraging consumers to develop a more resource-efficient way of consumption (Connolly and Prothero 2003), including the acceptance of consumption reduction, responsible consumption, and sustainable lifestyle (Peattie and Peattie 2009).

More precisely, the *marketing sustainability orientation* is based on three dimensions (Sung and Lee 2011). First one is the social dimension, related to the company's impact on society and the human well-being, including social equity and community relations. Second dimension is the environmental component that focuses on the company's activities relative to natural resources, as well as the company contribution to environmental sustainability. And the last dimension is the economic one, which refers to the value generation and financial performance of companies (Simpson and Radford 2014). The most relevant environmental-conscious marketing orientations are shown in Table 1.

**Table 1** Environmental-conscious marketing orientations

| Marketing orientations | Actions |
|---|---|
| Green marketing | Reduced packaging, energy efficient production. Focuses on the marketing mix and the minimization of its negative environmental impact |
| Social marketing | Marketing activities should take into consideration the welfare of society. Consumers should live an ecological lifestyle, demand of ecological products and local food produce |
| Critical marketing | Critical viewpoint to the marketing theory and practice<br>Marketing should increase customer satisfaction and quality of life |
| Sustainable marketing | Integration of environmental, social, and economic concerns and practices into the strategic and marketing activities<br>Sustainable production and consumption. Reduction of consumption and wastage. Development of resource-efficient ways of consumption |

*Source*: Own elaboration from Peattie and Peattie (2009) and Gordon et al. (2011)

## 2.2 *New Ways of Consumption*

Even though there is a general consensus on the need to reduce resource utilization and energy or reduce carbon emissions as part of sustainable development, a lack of consensus exists on whether consumption should be reduced or just changed and whether individual consumers have the capability to significantly contribute to resource conservation (Banbury et al. 2012). However, major changes in current consumption patterns are required to solve the global environmental issues, such as climate change (Perry 2006; Robinson et al. 2006).

### 2.2.1 Sustainable Consumption

Many consumers find it difficult to consume sustainably, because the acts of consuming and sustaining are contradictory to each other. Therefore, the concept of *sustainable consumption* itself is a problematic concept, and the reason is that *to consume* something means to use it up or to destroy it. In this context, consumption is being increasingly challenged by consumerist and anti-consumption movements (Lim 2017), and a new perspective of consumption is required. This new perspective could be the so-called sustainable consumption, which is a concept that goes beyond the traditional understanding of consumption.

According to the World Commission on Environment and Development (1987), sustainable consumption should meet current needs and wants without impoverishing future generations and the planet's ability to meet these needs and wants. In this vein, the European Environmental Agency (2005) defined sustainable consumption as "the use of goods and services that respond to basic needs and bring better quality of life, while minimizing the use of natural resources, toxic materials and emissions of waste and pollutants over the life cycle, so as not to jeopardize the needs of future generations." So, in general terms, the *sustainable consumption* emphasizes the need for a reduction in overall resource consumption and the assumption that changes in consumption levels and patterns are needed in order to achieve sustainability.

Following Cummins et al. (2014), *sustainable consumption* refers to the purchase and use of more efficiently produced goods, and environmentally friendly consumption practices, to address the resource needs of current and future generations. Likewise, *sustainable consumption* includes meeting needs, improving quality of life, improving efficiency, minimizing waste, and taking a life cycle perspective while contributing to reducing environmental damage (Manoochehri 2001). So, *sustainable consumption* is a way of life that rejects consumerism, high consumption, and materialistic lifestyles.

### 2.2.2 Responsible Consumption

The seminal work of Fisk (1973) explored the potential role of consumption reduction in his *Theory of Responsible Consumption* and referred to responsible consumption as the "rational and efficient use of resources with respect to the global human population"; similarly, focusing on the demand side, responsible consumption could be considered as a consumer behavior. According to Mohr et al. (2001) the *responsible consumption* means the acquisition, usage, and disposition decisions based on the desire to minimize or eliminate any harmful effects and maximize the long-term beneficial impact on the society and the environment. Other definitions of responsible consumption report that ethical, social, and environmental dimensions should be incorporated into the term of responsible consumption (Lim 2017).

### 2.2.3 Anti-consumption

The anti-consumption movements highlight the excessive unsustainable consumption.

Further, the anti-consumption orientation makes some propositions such as "we need to consume less," "you don't buy happiness in a shop," "new is not always better," or "maintaining and repairing products is smart" (Peattie and Peattie 2009), seeking to move individuals away from an intensive form of consumption.

More precisely, this new way of consumption considers three dimensions. In the first place, it considers the importance of *non-purchase* elements of consumer behavior including product use and disposal. Second, it considers the potential importance of non-purchase-*based behaviors* as the means for consumers to meet their needs and achieve satisfaction. Finally, this way of consumption considers the social and environmental concern to encourage a *reduction in the total level of consumption* (Peattie and Peattie 2009).

Numerous authors defined the concept of anti-consumption. Zevestoski (2002) refers to anti-consumption as an act of resistance to, distaste of, or even resentment of consumption. Later, Lim (2017) conceptualized anti-consumption as a subjective consumer practice motivated by personal interests or social or environmental concerns, in which the consumer rejects the consumption of a particular product/brand on both a personal and societal level. Similarly, according to Lee et al. (2011), anti-consumption includes rejection, restriction, and reclamation. First, in the process of rejecting, consumers intentionally exclude particular products/brands from their consumption. When total anti-consumption is not possible, consumers can chose to restrict consumption of particular products/brands. Then, reclamation represents an ideological shift to a holistic process that includes acquisition, use, and dispossession.

In addition, the anti-consumption orientation creates savings rather than expenses. So, the costs of changing behavior into anti-consumption are not financial, and costs may be in terms of psychological barriers, related with not keeping up with

**Table 2** Alternative ways of consumption

| | Alternative ways of consumption |
|---|---|
| Sustainable consumption | Consumers concern about the environment, social, and economic issues<br>Consumers' evaluations regarding companies sustainability practices |
| Responsible consumption | Consumers are aware of the negative effects of consumption. Social, environmental, and ethical concerns can be translated into consumers' decisions |
| Anti-consumption | Increase of non-purchase behaviors and non-purchase elements. Reduce the total level of consumption. Alternative consumption choices include restriction and reclamation |
| Mindful consumption | Great consumer awareness for oneself and the entire ecosystem, which may dampen the effects of unsustainable practices |

*Source*: Own elaboration from Lee et al. (2011) and Lim (2017)

fashion or trends or matching the consumption patterns and levels of peers (Peattie and Peattie 2009). However, it is difficult to make consumption reduction appealing to consumers, and in developed countries, promoting consumption reduction goes against the highly consumption-orientated dominant social paradigm (Kilbourne and Beckman 1998). Further, consumption is far easier to portray and communicate than frugality or a simplified lifestyle.

#### 2.2.4 Mindful Consumption

Mindfulness could be conceptualized as the open and receptive attention to and awareness of what is occurring in the present moment (Brown and Ryan 2004). Similarly, the seminal work of Sheth et al. (2011) offers the first extensive conceptual integration of the notion of mindfulness with consumption. More specifically, these authors define *mindfulness consumption* as the behavior of the consumer who guides his/her consumption behavior on whether to consume sustainably or unsustainably. Later, authors like Lim (2017) suggested that the core attribute of mindfulness consumption is consumer temperance in consumption, with the goal of enhancing personal well-being in a way that is consistent with personal values. So, consumers who engage in mindfulness consumption make conscious choices fully understanding the consequences of their consumption choices and practices.

These alternative new ways of consumption are shown in Table 2.

## 3 How the Sustainable Marketing Orientation Helps Reducing Climate Change

Any act of consumption will inevitably have some impact on the environment. In this vein, marketing could be used to change consumers' attitudes and activities: if consumers do not change their consumption behavior to become more sustainable, then little will be achieved. Likewise, marketing could help in shifting the cultural values of consumption to address environmental issues, making consumers understand and become aware about the environmental impact of consumption, while promoting sustainable ways of consumption.

However, the relationship of marketing to sustainability is complex. On one hand, marketing has developed a role in stimulating unsustainable levels of demand and unsustainable consumption patterns. And on the other hand, marketing has developed mechanisms to tackle environmental issues (Peattie and Peattie 2009). More precisely, through *sustainable marketing*, companies could introduce sustainability into the core marketing strategies and actions, ensuring that product design and development, manufacturing, distribution, and promotion are made sustainable and, further, reduce consumption and waste (Gordon et al. 2011).

This chapter supports that marketing holds many potential solutions to some of the challenges around sustainability faced today and that marketing actions and strategies could help companies to face the new environmental challenges.

## *3.1 Slow Food and Slow Fashion Movements*

### 3.1.1 Slow Food

Environmental problems are closely related to the dominant food production practices, and there are great links between food consumption practices and climate change. In fact, a large proportion of greenhouse gas emissions are caused by food production through the use of chemical fertilizers and pesticides, as well as by consumer habits (IFOAM 2009). Additionally, the localization of food production affects climate change and, more specifically, the long physical distance between the food production and the food consumption in the global food system (Soler 2012).

In this context, the American Public Health Association (2007) defines sustainable food production as "one that provides healthy food to meet current food needs while maintaining healthy ecosystems that can also provide food for generations to come with minimal negative impact on the environment, while encouraging local food production." Likewise, sustainable food is related to food that is minimally processed, organically produced, regional and seasonal, and fairly traded and packed in an environmentally friendly way, entailing the recycling of soil nutrients and the circulation of energy (Soler 2012).

Nowadays, the dominant food production system in agriculture is the conventional method of production and distribution that relies heavily upon high inputs of fertilizers, pesticides, and fossil fuels to cultivate large monocultures. Further, it could be considered a large-scale industrial food production system dependent on inputs of energy and nutrients, globally sourced and nonseasonal (Soler 2012). However, as a result of the increasingly negative impacts of conventional agriculture on the environment, alternative agriculture movements have emerged that propose alternative methods of production, distribution, and consumption. One of these movements is the *slow food movement*.

The main concern of the *slow food* movement is that agribusiness and food industries are standardizing taste and leading to the annihilation of thousands of food varieties (Stille 2001). As a consequence, this movement promotes food quality, environmental and cultural sustainability, and biodiversity (Sassatelli and Davolio 2010), through unique local foods prepared with locally grown ingredients, from local food producers and suppliers. More precisely, local food production is based on small-scale food production of seasonal food for the local market, following local-regional food production. Finally, slow food must be produced or harvested sustainably, without genetically modified plants or transgenic breeds. Finally, one major issue is that consumers cannot directly perceive the direct impact of climate change or global warming of unsustainably produced food, which makes the marketing of sustainably produced food more complex in terms of credibility (Soler 2012).

### 3.1.2 Slow Fashion

The fashion industry is considered as not a sustainable industry, with business activities creating pollution, gas emissions, and hazardous waste. The deep environmental impact of fashion activities is due to the volume of processes dependent on the massive use of natural resources. In addition, the fashion industry promotes overconsumption and fashion waste, mainly due to the short and seasonal product life cycle, the high volatility and low predictability of product's demand, and the high purchasing impulse, since many fashion purchase decisions are influenced by emotional factors (Fletcher 2007). Overconsumption and fashion waste have become important environmental issues in the fashion industry. More specifically, the *fast fashion* industry, focused on the capability to quickly respond to fast-changing fashion trends offering low prices, enhances overconsumption, which is a direct cause of unsustainability and climate change. Similarly, the *fast fashion* industry makes consumers purchase clothes at a certain time and discard them not long after by shortening the life span of clothing deliberately, resulting in huge resource consumption and fashion waste (Fletcher 2010).

In this context, and being in line with the slow food movement, Fletcher (2007) first introduced the concept *of slow fashion* which broadens the sustainability perspective to include the pace of fashion cycles, emerging as an alternative to environmentally unsustainable practices resulting from the fast fashion. *Slow fashion* products are manufactured slowly and in small quantities, reducing the consumption of natural resources and the amount of waste (Cline 2012) and emphasizing the product quality through a slower production and consumption cycle. Consequently, the prices of *slow fashion* products are higher in general terms than fast fashion products—which are produced through mass production systems (Jung and Jin 2016). Likewise, slow fashion encourages consumers to buy less products at a higher price and more durable quality (Fletcher 2007), offering designs that are less influenced by fashion trends, so consumers can wear the garments for a long time and keep the garment longer, rather than discarding it shortly after purchase (Fletcher 2010; Jung and Jin 2016).

Therefore, slow fashion makes individuals to buy less at a high quality, underlying a shift in the consumer mind-set from quantity to quality and shifting consumption patterns to reduce the consumption levels. However, even the slow fashion sustainable production can become unsustainable, when garments are worn only a few times and then discarded quickly.

# 4 Methodology

The methodology developed is the case study method. Case studies illustrating practice from the world of business are widely used and have become an integral part of the methodology for business studies that draw on reality. More specifically, case studies are useful when a case represents a special set of circumstances or phenomena that warrant intensive study or when researchers do not have sufficient knowledge of a case to place it in theoretical perspective (Bradshaw and Wallace 1991).

This case study analysis focuses on environmental aspects. The companies were selected for the study according to their innovation and originality in reducing climate change and according to their special circumstances that deserve analysis. Likewise, the companies selected may help inform general theory and explain conditions that deviate from conventional theoretical foundations, since the selected cases deviate from conventional marketing strategies. Consequently, in this study some companies belonging to different sectors have been selected to examine their contribution to climate change reduction, through a sustainable marketing orientation.

# 5 Companies Developing Sustainable Marketing Actions

## *5.1 Panera: Reduction of Food Waste Through Donations*

The bakery company Panera addresses the challenge of food waste through the donation of usable food that remains on their shelves at the end of the day. As a bakery-cafe, this company addresses two main types of food disposal, packaging and food waste, which are generated in the guest areas and back of house. Regarding packaging, waste ranges from paperboard, plastic, and aluminum, all of which are recyclable. The other main issue is the food waste generated in the bakery-cafes. To address this issue, Panera created the *Day-End Dough-Nation* program, which provides leftover bread and other foods to local charities, through the donation of unsold baked goods—that otherwise would go to waste—to food banks, organizations, and other charities.

Later, Panera developed the "Food For Thought Outreach" program with the goal of providing wholesome food to the Walton Academy in Florida—a charter school with a large at-risk student population. In this school, 90% of the students are eligible for free breakfast, since they do not get enough food to eat at home, arriving at school hungry. However, this school did not have a cafeteria, making it unable to meet this need. Today, this school partners with Panera, providing breakfast to 100% of the most at-risk children they attend: Panera provides breakfast food and snacks every week, donating much-needed nourishment to these kids and making a significant difference in their lives at school. So, in 2014, Panera donated a retail value of approximately $100 million worth of unsold baked goods to those in need, food that otherwise would go wasted.

## 5.2 *Patagonia: "Everything We Make Costs the Planet More Than It Gives Back"*

Patagonia developed downstream consumption, intending to influence consumer behavior to lower the environmental strain from the growing consumption levels. This outdoor clothing company aims to reduce the environmental and social impact of disposable fashion, developing an *alternative marketing* approach emphasizing fashion maintenance, repair, and recycling, thus discouraging consumption and the disposability of clothes. More precisely, Patagonia is putting sustainability ahead of profit and actively asking and encouraging consumers to *buy less*.

The company soon realized that their business has potentially harmful impacts on the environment, since Patagonia jackets come with an environmental cost higher than their price—one jacket requires 135 l of water to be manufactured and generates 20 pounds of carbon dioxide. So, the company believes that "we design and sell things made to last and be useful. In addition, we ask customers not to buy from us what they do not need or cannot really use, since everything we make costs the planet more than it gives back." For this reason, when the jacket comes to the end of its life cycle, the company offers customers to take it back to recycle into a product of equal value. To put the idea into action, Patagonia is partnering with eBay to provide consumers a way to resell their used Patagonia apparel. So, the Patagonia's message is that consumers should buy high-quality apparel that will last a very long time and such apparel should command a premium price, relative to lower-quality garments.

More surprisingly, on Black Friday, Patagonia launched an advertising campaign "Don't buy this jacket" (Fig. 1) asking consumers to rethink their purchasing decision if they really needed this jacket and encouraging them to repair their jackets instead. The advertisement was part of an initiative communicating to consumers: "Reduce what you buy, repair what you can, reuse what you no longer need, recycle what's worn out," attempting to close the loop of the product life cycle.

**Fig. 1** Patagonia advertising campaign

## 5.3 *Levi Strauss & Co.: Reducing Water Consumption*

The production of jeans comes with a considerable environmental footprint. In fact, some of the most significant environmental impacts of making jeans come from fabric mills, which use a great deal of water, chemicals, and energy during the dyeing and finishing process.

In this context, the fashion company Levi Strauss & Co. introduced various environmental strategies to reduce the negative impact of this fashion industry on the marketplace. In the first place, Levi Strauss & Co. developed a green strategy which consisted of a line of jeans crafted with organic cotton. In year 2007, Levi's began to deeply research the overall sustainability of jeans, conducting the *first life cycle assessment* to better understand the environmental impact of a pair of jeans (Fig. 2). Many important aspects of the life cycle and ripple effects of the production of Levi's denim jeans were examined, focusing on the effects of resource extraction, production, shipping, packaging, and afterlife of jeans when they are thrown away as waste. The company soon realized that every step of the process to create a pair of jeans made of cotton requires water—from the cotton fields to manufacturing facilities. More precisely, making one pair of jeans required almost 920 gallons of water—which means an average of 12.6 l of water per pair of jeans—and expelled 32 kg of carbon dioxide.

Surprisingly, their analysis showed that while the overall water consumption in a jean's life cycle is largely accounted for in the cotton growing stage (49% of water consumption), the consumer phase comes representing the 43% of water consumption in the jean's life cycle. Moreover, the *consumer phase* could have the greatest overall impact on climate change, compromising 58% of its total estimated environmental impact, based on an average life span of 2 years with one washing per week.

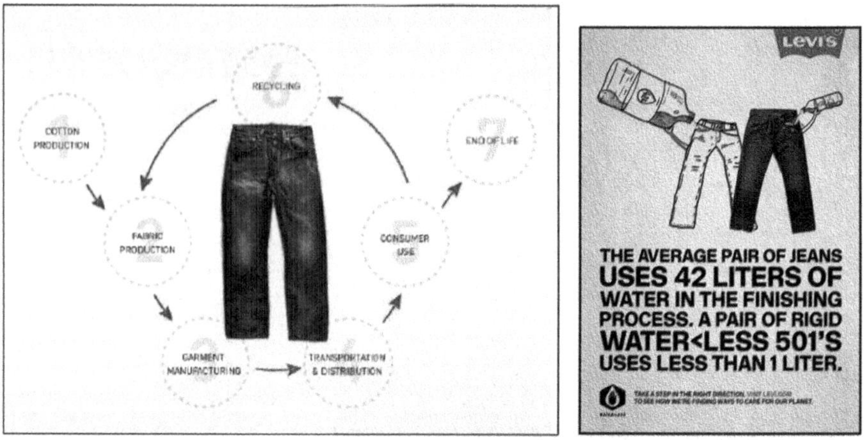


**Fig. 2** Life cycle of a pair of jeans and Levi's waterless advertising

Since then, the company Levi's tried to reduce the water consumption and its environmental impact in three ways. First, the company creates the *WaterLess*™ process that reduces water in the finishing process of denim jeans by up to 96%. This manufacturing process is based on the addition of a chemical which causes the dye to more readily fix to the fabric, resulting in less water needed for rinsing the fabric. The process results in 75% water savings and up to a 96% compared to traditional jeans' dyeing (Fig. 2).

Second, the company implemented a marketing campaign "Care Tag for the Planet" to teach and educate consumers about caring for their jeans in a more responsible way in order to reduce their footprint. The main purpose of this marketing campaign was to educate consumers in not washing jeans so frequently, given that by decreasing the number of times people wash their jeans, they can significantly reduce the consumer impact on environment. Finally and third, the company joined the "Better Cotton Initiative" to educate farmers on water efficiencies in order to reduce water consumption in the production process.

## *5.4 Ecoalf: "Because There Is No Planet B"*

Ecoalf is a pioneer business into the high-end fashion market. This company production process is based on *reverse logistics*, since the manufacturing process is based on collecting wastage and recycled plastic to offer fashion garments and accessories. So, this company is strongly focused in sustainability and technological innovation, as well as on reducing resources, consumption, and waste. More precisely, Ecoalf manufactures products such as clothing, handbags, luggage, and accessories exclusively from recycled materials that include PET plastic bottles,

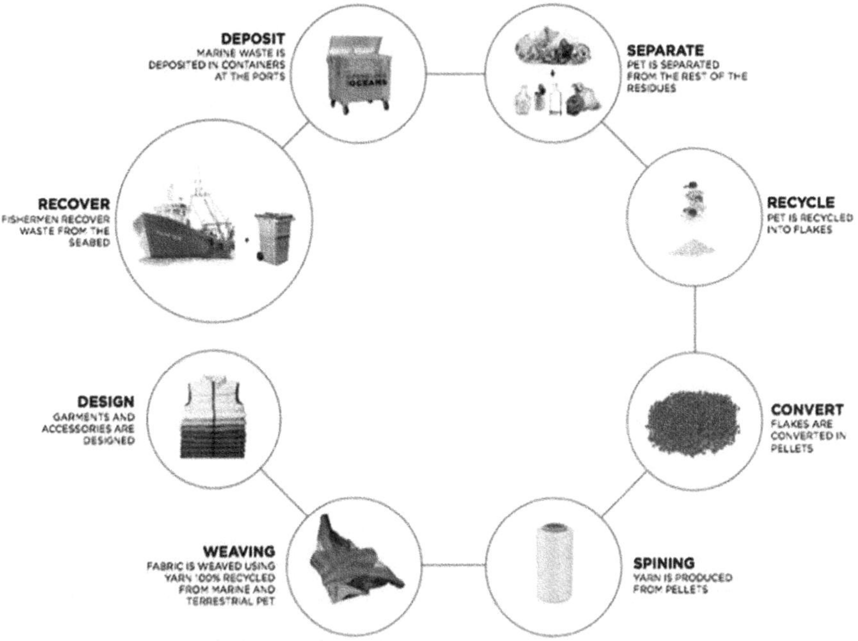


**Fig. 3** Ecoalf recycling process

fishing nets, or rubber tires, among other materials upcycled or reused. Consequently, this company helps in reducing plastic waste and cleaning the sea.

What is going on in the oceans and seas? Over 500,000 million plastic bags are used every year, 200 million plastic bottles are not collected every year around the world ending in the oceans, and 650,000 tons of plastic fishing nets is left at the bottom of the oceans. Unfortunately, fishermen have to change their nets every 5–6 years, and they need to pay to leave those nets in the ports; so often nets are thrown in the ocean creating a lot of damage to marine life. In addition, this waste is growing and it is very difficult to recycle.

In this context, Ecoalf founded "Upcycling the Oceans," a revolutionary project to collect the discarded plastic bottles that are harming the Mediterranean Sea and turn it into top-quality thread. Using sophisticated research and design processes to recycle bottles from the ocean floor, the project's main objective is to develop production technologies that will allow the company to create textile products made with marine debris. Fortunately, the fishermen agreed to participate in the project and place plastic containers in the ports where they could put all the waste they caught in their nets.

Nowadays, the company recycles discarded fishing nets, plastic bottles, used tires, post-consumer coffee, postindustrial cotton, and postindustrial wool and has developed over 98 different recycled fabrics through an innovative recycling process (Fig. 3). Today, the company needs 80 plastic bottles to make one their our jackets.

**Fig. 4** Ecoalf advertising campaigns

Ecoalf was born with the idea of creating a truly sustainable fashion company that wanted to develop the "trashion" concept, since each recycled material is processed and turned into new fashion products. The company believes that the most sustainable thing to do was to not use natural resources, so it focused on recycling to create fashion products. More interestingly, the company aims to collect the plastic before it enters the oceans and is converted into micro-plastic. This way, the company could demonstrate that it is possible to transform what other people call waste into amazing products, so that the company thinks of waste of something that has great value (Fig. 4).

## 5.5 *Tide: Laundry Washing in Cold Water*

Promoting consumer behaviors that can decrease current greenhouse gas emissions can provide great benefit for the environment and decrease consequences of climate change. Nowadays, laundry is becoming a more sustainable practice, due to a variety of innovations in products, and the increasing consumer awareness has also led to the introduction of detergents that are readily biodegradable, phosphate-free, or made from vegetable-based ingredients, instead of being petroleum-based. In this context, a range of products have emerged that market themselves specifically based on the absence of harmful chemicals. The success of Seventh Generation or Method Laundry illustrates the rise of detergents with new formulations that represent sustainable alternatives that are playing their part in reducing the environmental impact of each laundry cycle.

However, many consumers may not be aware that washing clothes is responsible for 341 kWh of electricity consumption and 0.24 metric tons of greenhouse gas emission per household per year. More precisely, if just one load of laundry per week

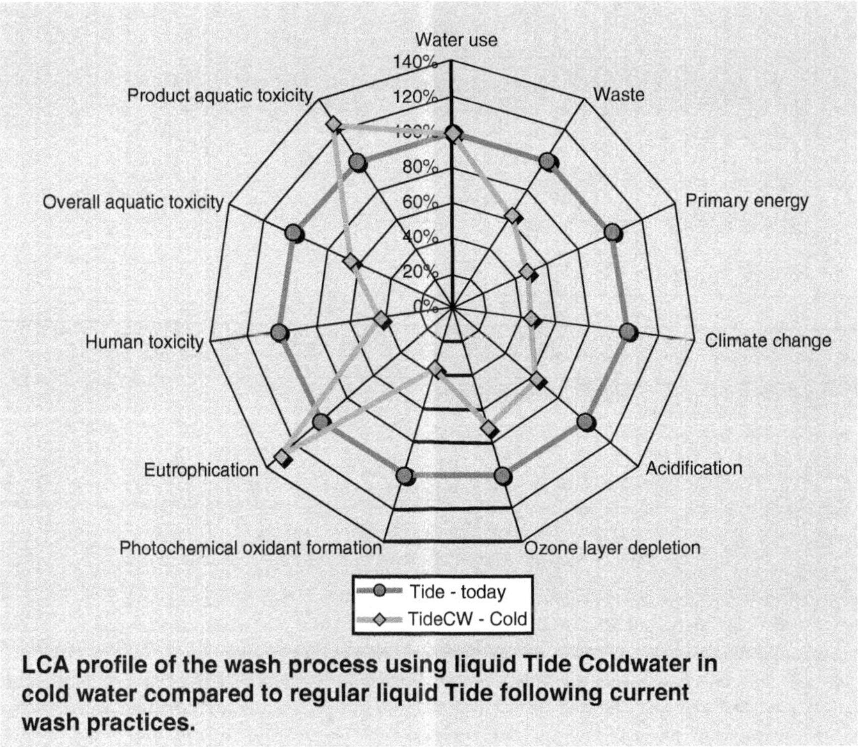


**Fig. 5** Life cycle assessment profiles of conventional versus cold-water laundry soap

was shed on *cold instead of hot or warm* cycles, over the course of a year, 2007 million kWh of electricity, 166 million tons of natural gas, and 2.3 million metric ton of greenhouse gas emissions could be averted (Fig. 5). Therefore, small changes in wash temperatures are able to generate significant savings in energy consumption. Further, changes in washing machines, particularly in making cold the default for most cycles, will contribute to greater sustainability. When looking at laundry process, the choice of the machine wash temperature is the single biggest driver of the environmental footprint, since the most important impact is from the electricity used to heat the wash water. In this vein, public campaigns in some European countries encouraged the reduction of wash cycle temperatures among consumers.

In this context, DuPont and Procter & Gamble designed and launched a washing powder to work at cold-water and low water temperatures with less harmful chemical additives through the inclusion of enzymes that improved detergent cleaning performance: "Tide ColdWater" (Fig. 6). So, with an adequate cold-water detergent, it is easier for consumers to make a small change in their habits to care for the environment, through energy saving, and a transition to lower temperature washing could be enabled by laundry product innovations.

**Fig. 6** Tide ColdWater advertising

## *5.6 Carrefour: Removal of Disposable Plastic Bags*

Single-use disposable plastic bags used to carry goods from supermarkets and other shops—these bags are usually only used for one shopping trip—and often provided free of charge. So, plastic bags have facilitated a more convenient shopping experience for consumers, since plastic bags are light and inexpensive.

However, the use of disposable plastic bags entails negative environmental externalities that are not taken into account in the prices paid by retailers or end users. First, plastic bags in landfill can take decades or longer to degrade, since depending on the type of bag, they persist in the environment for a long time—biodegradable plastic bags do not last as long in the environment as non-biodegradable bags; however, they will only degrade within a reasonable time if disposed of in appropriate conditions. Second, additives in plastic bags can contaminate soil and waterways and, if ingested by animals, can enter the food chain. Many species accidentally ingest plastic bags because they confuse them with prey species, and to some extent, ingestion by marine mammals may occur indirectly as a result of ingesting fish that have eaten plastic. Other particular concern emerging in the marine environment is the giant masses of plastic waste known as *plastic soup*

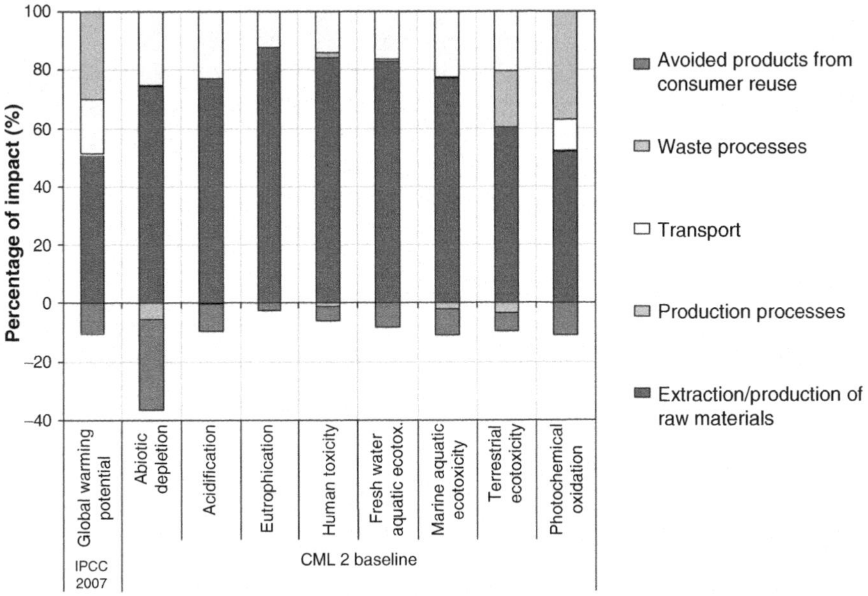


**Fig. 7** The life cycle impact of starch-polyester plastic bags

that have been discovered in the oceans (European Commission 2011), and substantial quantities of plastic bags are accumulating in natural habitats worldwide. Finally, other environmental issue is that plastic bags are likely to end up as litter, and once littered they are visually intrusive and persistent, due to their lightness and mobility. Therefore, the reduction in the use of single-use plastic bags would have a highly positive influence on all environmental indicators, including energy use and greenhouse gas emissions (Fig. 7).

But how is the scale of this environmental issue? The number of plastic bags used in trade in year 2010 is about 98.6 billion plastic bags, of which 89% were the single-use type, the vast majority non-biodegradable. Plastic bags have a high calorific value, and this energy can be recovered in waste-to-energy plants for use in heating and electricity generation; however, the share of plastic bags going to energy recovery in the EU is only the 39% (European Commission 2011).

In this context, many countries have introduced strong policies to reduce single-use plastic bag use, and such initiatives have proved popular. The fact that many retailers still distribute plastic bags for free is the main driver behind their use and thus an important lever for change. So, pricing measures requiring that a price be placed on all plastic bags would result in both a reduction in their use and a shift from single-use to multiple-use plastic bags, thus having an almost immediate effect on consumer behavior and being a highly effective measure (European Commission 2011). In addition, other policies to reduce the use of plastic bags would be to switch to multiple-use plastic bags, since many consumers prefer to use multiple-use bags made of plastic or other materials, instead of paying for single-use plastic bags.

Nevertheless, whatever type of bag is used, the key to reduce its environmental impact is to reuse it as many times as possible.

In 2008, the Carrefour Group decided to take action and adopt an effective strategy aimed at identifying and reducing their environmental impact. This retailer group aimed to reduce $CO_2$ emissions from its stores by 40%. Later, in year 2012, Carrefour decided to completely stop giving out free disposable plastic checkout bags in its stores. This has come into effect in hypermarkets and supermarkets in France, Belgium, Spain, Greece, or Argentina, and in these countries, the company now offers alternative solutions to their customers, such as reusable shopping bags. This is a pioneering initiative that has been accompanied by a major customer awareness-raising campaign called "No more plastic bags" in partnership with local authorities.

## *5.7 Bulk Purchase: The Removal of Food Packages in Grocery Shopping*

Consumers are increasingly looking for *convenience foods* that reduce preparation and cooking time, such as fresh produce that is prepacked, for example, cut and washed lettuce leaves, as well as foods that are ready to eat, for example, frozen meals, and this trend is expected to continue. However, food packaging has a great environmental impact, increasing the amounts of packaging waste requiring disposal or recycling at the household level and being a great source of pollution.

Food packaging is wasteful and mostly unnecessary, and consumers may probably love the idea of a package-free grocery store. In this context a recent trend has emerged, the *bulk purchasing*, which consists on the removal of food packaging by offering some food product categories in large containers or bins through bulk purchase. These bulk bins usually offer dry foods from dried fruit to nuts, dried beans, grains, teas and coffees, or spices.

Bulk food requires less packaging than individual serving packaged food items, resulting in less waste; and consequently when consumers purchase bulk items, they are reducing their carbon footprint. Therefore, buying in bulk eliminates the need for wasteful packaging and lessens the amount of trash, and the transportation of food products is simplified, easing the burden of $CO_2$ emissions. In addition, bulk purchase also reduces food waste, since this method of purchasing allows the portion control by customers, who will only purchase the needed quantity, decreasing the quantity of food wasted when using prepackaged products.

The store will offer many items from gravity bins, which do the work of dispensing the food (Fig. 8). Customers will often have plastic and/or paper bags available to them for transporting their food bulk items home. However, using these bags will produce a tremendous amount of waste, negating one of the benefits of bulk purchasing. Therefore, most of these grocery retailers allow customers bring their own containers or reusable bags.

**Fig. 8** Examples of bulk purchase in grocery food retailers

*Ingredients*, in Austin (Texas), was the first store to apply this concept, offering local food and beverages that were filled in customers' own containers. Today, some retailers such as *Whole Foods* and *Auchan* offer bulk purchase, covering the walls in the store with clear bins stocked with grains, beans, and dried goods.

# 6 Conclusion

In general terms marketing focuses on encouraging increased and unsustainable consumption. However, marketing can influence and change consumers' behavior, helping in the shift of cultural values of consumption to address environmental issues, making consumers become aware about the environmental impact of consumption, and promoting sustainable ways of consumption. Similarly, there is a general consensus on the need to reduce resource utilization, energy consumption, and carbon emissions, but a lack of consensus exists on whether consumption should be reduced or just changed and whether consumers have the capability to contribute to resource conservation. In this context, the sustainability marketing orientation could be considered an alternative approach to marketing theory and practice. Through *sustainable marketing*, companies introduce sustainability into the core marketing strategies and actions to ensure the reduction of consumption and waste and to encourage consumers in developing a resource-efficient way of consumption, which includes the acceptance of consumption reduction.

The present study supports that marketing can provide strategies and actions for radical changes to the unsustainable consumption system. Therefore, marketing holds potential solutions to some of the challenges of sustainability faced today, such as the climate change.

## References

American Public Health Association. (2007). *Toward a healthy, sustainable food system* (Policy number 200712). Washington, DC.

Banbury, C., Stinerock, R., & Subrahmanyan, S. (2012). Sustainable consumption: Introspecting across multiple lived cultures. *Journal of Business Research, 65*, 497–503.

Bradshaw, Y., & Wallace, M. (1991). Informing generality and explaining uniqueness: The place of case studies in comparative research. *International Journal of Comparative Sociology, 32*(1/2), 154–171.

Brown, K. W., & Ryan, R. M. (2004). Perils and promise in defining and measuring mindfulness: Observations from experience. *Clinical Psychology: Science and Practice, 11*(3), 242–248.

Cline, E. L. (2012). *Overdressed: The shockingly high cost of cheap fashion*. New York, NY: The Penguin Group.

Connolly, J., & Prothero, A. (2003). Sustainable consumption: Consumption, consumers and the commodity discourse. *Consumption, Markets and Culture, 6*(4), 275–291.

Cronin, J. J., Smith, J. S., Gleim, M. R., Ramirez, E., & Martínez, J. D. (2011). Green marketing strategies: An examination of stakeholders and the opportunities they present. *Journal of the Academy of Marketing Science, 39*(1), 158–174.

Cummins, S., Reilly, T. M., Carlson, L., Grove, S. J., & Dorsch, M. J. (2014). Investigating the portrayal and influence of sustainability claims in environmental advertising context. *Journal of Macromarketing, 34*(3), 332–348.

European Commission. (2011). *Assessment of impacts of options to reduce the use of single-use plastic carrier bags* (DG Environment. Bio Intelligence Service).

European Environmental Agency. (2005). *Household consumption and the environment* (EEA Report No. 11/2005). Copenhagen.

Fisk, G. (1973). Criteria for a theory of responsible consumption. *Journal of Marketing, 37*(2), 24–31.

Fletcher, K. (2007). Slow fashion. *Ecologist, 37*, 61.

Fletcher, K. (2010). Slow fashion: An invitation for systems change. *Fashion Practice, 2*, 259–266.

Gordon, R., Carrigan, M., & Hastings, G. (2011). A framework for sustainable marketing. *Marketing Theory, 11*(2), 143–163.

Hastings, G. (2009). *Critical social marketing*. In J. French, C. Blair-Stevens, D. McVey, & E. Merritt (Eds.), *Social marketing and public health: Theory and practice*. Oxford: Oxford University Press.

International Federation of Organic Agriculture Movements. (2009). *The contribution of organic agriculture to climate change mitigation*. Bonn: IFOAM.

Jung, S., & Jin, B. (2016). From quantity to quality: Understanding slow fashion consumers for sustainability and consumer education. *International Journal of Consumer Studies, 40*, 410–421.

Kilbourne, W. E., & Beckman, S. (1998). Review and critical assessment of research on marketing and the environment. *Journal of Marketing Management, 14*(6), 153–532.

Kotler, P., Roberto, N., & Lee, N. (2002). *Social Marketing: Improving the quality of life*. Thousand Oaks, CA: Sage.

Kotler, P., & Zaltman, G. (1971). Social marketing: An approach to planned social change. *Journal of Marketing, 35*(1), 3–12.

Lazer, W., & Kelley, E. J. (1973). *Social marketing: Perspectives and viewpoints*. Homewood, IL: Richard D. Irwin.

Lee, M., Roux, D., Cherrier, H., & Cova, B. (2011). Anti-consumption and consumer resistance: Concepts, concerns, conflicts and convergence. *European Journal of Marketing, 45*(11/12), 1.

Lim, W. M. (2017). Inside the sustainable consumption theoretical toolbox: Critical concepts for sustainability, consumption and marketing. *Journal of Business Research, 78*, 69–80.

Manoochehri, J. (2001). *Consumption opportunities: Strategies for change*. Paris: United Nations Environment Programme.

Mohr, J. A., Webb, D. J., & Harris, K. E. (2001). Do consumers expect companies to be socially responsible? The impact of corporate social responsibility on buying behavior. *Journal of Consumer Affairs, 35*(1), 45–72.

OECD. (2002). *Towards sustainable household consumption? Trends and policies in OECD Countries*. Paris: OECD.

Papadas, K. K., Avlonitis, G. J., & Carrigan, M. (2017). Green marketing orientation: Conceptualization, scale development and validation. *Journal of Business Research, 80*, 236–246.

Peattie, K. (1995). *Environmental marketing management*. London: Pitman.

Peattie, K. (2007). *Sustainable marketing: Marketing re-though, re-mixed and re-tooled*. In M. Saren, P. Maclaran, C. Goulding, R. Elliot, A. Shankar, & M. Catterall (Eds.), *Critical marketing: Defining the field*. London: Butterworth-Heineman.

Peattie, K., & Peattie, S. (2009). Social marketing: A pathway to consumption reduction? *Journal of Business Research, 62*(2), 260–268.

Perry, A. (2006). Will predicted climate change compromise the sustainability of Mediterranean tourism? *Journal of Sustainable Tourism, 14*(4), 367–375.

Prothero, A. (1990). Green consumerism and the societal marketing concept: Marketing strategies for the 1990s. *Journal of Marketing Management, 6*(2), 87–104.

Robinson, J., Bradley, A., Busby, P., Connor, D., Murray, A., & Sampson, B. (2006). Climate change and sustainable development: Realizing the opportunity. *Journal of the Human Environment, 35*(1), 2–8.

Saha, M., & Darnton, G. (2005). Green companies or green con-panies: Are companies really green or are they pretending to be? *Business and Society Review, 110*(2), 117–157.

Sassatelli, R., & Davolio, F. (2010). Consumption, pleasure and politics. *Journal of Consumer Culture, 10*(2), 202–232.

Schor, J. B. (1998). *The overspent American: Why we want what we don't need*. New York, NY: Harper Perennial.

Sheth, J. N., Sethia, N. K., & Srinivas, S. (2011). Mindful consumption: A customer-centric approach to sustainability. *Journal of the Academy of Marketing Science, 39*(1), 21–39.

Simpson, B. J. K., & Radford, S. K. (2014). Situational variables and sustainability in multi-attribute decision making. *European Journal of Marketing, 48*(5/6), 1046–1069.

Soler, C. (2012). conceptualizing sustainably produced food for promotional purposes: A sustainable marketing approach. *Sustainability, 4*, 294–340.

Stille, A. (2001). Slow food. *The Nation, 273*(6), 11–16.

Sung, H., & Lee, J. (2011). Environmental management portfolio of Korean fashion brands. *Journal of Global Fashion Marketing, 2*, 44–54.

Unitied Nations Environment Programme (UNEP). (2005). *Integrated assessment and planning for sustainable development: Key features, steps and tools*. Division of Technology, Industry and Economics. UN Global Compact (UNGC). UNEP/ETB, Geneva.

Zevestoski, S. (2002). The socio-psychological bases of anti-consumption attitudes. *Psychology & Marketing, 19*(2), 149–158.

# Madeira Island: Tourism, Natural Disasters and Destination Image

António Manuel Martins de Almeida and Luiz Pinto Machado

## Contents

1 Introduction ........ 286
2 Contextual Setting ........ 288
3 Literature Review ........ 289
4 The 2010 Natural Disaster ........ 292
5 Suggestions for an Action Plan ........ 296
6 Conclusion ........ 298
References ........ 299

**Abstract** There is no doubt that globalization and technology have in recent years been responsible for the economic and social progress of the four corners of the world. Tourism has also been "affected" by these phenomena and is generally referred to as the greatest expression of globalization. Today, the business volume of tourism equals or even surpasses that of oil exports, food products or automobiles. Tourism has become one of the major players in international commerce and represents at the same time one of the main income sources for many developing countries. This growth goes hand in hand with an increasing diversification and competition among destinations (UNWTO). In the service area, tourism distinguishes itself as the main industry for technological use and innovation. But as tourism grows, climate risks increase. Storms, hurricanes, torrents of water, earthquakes and general natural disasters arrive without warning!

Last decade can be remembered for the multiplicity of negative events including natural disasters, terrorist attacks and bird flu that have severely impacted tourist destinations. Whether the incidence of disasters or crises, both natural and man-made, is increasing, people have become more concerned about their safety,

A. M. M. de Almeida (✉)
Universidade da Madeira, Funchal, Portugal
e-mail: antonioa@staff.uma.pt

L. P. Machado
Universidade da Madeira, Funchal, Portugal

CEFAGE Center for Advanced Studies in Management and Economics, Universidade de Èvora, Èvora, Portugal

 
T. Sequeira, L. Reis (eds.), *Climate Change and Global Development*, Contributions to Economics, https://doi.org/10.1007/978-3-030-02662-2_14

particularly when they decide to go travelling (Machado and Almeida, *Natural disasters: Prevention, risk factors and management.* Nova Science Publishers, 2012). The Atlantic islands, fragile by nature, have been plagued in recent decades by relatively frequent events, leaving tracks of destruction and extremely damaging image of fate.

Madeira Island was no exception; after the storm of 2010, which left marks still visible today, the island once again suffered a fire of unprecedented dimensions in August 2016, terrorizing the population and tourists who quickly helped spread the news around the world. The strong winds and very high temperatures far above normal pushed the flames to the city, populated by old buildings and badly treated pieces of land that worked as a fuse to make the city a hell of flames. The initial under-assessment of risk and clear lack of preparation for such events helped, and chaos settled for several days. The economic impact of the fires, the damage to the local entrepreneurs and the recovery of the destination image are something that can take years to recover. This chapter intends to deepen the damage caused to the island tourism sector while suggesting some actions that can minimize the effect of a similar crisis in events that seem to arise more and more frequently!

**Keywords** Safety and security crisis · Crisis in Atlantic islands · Managing crisis · Place image

## 1 Introduction

Islands offer several of the most beautiful and unforgiving places on earth. Superb landscapes, a diversity of natural habitats, crystalline waters and golden sands attract thousands of tourists every year. Tourism is traditionally considered the almost ideal instrument to boost the islands' economies (Machado and Almeida 2012). Not all is rosy, however, in this regard. The rising trend in natural disasters in small islands is well evident, and the global warming will lead to a dramatic increase in the vulnerability of low-lying coastal and tourism infrastructures (Okuyama 2018). For several decades, short-term conservationist concerns have been seen as primary over long-term concerns. Small islands possess high levels of coastal and marine biodiversity. In this regard, any detrimental activity would lead to unsustainable losses in terms of the quality of the ecological systems, followed by a reduced number of arrivals and a deteriorated image abroad. Nowadays, however, in a growing number of cases, risks relate mainly to major natural disasters and episodes of extreme weather, which are proportionately more difficult to prevent and contain than human-made detrimental and pollution activities. The increasing number and intensity of natural disasters highlight the importance of studying tourists' post-disaster behaviour in order to enhance and promote risk management practices.

Research on tourist behaviour in post-disaster timeframes is rather limited (Miller et al. 2017; Jiang and Ritchie 2017). What is known, however, is that destination image impacts profoundly travel decision-making processes at the individual level.

Destination image is a critical selection factor. As "connotations" derived from the destination image are constantly mixed up with travel's expectations from past experiences at the destination and insights from meaningful events broadcasted by the media, meaningful events such as natural disaster may lead to a downgrade of the image. Destination images are shaped in visitors' minds through a synthesis and analyses of the information gathered over time (Letho et al. 2010, p. 32). An increasing number of studies on the subject on crisis management have been dealing with the subject from a supply side perspective. These efforts have been instrumental in helping regions and islands to proactive planning schemes and systems of reactive crisis management (Prideaux et al. 2003; Prideaux 2003; Coles 2003). Nevertheless, a key component of any crisis management scheme lies in visitors' perceptions and "perception" changes. In this respect, the analysis of the destination image is worthwhile and a matter of concern because natural disasters may have a profound impact on the psyche of tourists with far-reaching effects on their behaviour.

In line with the trend observed worldwide, Madeira is increasingly vulnerable to natural and man-made disasters (EM-DAT CRED 2012). Faced with such challenges, the local government has considered to direct its focus to the development of planning systems aiming at the identification and assessment of potential risks. Good reasons abound to follow the "natural disaster agenda" closely. The region suffered a string of natural disasters and human-made events since 2010, with far-reaching consequences, notably in the case of the tourism industry. Tourism plays a crucial role in generating employment and added-value in Madeira. This high level of dependency on tourism leaves the region's economy highly vulnerable external shocks in the form of "erratic and uncertain movements of tourism" (Ismeri 2011, p. 128). The preference for islands seems to be associated with factors such as the diversity of natural landscapes and habitats and a minimum or total absence of risks and failures in terms of life-threatening events, disruptive weather, poor quality of services, etc. The study analyses the recent experience of the tourism sector in Madeira in dealing with the aftermath of a major natural disaster. Based on qualitative content analysis of the local press headlines and news and reports in leading international newspapers, plus interviews with leading officials, we analyse the efforts pursued by the local authorities to restore normality as soon as possible, in order to improve the positive image of the region shared by a number of potential tourists. Section 2 offers a brief overview of the tourism industry in Madeira. Section 3 reviews the literature pertinent to the topic under analysis, and Sect. 4 provides a detail of the 2010 natural disaster and looks at local government response to the crisis in order to mitigate the magnitude of most severe potential impacts. Section 5 summarizes the study and presents conclusions.

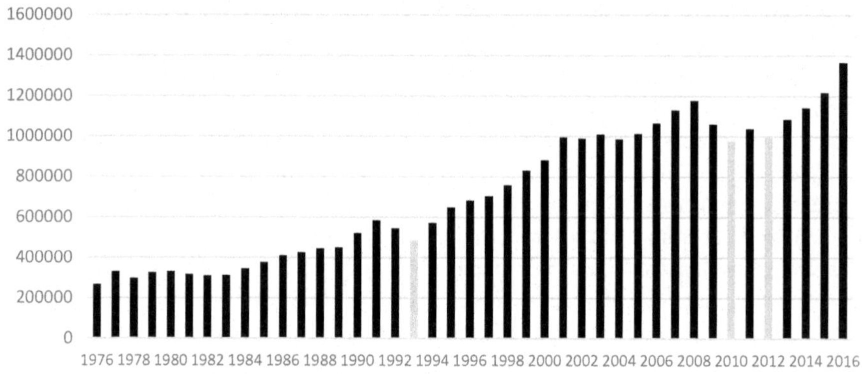


Source: Direção Regional de Estatística da Madeira (Madeira Statistical Office)

**Fig. 1** Annual arrivals (guests) from 1976 to 2016. *Source*: Direção Regional de Estatística da Madeira (Madeira Statistical Office)

## 2 Contextual Setting

The tourism is the dominant sector of the economy in Madeira. The sector benefits from a "centenary tradition" (Madeira is one of the oldest touristic destinations in the world) (Ismeri 2011, p. 135). Ismeri (2011, p. 139) places the sector in an intermediate stage of development: "far from the mass offer of the Canary Islands but also far from the scarce hotel capacities of the Azores: far from the mass approach of the former but far from the 'niche' approach of the latter".

By taking into account direct and indirect impacts, Ismeri (2011) indicates the following figures in terms of GDP and employment: 21% for GDP and 14% for employment. The accommodation and hotel sector employs 6205 persons, accounting for 5.8% of the total employment. The sector has been growing consistently in the recent decades with intermittent periods of stagnation and slower growth, as indicated by the yellow-coloured bars in Fig. 1. In 2016, the region counted 7,368,095 overnight stays. The number of arrivals grew by 13% in 2015 to 1,162,621, and, in 2016, the sector generated 377.75 million euros in earnings. Forecasts for 2017 point out to another sizeable increase in terms of arrivals. The sector operates 29,118 beds and welcomed 1,162,621 visitors in 2016. By 2016, Madeira had 156 hotels providing 13,914 rooms. The average occupancy rate was 69.2% in 2016, 4.9 percentage points higher than in 2015.

As mentioned above, the region experienced just two major upheavals since 1976: in 1993 and then in 2010–2012. The 1993 crisis is largely attributable to the effects of the financial crisis at the EU level, and the number of tourists had already been faltering since the last quarter of 1991. It should be noted that in October 29, 1993, a flash flood, thought to be one of the most violent episodes to be recorded in the city of Funchal history left a trail of destruction and death. The tragedy lead to high losses in terms of public infrastructure and private sector facilities, with more than 400 evacuees and 5 deaths. There is not much, however, to indicate that the tourism sector was adversely affected by this event. The sector was much more affected by the 2010 natural disaster. The total number of visitors in 2010 fell by 7.8%, although the number of guests increased by 6.2% in 2011. The sector fallen again by 4.1% in 2012 (compared with the previous year), but since then, the sector has experienced continuous growth.

The South Coast in general and Funchal in particular are particularly affected by flash floods with most deaths recorded in and around Funchal. Funchal is an extremely popular "vacation" destination hosting about 2/3 of the region's accommodation capacity (more than 10,000 tourists per day). Madeira is densely populated, with 254,876 inhabitants as of 2016. Not surprisingly, significant negative environmental loss of infrastructure or health impacts are likely to occur in a such densely populated area in the event of a major disaster. It is worth to mention that around 104,813 inhabitants (approximately 41.1% of the total population) live in the city capital Funchal, with most tourists staying or visiting the capital city every single day.

## 3 Literature Review

Tourism emerges as one of the most vulnerable industries to natural disasters and climate change. On islands, the tourism industry is prone to a myriad of risks including traditional natural disasters, outbreaks of tropical diseases, climate change-related events and political instability. In addition, there is a significant trend in unpredictable and erratic variations in demands patterns for a wide variety of reasons. For example, cancelled flight schedules caused by weather events and localized political and economic crises may explain changes in market demand. Natural disasters had been defined as "an event caused by forces of nature such as weather which leads to a significant damage to the physical environment and may threaten human safety" (Henderson 2007, p. 100). On the subject of safety, Beattie (1992) refers that tourism and natural disasters conjure up radically different images and emotions in visitors' minds. When thinking of tourism, visitors dream of "fun, relaxation, sightseeing, and beautiful surroundings" (Beattie 1992, p. 7; Mazursky and Jacoby 1986). In a similar vein, Santana (2003) refers that it is unusual to think about tourism and natural disaster simultaneously. Tourism evokes feelings of "enjoyment, pleasure, relaxation, safety", while natural disaster relates to "distress, death, fear, anxiety, trauma, panic; contrariwise, natural disasters 'bring to mind' images of "destruction, death, and tragedy" (Santana 2003, p. 230). Therefore, disaster-hit areas do not appear to be regarded as ideal vacation settings. As pointed out by Beattie (1992), tourism and natural disasters are located on the opposite ends of a spectrum, being "mutually exclusive rather than complementary". It is for that reason that, in a number of cases, DMOs in hazard-prone regions may be tempted to discuss the topic in a cautious approach in order to avoid giving rise of concern among potential visitors. However, Roger Yates in The Guardian claims that "it has been (in 2016) a busy year for disasters: from the Haitian earthquake to Chinese landslides, food insecurity in Niger to the current floods in Pakistan, it has seemed like a constant barrage", which suggests there is no point in avoiding a frank discussion of the subject. If countless tourists are affected in one way or another by natural disaster while on vacation, the negative impact on the image conveyed abroad may be quite serious and hard to manage and restore.

In the aftermath of a natural disaster, destinations are urged to rebuild infrastructure, facilities and communities. However, in the post-recovery stage, marketing communication aiming at "regaining market shares" is key. In this case, marketing campaigns are critical to support the region's efforts to recover and change any customers' misperceptions. In this regard, an assessment of visitors' profile in terms of attitudinal and affective responses must be carried out as soon as possible to gather accurate and updated information. It is essential to take into account that the tourism industry relies on perceptions of safety, smooth operational systems and predictability (Ritchie 2008). Risk influences tourist's behaviours and expectations. Natural disasters affect the levels of perceived security, which are a major reason to outline appropriate risk control measures. The key reason to control risks and plan in advance mitigation measures lies in capability to manage the crisis aftermath. In fact, the magnitude of the impact of a major natural disaster on the tourism industry is largely determined by tourists' responses. Tourists are entitled to a large degree of autonomy and flexibility in deciding their next move: whether to visit or not a disaster-hit destination, as well as their travel arrangements and timing of visit and most importantly, word-of-mouth. Therefore, in a significant and surprisingly higher number of cases, it is possible to mitigate to a large extent the impact of a major disaster, no matter how catastrophic it may appear at first.

Since September 11, 2001, terrorism and insecurity are perceived as a critical factor affecting the tourism industry in a negative way. The September 11, 2001, terrorist attack is seen as a major turning point in international travellers' point of view on issue of security. In more recent years, travellers' fears were pushed to the limit by the Indian Ocean Tsunami in 2004, the Hurricane Katrina in 2005, the 2010 eruptions of Eyjafjallajökull and the 2011 Tohoku earthquake and tsunami. These events had a major impact on the tourism industry (Orchiston 2012; Tsai and Chen 2010; Yang et al. 2011). In the aftermath of such events, a number of studies have been carried out to analyse post-disaster recovery and planning strategies (Huang and Min 2002; Ritchie 2008). It became apparent the key role of the media as the primary actor in shaping the disaster-hit area image abroad (Sharpley 2005). Media coverage, by increasing the public awareness, can negatively affect the post-disaster phase. Roger Yates, already mentioned above, highlights the key role played by the media. A balanced and informative coverage from the very beginning helps the local authorities to manage the aftermath. Not every natural disaster is newsworthy. Normally, high death toll incidents tend to lead the audiences and media coverage, which offers the opportunity to escape unharmed. Quite ironically, it is also possible to turn an out-of-control flow of events into an opportunity to attract new audiences. As observed by McClure (2012, p. 2) "in an ironic signal of the supremacy of the technological psychosis, 24 h cable news now transforms all major natural disasters into profitable spectacles of destruction, complete with an endless line of supernumerary victims—'the CNN syndrome'—while screening out the politics of disaster management policy".

In any event, to study visitors' reaction is essential in order to identify patterns of behaviour. Itzhaky et al. (2016) based on a qualitative analysis of the experiences felt by a group of backpackers, who were travelling in Nepal at the time of the earthquake, identified four main themes: emotional turmoil, quick recovery, springing into action and connection to the army. While most interviewees understood the crisis in an emotional manner at first, in a second stage, respondents searched for ways to cope effectively with the situation in the field based on problem-solving strategies (community of action, analysis of past experiences). A positive assessment of the region's image is instrumental in this regard.

Image has been defined as an attitudinal construct relating to the individual's beliefs, feelings and general impressions about an object or tourism destination (Letho et al. 2010: Crompton 1979; Echtner and Ritchie 1991). Lawson and Baud Bovy (1977) define destination image as "the expression of all objective knowledge, prejudices, imagination and emotional" that contribute to define the perceptions, attitudes and views of an individual about a particular destination. Bigné et al. (2001) consider that an individual's destination image is subjective, ambiguous and immaterial by nature because closely linked to tourists's individual interpretation of "reality" made by the tourist. The image of a destination in a visitor' mind is subjective to a great extent because it is based primarily on perceptions rather than on attributes, which is especially true when no visit has actually taken place. There is a widespread consensus that image is compounded by cognitive and affective components. Baloglu and McCleary (1999a) assert that the cognitive component relates to beliefs or previous knowledge about the destination' attributes, while the affective component relates to feelings towards or an emotional attachment to salient characteristics of the destination. A number of authors admit (Baloglu and Brinberg 1997; Walmsley and Young 1998; Baloglu and McCleary 1999a; Lin et al. 2007) that the development of the image of a tourist destination in visitor's minds is governed by principles of rationality and emotionality, along with the above-mentioned components or dimensions: cognitive and affective. It is worth to mention the cognitive component of the image impacts to a large extent the affective component (Lin et al. 2007; Ryan and Cove 2007; Baloglu and McCleary 1999b). However, there are other factors at play. Beerli and Martín (2004) found out that the socio-demographic background of the tourists exerts a great influence on the individuals' cognitive and affective assessment of the overall image. In all, the overall image of the destination being formed in visitors' minds results from a combination of cognitive and affective aspects, with past experiences and previous visits impacting the destination image being formed on visitors' minds (Beerli and Martín 2004). The process of image formation is intricately entwined with perceptual and cognitive process, which is, in turn, conditioned by several sources of information (e.g. reference groups, group membership, media, etc.). Under these circumstances, any potential visitor person can build a mental image of any territory without having experienced it for himself. The image of the destination is conditioned by information relating to historical, political, economic and social facts and also natural and man-made disasters, which will lead to a reappraisal of the initial image being stored in his/her mind (Echtner and Ritchie 1991). Furthermore, the values (and misconceptions and stereotypes) shared each potential visitor will influence the overall image of any tourist destination, by offering a selective attention filter (Moutinho 1987). Moreover, the country of origin of the visitor has been found to influence the image in visitors' mind (Bonn et al. 2005; Stern and Krakover 1993).

Traditionally, it was considered that tourist's images were rather resistant to change and relatively persistent. However, major changes in perception can occur after a major natural disaster, especially in cases of severe damage to the environment and loss of life. In this case the overall impact in terms of a negative image can be "serious and long-lasting" (Obasi and Frangialli 1998).

## 4 The 2010 Natural Disaster

On February 20, 2010, the region was hit by a natural disaster in the form of torrential rainfall. The episode of torrential rainfall triggered "catastrophic flash-floods" (Fragoso et al. 2012, p. 715). When the rain finally stopped, at least 42 persons were recorded as dead, with 13 persons still missing 1 week later. Around half of the victims were found in Funchal. As specified by Fragoso et al. (2012, p. 715), "it was the deadliest hydro-meteorological catastrophe in the Portuguese territory in the last four decades". Madeira usually enjoys a mild attractive weather all year round. However, in 2010, the weather stations recorded the rainfall of 426 mm during the first 22 days of February, which was equivalent to approximately 70% of the total annual average.

The economic damage was estimated at 1.1 billion euros. Graphic images of rivers of mud and rocks ripping through the city' streets were pictured in the world's press. The island's topography contributed to the disaster. The island orography is characterized by deep valleys and steeped slopes and scarps (Fragoso et al. 2012, p. 715).

As mentioned earlier, the region boasts a "Mediterranean type climate moderated by the Atlantic Ocean" (Fragoso et al. 2012, p. 715), but it is not immune to episodes of extreme weather. In terms of natural disasters, it is worth to mention that the region is not affected by major volcanic activity with only relatively modest earthquake incidents being recorded over the recent decades (Fragoso et al. 2012). However, rain-induced natural hazards are a major reason for concern. As mentioned by Fragoso et al. (2012, p. 716), "flash-floods, landslides, debris flows and less frequently tsunamis, which can be caused by coastal rock slides", have been recorded since the nineteenth century (Rodrigues and Ayala-Carcedo 2003; Fragoso et al. 2012). Flash floods are understood as the most dangerous and dramatic natural event that occurs in the island. A particularly dangerous type of flash floods, locally known as "aluvião", takes the forms of "flash-floods affecting streams whose discharges and energy increase dramatically after intense rainfall episodes, dragging mud, blocks and debris, and producing a powerful and destructive current" (Fragoso

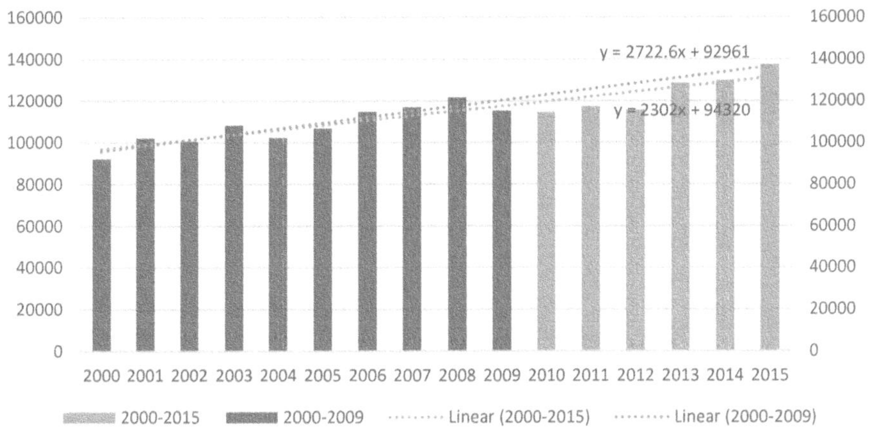


**Fig. 2** Arrivals' highest value recorded each year (2000–2015). *Source*: Direção Regional de Estatística da Madeira (Madeira Statistical Office)

et al. 2012, p. 715). Quintal (1999) produced an inventory of flash flood events leading to deaths or severe damage. Thirty major events were recorded since the historical aluvião of the October 9, 1803, with more than 800 residents killed, most of them in Funchal. A recent compilation produced by the SRES points to nine major events over a period of 210 years running from 1800 to 2009 (SRES 2010).

The 2010 natural disaster damaged to a certain extent inbound tourism in Madeira. The total number of arrivals in 2010 drop by 7.8% to 968,000, compared to 1.05 million tourism arrivals recorded in the previous year. The annual growth rate in terms of arrivals recovered somewhat during the first three quarters of 2011 (annual growth rate of 6.2%), but minor losses were reported again in 2012 (−4.1% compared with the previous year). By considering the highest value recorded each year, it is possible to notice that the region recovered the previous dynamics by 2013–2015. The values recorded in 2015 are in line with the 2000–2009 trend, a clear understanding that the sector has successfully overcome the effects of the 2010 economic crisis (see Fig. 2). We provide a closer look at the figures available now. It is worth to mention that the recovery process of different markets exhibits different patterns, with Portuguese nationals to recover quickly. A further relevant aspect can be examined by looking at monthly data covering the period 2009–2013. Figure 3 shows that the number of arrival in January 2010 is up to 1.8% compared with the homologous period in 2009. However, throughout the remainder of the year, but October, Madeira recorded losses in terms of number of arrivals amounting to 83,175 visitors. By 2011, monthly arrivals were already above the values observed in 2010, except for the winter months (October to February). The same pattern can be observed in 2012 (October to March) and, to some extent, in 2013 (just for January and February). The "long-term" impact of the 2010 natural disaster appears to die out by January 2014, less than 4 years after the disaster.

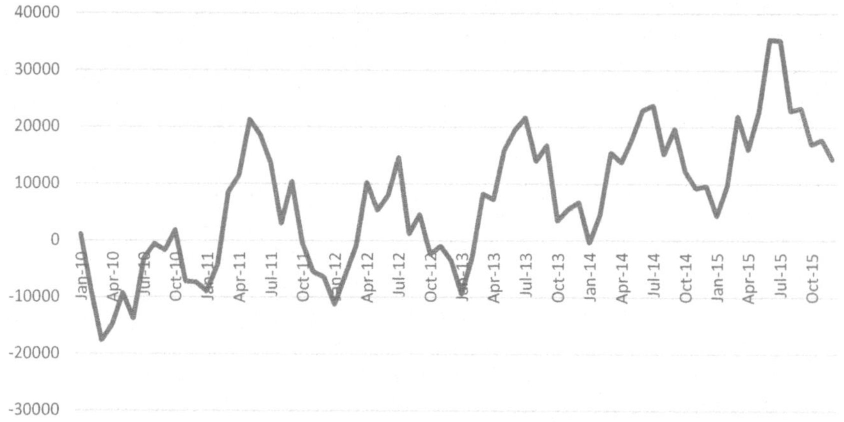


**Fig. 3** Arrivals: differences in monthly data from January 2010 through December 2015 (Data for 2010 refers to the monthly difference compared to the previous year; and data for 2011–2015 refers to the monthly differences compared to 2010)

The 2010 natural disaster received worldwide attention. The 2010 Madeira Island floods and mudslides were classified as an extreme weather event by the press, with an abundance of words and images conveyed abroad. The media impact in Portugal and in the UK was greater than in other European countries, with a number of officials expressing their deepest sympathy to the victims' families. In Portugal, the civil society, in general, and previous visitors expressed their sympathy for the region. At least 250 persons were injured as a result of the mudslides, and about 650 residents lost their houses.

In the aftermath of the natural disaster, the authorities work hard to restore and rebuilt key facilities and infrastructure. A few days later, the local government launched an extensive marketing campaign to mitigate the negative impact on the public opinion. According to the BBC website, the president of the regional government, Joao Jardim, urged the economic operators (more specifically shop owners) to reopen as soon as possible, in order to give an appearance of "normality". The president went to quite extraordinary lengths to protect the tourism industry, namely, by downplaying the scale of the tragedy: "We don't know how much it will affect the tourism, but there is no point in dramatizing the situation too much". Key players urged tourists planning their vacations in Madeira to not cancel their travel arrangements, by making an appeal to international solidarity. Michael Blandy, chairman of the Blandy group, and one of the leading operators in the sector with five hotels targeting the British market, complained that as a result of the "dramatic images of the devastation", the number of cancellations went out of control. This leading hotelier was hopeful that "the impact would be quick to pass", with life totally back to normal "hopefully within days" (quoted from The Telegraph, 23th of February, 2010). The local authorities try to convey the message only a "very small area that has been affected" (The Telegraph, 23th 2010). In the end, they succeeded in instilling a sense of normalcy. The streets were thoroughly cleared of mud and debris in a few days. In the following day, the recovery and cleaning operations were underway thanks to a 2000 strong army of public servants, private sector employees, army personnel and volunteers from the civil society. The local government succeeded in restoring normal living conditions as soon as possible, in most areas. A few days later, essential services such as electricity and water services were restored, as well as telecommunications services (mobile phone and the Internet). At the very same time, the oldest and most emblematic alley in Funchal was fully operational, just a week after, with shops and coffees open to welcome 4500 cruiser's passengers. It is worth to mention that most hotels were unaffected and were fully operational during the entire crisis. The airport, the harbour and the main tourist areas remain operational and ready to welcome visitors, and most areas in the island suffered only minor damage.

Aiming at attracting tourists back, the local government implemented a series of countermeasures to revitalize the tourism industry. One of the leading hoteliers recommended aggressive marketing campaigns based both on price and normal advertising campaigns to bring traditional visitors back. Other recommended stepping up decisive action to reduce the airport fares. It must be highlighted that a sizeable number of tourists opted for remaining in the area. Most were genuinely surprised with the recovery efforts and the street cleaning process, and a number of them interviewed by leading national newspapers were rather pleased of being able to resist calls from home to go back. They claim that the alarmist and often factually incorrect reporting by some journalists did not do justice to the strenuous efforts being undertaken by the local authorities and operators to offer suitable replacement activities. Dozens of celebrities were filmed telling their intention to visit the area in 2010.

To enhance the competitiveness of the tourism industry, the local government has undertaken a number of initiatives to attract and enhance the visitor's experiences and to meet visitor's need. For example, the tourism sector was able to offer alternative visitor itineraries and original ways of exploring the region through the region's roads without being forced to contact with areas severely damaged. But, the impact in terms of daily prices was quite heavy. As depicted in Fig. 4, RevPar figures are below the 2009 ones, with the biggest impacts felt in the following months (March to June of 2010) and then in the peak of the low season (November 2010 to February 2011). A notable recovery in terms of the RevPar figures can be seen in 2011, except for the low season (November 2011 to February 2012). Once again, it can be concluded that the 2010 natural disaster was interpreted by visitors as indicating a higher degree of risk in the winter periods. However, the negative impacts were concentrated mainly in the low season (winter months) and were particularly strong in the following months and in the rain season of 2010/2011. By January 2013, no signs of price cuts are visible. From the end of 2012, demand has recovered to the level before the natural disaster. However, more recently, Madeira has been increasingly featured in the news for reasons other than the 2010 natural disaster. For example, the 2011 forest fires made the headlines, and more recently, the fall of 200-year-old tree killed 13 people and injured dozens during the Celebration of Our Lady of the Mount festival near Funchal.

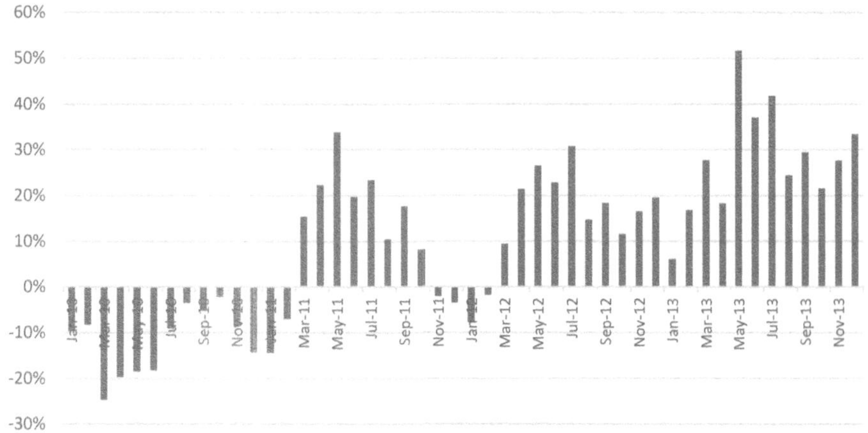


**Fig. 4** Monthly differences in RevPar. *Source*: Direção Regional de Estatística da Madeira (Madeira Statistical Office)

## 5 Suggestions for an Action Plan

Avraham (2016) identified a number of strategies employed by the Egyptian authorities in the aftermath of the Arab Spring events. Basically, the local authorities adopted a dual approach: (a) either by cooperating with the media, based on interviews, press conferences and readiness and willingness to respond to any of the journalists' questions and to provide information in "upcoming events" or (b) by resorting to strategies of information control to downplay the seriousness of the events taking place. We will focus our attention on information control strategies. The first approach lies in (ba) ignoring the crisis (by pretending that "nothing happened") and/or (bb) by conveying abroad the message of "business as usual". A variant of this approach lies in admitting a fortuitous combination of unpropitious events followed by the announcement that the life goes on as usual. When it is not possible and convenient to deny the occurrence of a negative event, local authorities may resort to (bba) downplaying the importance of the events. In this case, DMOs attempt to mitigate or limit the scale of the crisis. In this case, government officials develop efforts to explicitly delimit the geographical boundaries of the crisis, its severity and impacts in terms of disruption on daily life activities and the degree of impact on tourism activities and tourists' wellbeing (Benoit 1997, 2015). As reported above, the local authorities in Madeira fully acknowledged the extent of the tragedy, at domestic level, but attempt to divert the focus towards a few areas in Funchal while pointing out the efforts being deployed to clean and secure all areas around the key tourism assets in order to downplay the problem abroad. The local authorities claimed that less than one tenth of the city area were affected by disturbances and that, as opposed to Funchal, other parts of the islands were ready to welcome tourists. Another approach to handle the crisis lies in downplaying and lessening the impact of the crisis and even promoting an early recovery by depicting the event as "insignificant or marginal" in terms of damage, loss of life and injuries. In a few cases, such as in Madeira in 2010, it is not possible to ignore or limit the importance of the crisis to the tourism industry. Neither it was possible to portray the events as "as insignificant, marginal or irrelevant for foreign tourists" (Avraham 2016, p. 45). Damage control measures as RP tools are often employed by officials confronted with journalists bombarding the government with annoying issues (Bonn et al. 2005). Another variant of the downplaying approach lies (bbc) in portraying the crisis as irrelevant for tourists, with no risk or unattended consequences whatsoever to foreign nationals and tourists alike (Avraham 2016). The local government resorted to this approach to a certain extent. Still, in the field of information control techniques, acknowledging the negative nature of the unfold events in terms of destination image may offer interesting avenues to gain the upper hand and turn the game around. Admitting to the full extent of the crisis helps to build an image of transparency and honesty (Avraham 2013), which was the case in Madeira. In this regard, destinations may try to sell the idea of better future and the beginning of the new era. The basic premise of this communication strategy lies in conveying abroad information on the changes and initiatives being pursued to minimize all the difficulties felt in the past along with the steps taken to consolidate the beginning of a new era and a bright future ahead based on urban planning, risk management measures, investment in infrastructures aiming at minimizing the impact of future natural hazards, etc. (Avraham 2015, 2016; Paraskevas and Arendell 2007). In other cases, the local authorities may resort to "spotlight events" in order to create positive news and shift international media attention from a negative to a positive portrayal of the place (Avraham and Ketter 2008, p. 145).

Avraham (2014, 2016) provides an example of turning a liability into an asset, by transforming the chaos, danger and instability into a political experience to be participated by visitors interested in witnessing the very beginning of new era of democracy (Avraham 2013; Kwok and Chew 2017). In the case of a wider geographical area, it is possible to promote a "geographic disassociation" of the destination under analysis from problematic areas in the vicinity. In the case of destinations suffering from a persistent image problem and negative media views expressed frequently in the press, the association with celebrities and brands can be advisable. Opinion leaders, representatives of civil society in selected countries, business world moguls and academic researchers may encourage potential visitors to reach a more balanced opinion (Mair et al. 2014; Baloglu and Mangaloglu 2001). In a number of cases, image problems derive from preconceptions, generalizations and stereotypes, which may be unworkable to redress because images tend to be persistent (Manheim and Albritton 1984; Mansfeld and Winckler 2015). The local authorities can also choose to target specific audiences to increase the number of arrivals instead of opting for damage control strategies. In this case, the DMO attempts to improve the destination's image based on specific targets and market segments (Cooper and Momani 2009; Okuyama 2018). For example, Avraham (2016) suggests emphasizing similarities, resemblances and relevance to specific market segments to attract their attention. The basic idea lies in connecting specific audiences' sharing interests, values and political ideologies or holding in high esteem a key feature and characteristic of the region, by several potential visitors to the region's/destination background in terms of history, past historical events and values. Such audiences may be attracted through "solidarity tourism" initiatives (Tucker 2016; Avraham and Ketter 2016; Walters and Mair 2012). A more generic approach lies in developing from scratch new market niches. In this case, efforts are directed towards the promotion of new market segments such as golf, nature, rural tourism, eco-tourism and any other sector being neglected in the past. Market segments less sensitive to natural disasters and safety issues are of special value. There is a consensus that cultural tourism is, to a certain extent, less affected by local crisis. In Madeira, aiming at increasing the availability of events and the number of reasons to visit, instead of promoting the traditionally over-represented aspects of nature and landscapes, the local government started to promote and stage a series of cultural events. Successful campaigns were designed to promote the region's culture, because it is believed that region's culture is not merely a green hotspot in the Atlantic.

In the end, the region succeeded in avoiding the most irksome consequences of the 2010 natural disaster. The public relations and image restoration campaign were able to counteract to a large extent any negative image being conveyed abroad by the media. By 2014 most negative impacts in terms of losses in arrivals and RevPar had just died out.

## 6 Conclusion

Islands are under pressure to cope with an increasing number of external threats in the form of natural disasters, climate change-related events and other risks such as outbreaks of tropical diseases. Madeira provides an instructive example in this regard. After years of continued growth in terms of arrivals, overnights and revenue, the industry was forced to handle complex issues resulting from a major natural disaster. Owing to the quantity and quality of the sector's competitive advantages, the industry recovered quite rapidly, and the demand's levels quickly returned to its 2010 levels, which can be dangerous because it effectively led to a reduced level of risk. Operators need to incorporate into the decision-making process and strategic planning lessons learnt from the 2010 natural disaster. As described in Sect. 5, a number of strategies can be successfully deployed if the impact of a natural disaster lies with an acceptable range. Disruptive disaster will demand further measures, which should prompt operators in the regions to adopt a "disaster-ready" strategy.

## References

Avraham, E. (2013). Battling stereotypes of terror and wars: Media strategies for marketing tourism to Middle Eastern countries. *American Behavioral Scientist, 57*, 1350–1367.

Avraham, E. (2014). Spinning liabilities into assets in place marketing: Towards a new typology. *Place Branding and Public Diplomacy, 10*(3), 174–185.

Avraham, E. (2015). Destination image repair during crisis: Attracting tourism during the Arab Spring uprisings. *Tourism Management, 47*, 224–232.

Avraham, E. (2016). Destination marketing and image repair during tourism crises: The case of Egypt. *Journal of Hospitality and Tourism Management, 28*, 41–48.

Avraham, E., & Ketter, E. (2008). *Media strategies for marketing places in crises: Improving the image of cities, countries, and tourist destinations*. Oxford: Butterworth Heinemann.

Avraham, E., & Ketter, E. (2016). *Marketing tourism for developing countries: Battling stereotypes and crises in Asia, Africa and the Middle East*. London: Palgrave-Macmillan.

Baloglu, S., & Brinberg, D. (1997). Affective images of tourism destination. *Journal of Travel Research, 35*(4), 11–15.

Baloglu, S., & Mangaloglu, M. (2001). Tourism destination images of Turkey, Egypt, Greece, and Italy as perceived by US-based tour operators and travel agents. *Tourism Management, 22*(1), 1–9.

Baloglu, S., & McCleary, K. (1999a). A model of destination image formation. *Annals of Tourism Research, 35*(4), 11–15.

Baloglu, S., & McCleary, K. (1999b). US international pleasure travellers' images of four Mediterranean destinations: A comparison of visitors and nonvisitors. *Journal of Travel Research, 38* (2), 114–129.

Beattie, M. (1992). *The effect of natural disasters on tourism a study of Mount Saint Helens and Yellowstone National Park*. Master thesis. http://scholarworks.rit.edu/theses/7428/

Beerli, A., & Martín, J. (2004). Tourists' characteristics and the perceived image of tourist destinations: A quantitative analysis – A case study of Lanzarote, Spain. *Tourism Management, 25*, 623–636.

Benoit, W. (1997). Image repair discourse and crisis communication. *Public Relations Review, 23* (2), 177–187.

Benoit, W. (2015). *Accounts, excuses and apologies: Image repair theory and research*. Albany: SUNY Press.

Bigné, E., Sánchez, M., & Sánchez, J. (2001). Tourism image, evaluation variables and after purchase behaviour: Inter-relationship. *Tourism Management, 22*, 607–616.

Bonn, M., Joseph, S., & Dai, M. (2005). International versus domestic visitors: An examination of destination image perceptions. *Journal of Travel Research, 43*(3), 294–301.

Coles, T. (2003). A local reading of a global disaster: Some lessons on tourism management from an Annus Horribilis in South West England. *Journal of Travel & Tourism Marketing, 15*(2/3), 173–197.

Cooper, A., & Momani, B. (2009). The challenge of re-branding progressive countries in the Gulf and Middle East: Opportunities through new networked engagements versus constraints of embedded negative images. *Place Branding and Public Diplomacy, 5*(2), 103–117.

Crompton, J. (1979). Motivations for pleasure vacations. *Annals of Tourism Research, 6*(4), 408–424.

Echtner, C., & Ritchie, J. (1991). The measuring and measurement of destination image. *The Journal of Tourism Studies, 2*(2), 2–12.

EM-DAT CRED. (2012). *EM-DAT: The OFDA/CRED international disaster database*. Brussels: Universit'e Catholique de Louvain. www.nat-hazards-earth-syst-sci.net/12/715/2012/

Fragoso, M., Trigo, R., Pinto, J., Lopes, S., Lopes, A., Ulbrich, S., et al. (2012). The 20 February 2010 Madeira flash-floods: Synoptic analysis and extreme rainfall assessment. *Natural Hazards and Earth System Sciences, 12*, 715–730.

Henderson, J. (2007). *Tourism crises: Causes, consequences and management*. Oxford: Elsevier/ Butterworth-Heinemann.

Huang, J., & Min, J. (2002). Earthquake devastation and recovery in tourism: The Taiwan case. *Tourism Management, 23*, 145–154.

Ismeri, E. (2011). *Growth factors in the outermost regions*, Final report Vol. II, European Commission

Itzhaky, H., Kissil, K., & Weiss-Dagan, S. (2016). International tourists' reactions to a natural disaster: Experiences of the 2015 earthquake in Nepal among Israeli travelers. *Journal of Trauma Stress, 29*(6), 522–529.

Jiang, Y., & Ritchie, B. (2017). Disaster collaboration in tourism: Motives, impediments and success factors. *Journal of Hospitality and Tourism Management, 31*, 70–82.

Kwok, A., & Chew, E. (2017). The valley of dry bones: A city's revival for tourism. *Annals of Tourism Research, 66*, 183–215.

Lawson, F., & Baud Bovy, M. (1977). *Tourism and recreational development*. London: Architectural Press.

Letho, X., Douglas, A., & Park, J. (2010). Mediating effects of natural disasters on travel intention. In N. Scott, E. Laws, & B. Prideaux (Eds.), *Safety and security in tourism: Recovery marketing after crises*. London: Routledge.

Lin, C.-H., Duarte, B., Kerstetter, D., & Hou, J.-S. (2007). Examining the role of cognitive and affective image in predicting choice across natural, developed, and theme-park destinations. *Journal of Travel Research, 46*, 183–194.

Machado, L. P., & Almeida A. (2012). Natural disasters in touristic destinations: The case of Portuguese Islands. In B. Raskovic & S. Mrdja (Eds.), *Natural disasters: Prevention, risk factors and management*. Hauppauge, NY: Nova Science Publishers. (Chapter ID: _9303_ 2012).

Mair, J., Ritchie, B., & Walters, G. (2014). Towards a research agenda for postdisaster and post-crisis recovery strategies for tourist destinations: A narrative review. *Current Issues in Tourism*, 1–26.

Manheim, J., & Albritton, R. (1984). Changing national images: International public relations and media agenda setting. *American Political Science Review, 78*, 641–657.

Mansfeld, Y., & Winckler, O. (2015). Can this be spring? Assessing the impact of the "Arab Spring" on the Arab tourism industry. *Tourism: znanstveno-strucni casopis, 63*(2), 205–223.

Mazursky, D., & Jacoby, J. (1986). Exploring the development of store images. *Journal of Retailing, 62*(2), 145–165.

McClure, K. (2012). Media coverage of natural disasters: Pentadic cartography and the case of the 1993 great flood of the Mississippi. *The Journal of the Kenneth Burke Society – KB Journal, 8* (1, Spring 2012 Special Issue), 7.

Miller, S., Gonzalez, C., & Hutter, M. (2017). Phoenix tourism within dark tourism: Rebirth, rebuilding and rebranding of tourist destinations following disasters. *Worldwide Hospitality and Tourism Themes, 9*(2), 196–215.

Moutinho, L. (1987). Consumer behaviour in tourism. *European Journal of Marketing, 21*(10), 5–44.

Obasi, B., & Frangialli, F. (1998). Preface. In World Tourism Organization & World Meteorological Organization (Eds.), *Handbook on natural disaster reduction in tourist areas*. Madrid: WTO.

Okuyama, T. (2018). Analysis of optimal timing of tourism demand recovery policies from natural disaster using the contingent behavior method. *Tourism Management, 64*, 37–54.

Orchiston, C. (2012). Seismic risk scenario planning and sustainable tourism management: Christchurch and the Alpine Fault zone, South Island, New Zealand. *Journal of Sustainable Tourism, 20*, 59–79.

Paraskevas, A., & Arendell, B. (2007). A strategic framework for terrorism prevention and mitigation in tourism destinations. *Tourism Management, 28*(6), 1560–1573.

Prideaux, B. (2003). The need to use disaster planning frameworks to respond to major tourism disasters: Analysis of Australia's Response to Tourism Disasters in 2001. *Journal of Travel Tourism Market, 15*(4), 281–298.

Prideaux, B., Laws, E., & Faulkner, B. (2003). Events in Indonesia: Exploring the limits to formal tourism trends forecasting methods in complex crisis situations. *Tourism Management, 24*(4), 475–487.

Quintal, R. (1999). Aluviões da Madeira. Séculos XIX e XX. *Territorium, Revista de Geografia Física aplicada ao Ordenamento do Território e Gestão de Riscos Naturais, 6*, 31–48.

Ritchie, B. (2008). Tourism disaster planning and management: From response and recovery to reduction and readiness. *Current Issues in Tourism, 11*, 315–348.

Rodrigues, D., & Ayala-Carcedo, F. (2003). Rain-induced landslides and debris flows in Madeira Island. *Landslide News, Journal of the Japan Landslide Society, 14*, 43–45.

Ryan, C., & Cove, J. (2007). Structuring destination image: A qualitative approach. *Journal of Travel Research, 44*(2), 143–150.

Santana, G. (2003). Crisis management and tourism: Beyond the rhetoric. *Journal of Travel and Tourism Marketing, 15*, 299–231.

Sharpley, R. (2005). The Tsunami and tourism: A comment. *Current Issues in Tourism, 8*, 344–349.

SRES (2010), *Estudo de Avaliacão do Risco de Aluviões da Ilha da Madeira – Relatório S'ıntese*. Instituto Superior Técnico, a Universidade da Madeira e o Laboratório Regional de Engenharia Civil.

Stern, E., & Krakover, S. (1993). The formation of a composite urban image. *Geographical Analysis, 25*(2), 130–146.

The Telegraph. (2010). *Madeira floods: Tourists urged not to cancel holidays*, from https://www.telegraph.co.uk/news/worldnews/europe/portugal/7301442/Madeira-floods-tourists-urged-not-to-cancel-holidays.html

Tsai, C., & Chen, C. (2010). An earthquake disaster management mechanism based on risk assessment information for the tourism industry – A case study from the island of Taiwan. *Tourism Management, 31*(4), 470–481.

Tucker, H. (2016, March). Empathy and tourism: Limits and possibilities. *Annals of Tourism Research, 57*, 31–43.

Walmsley, D., & Young, M. (1998). Evaluative images and tourism: The use of perceptual constructs to describe the structure of destination images. *Journal of Travel Research, 36*(3), 65–69.

Walters, G., & Mair, J. (2012). The effectiveness of post-disaster recovery marketing messages—The case of the 2009 Australian bushfires. *Journal of Travel & Tourism Marketing, 29*(1), 87–103.

Yang, W., Wang, D., & Chen, G. (2011). Reconstruction strategies after the Wenchuan Earthquake in Sichuan, China. *Tourism Management, 32*, 949–956.

# Conclusion

**Tiago Sequeira and Liliana Reis**

## Contents

1 Lessons from What Has Been Discussed in the Book ........ 304
2 Topics for Further Discussion and Research ........ 308
Bibliography ........ 309

**Abstract** A broad lesson from this book is the recognition that climate change is a global phenomenon, which should be accounted for in a multidisciplinary perspective. In fact, it is impossible to deal with climate change and promote its mitigation without taking into account the agents' choices and priorities in different countries and regions. Those choices determine the politics followed by decision-makers at every opportunity (through elections in democracies) and iterations between countries (and eventually have an effect on the well-being of populations) that should take into account the income level, the income distribution, and the health status of populations. Thus, the interplay between politics, international relations, and economics becomes a stage for the analysis made in the different chapters of this book. Additionally, climate change mitigation is crucially dependent on technological change, as is the economic development of societies.

A broad lesson from this book is the recognition that climate change is a global phenomenon, which should be accounted for in a multidisciplinary perspective. In fact, it is impossible to deal with climate change and promote its mitigation without taking into account the agents' choices and priorities in different countries and regions. Those choices determine the politics followed by decision-makers at every opportunity (through elections in democracies) and iterations between

T. Sequeira (✉)
Universidade da Beira Interior and CEFAGE-UBI, Covilhã, Portugal
e-mail: sequeira@ubi.pt

L. Reis
Universidade da Beira Interior, Covilhã, Portugal

Instituto Português de Relações Internacionais (IPRI-NOVA), Lisboa, Portugal



T. Sequeira, L. Reis (eds.), *Climate Change and Global Development*,
Contributions to Economics, https://doi.org/10.1007/978-3-030-02662-2_15

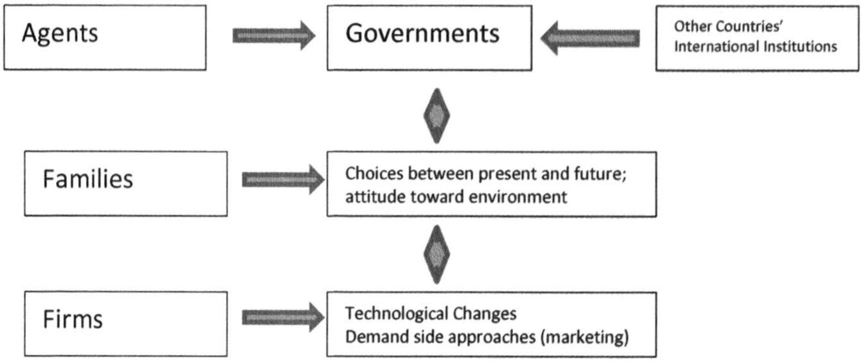


**Fig. 1** The schematic relationship between international relations, politics, economics, and management

countries (and eventually have an effect on the well-being of populations) that should take into account the income level, the income distribution, and the health status of populations. Thus, the interplay between politics, international relations, and economics becomes a stage for the analysis made in the different chapters of this book. Additionally, climate change mitigation is crucially dependent on technological change, as is the economic development of societies. If the agent-based decisions (both families and firms) and their political effects can bias technological change towards cleaner technologies that mitigate the effects of climate change, the future can provide more well-being for the human species (Fig. 1).

## 1 Lessons from What Has Been Discussed in the Book

This book brought together the contributions of authors from several countries with a variety of analyses on climate change. From the research on the chapters of the three parts of this book, we are able to draw some lessons and reveal some of the future research challenges regarding the implications of climate change for global development.

In the first part, the market economies' effects of climate change linked with civil society and states are analysed. For example, Marya Gubareva and Orlando Gomes argue that the main challenge that should be raised vis-à-vis agent-based modelling is how to develop more accurate and precise approaches to modelling and reflecting economy-climate interrelations. Perhaps one of the solutions can be found in incorporating into agent-based modelling the interrelations between climate change and social inequality. Using a game theory approach, Caleiro, Rocha de Sousa, and Oliveira have a very pessimistic view. In fact, they state that the *ecocide* might occur and we are on the way to it. As they argue, their parametrization of a sufficiently low parameter of no punishment for pollution (and thus a high level of defection reward in our agent-based simulations) makes a cooperative situation ending easily in a disastrous situation—the so-called ecocide. Empirical contributions in this part may suggest that heterogeneity between countries should be taken into account. Results from panel threshold regression indicate that both GHG emissions and $C0_2$ emissions affect real GDP growth negatively in the high per capita income regime and contribute to growth in the lower per capita income regime (see chapters from Qaiser Munir and Sook Chin Kok). An interdisciplinary lesson calls for an intergovernmental commitment to face those differences on economic outcome.

In the second part, dedicated to international actors, we realize that if the European Union has been perceived as a "green player" due to its leading role in the external action of climate change, it is also by the increasingly intense regulation of environmental issues. Ana Isabel Xavier analyses the historical background of the EU's environmental policies (from the Single European Act to the Europe 2020 Strategy and the Framework for Climate & Energy 2030) and policy and EU commitments towards global development and climate change, concluding that the EU plays an important role to promote, manage, and regulate climate change and sustainable development worldwide and has a significant degree of *actorness* that helps its presence in the international climate and environmental agenda, underlining that, given the strong intergovernmental nature of the EU climate regime that still has a significant impact in terms of an autonomous, coherent, and effective external diplomatic strategy, EU actors will then depend on the ability not only to contribute to global action to combat climate change but on an endogenous point of view that there are still a number of issues that need to be overcome and may have implications for the future of the union. The intergovernmental nature of most of this subject is also influenced by the idiosyncrasies of some of the countries in their neighbourhood, such as the Balkans. In fact, Teresa Cierco shows that the Balkans are facing very serious challenges in the field of climate change, due to the lack of legal framework for most environmental issues. But what the data presented suggest is that poor environmental governance in the Balkans may not only compromise environmental standards in these countries but also across Europe. However, with this situation environmental issues could also threaten the path of belonging to the EU of these countries. The question of the fragility of some states in responding to the need to adapt legislation also affects the mitigation of the effects of climate change. The continent of Africa, although contributing little to climate change, has been one of the most penalized by its effects, as underlined by Lynus Nyiwul in the chapter "Climate Change Mitigation and Adaptation in Africa: Strategies, Synergies and Constraints". Yet the political vulnerability of several states on this continent seems to compromise the adoption of policy environment necessary for climate-resilient development. Not only do the bureaucratic and legal frameworks require reforms to enhance efficiency, they will certainly need to be updated to account for environmental and more general climate change realities. However, achieving these synergies will require successful implementation of the mitigation and adaptation priorities. This successful implementation in Africa will require fundamental transformations in the institutional and policy environment. Following one of the ideas of the game theoretical approach of Caleiro, Rocha de Sousa, and Oliveira (first chapter, Part 1), Joana Pereira also argues that a climate change catastrophe may be closer than perceived in the public opinion, focusing on the Amazon forest and considering the recent deterioration of the Brazilian governance of the forest. That chapter explored the challenge of sustainable development in the forest, addressing the issues that have marked the federal politics and policies towards the region from 2005 to 2017. The future of this crucial biome is highly dependent on the development paradigm followed by Brazil as well as on the performance of Brazilian diplomacy in international environmental arenas.

With respect to the BRICs (Brazil, Russia, India, China, and South Africa), they claim to have a place in world leadership and indeed have enormous biodiversity richness, as mentioned by Marco Martins. However, these countries continue to present some of the greatest challenges to environmental issues. The struggle for growth and the guarantee of their status quo in the international system has compromised much of their own natural resources, with inevitable consequences for climate change. However, BRICS countries' environment ministers have reaffirmed their commitment to the implementation of the Paris climate deal despite the withdrawal of the USA from the accord.

There will probably be rearrangements in theoretical models of both international relations and economic theory. Classic theories, at least in isolation, appear inadequate to capture the current challenges posed by climate change to global development, which will certainly be greater in the future due to the challenges discussed in the various chapters of this book. Thus, the moment we are living through is a time of profound uncertainty about future political and economic choices among the many governance networks. Probably in the near future, it will certainly be compromise, policy, and right options at the legal level that will determine our ability to reduce or at least mitigate the effects of climate change, and not compromise global development, as Fig. 2 reveals.

From the point of view of international relations, the first part revealed that many of the idiosyncrasies of climate change mitigation in many countries are related to the multiple weaknesses in political institutions and the difficulty in consolidating the rule of law and thus the adoption of legislative measures necessary for the protection of the environment. The USA, which also seemed committed to environmental issues and climate change mitigation globally, suffered a change in the conduct of its foreign environmental policy with the election of Donald Trump.

Global governance has altered institutional architecture and the systemic and institutional conditions under which power is exercised, as well as the characteristics of the political system, the form of government, and the system of intermediation of interests. However, although it has surpassed the state's dimension of power, it created new interstate dimensions and new relations between powers (Vaz and Reis 2017, p. 13). The decentralization of governance to the supranational and subnational level has made it possible for multiple actors (e.g. civil society, NGO, cities) to assume a central role not only in governance but also in the management and sharing of responsibilities in resolving states' problems from an endogenous and exogenous point of view.

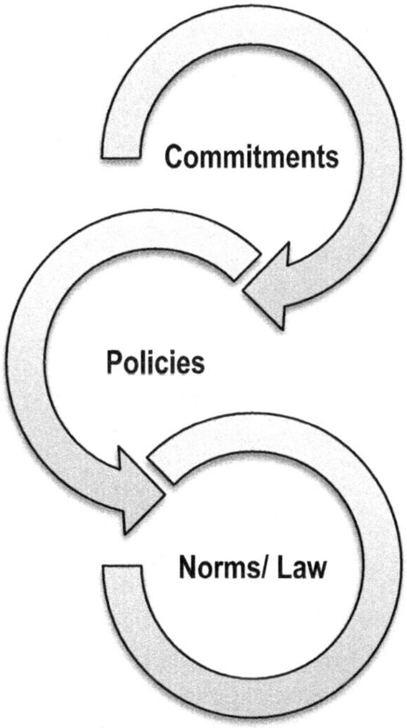


**Fig. 2** Climate change drivers

Keohane and Nye (1974, p. 41) used the term "transgovernmental" for the first time to describe interactions among "sub-units of governments" in response to the "greater complexity" of governance. Currently, threats to the international system and, particularly, to states are diffuse and interdependent. Indeed, the most recent World Economic Forum (2017) named environmental risks (natural disasters, biodiversity loss and ecosystem collapse, extreme weather events, failure of climate change mitigation and adaption, man-made environmental disasters) as a main threat to states' security. These broader and more diversified threats confront and go beyond the powers of the Westphalian state. They require global responses, given the globalized characteristics of threats and risks; and they require an investigation that considers the responses from "inside" states, within a subnational sphere, as we see in Fig. 3.

Part 3 continues the analysis of climate change issues related to climate change but from a more microeconomic perspective. There are three chapters focusing on micro-issues. Together they present a more optimistic view on climate change from a local or micro perspective. These show examples of how micro experiences and lessons, if they could be generalized, could help to avoid some of the climate change catastrophe highlighted above. For example, Cristina Porral analyses a set of marketing experiences that are both the response and the cause of changing attitudes of consumers towards more environment-friendly behaviours and choices. Bruno Pereira, Radu Godina, João Matias, and Susana Azevedo compare the environmental performance of diesel and hybrid automobile engines and conclude for a better

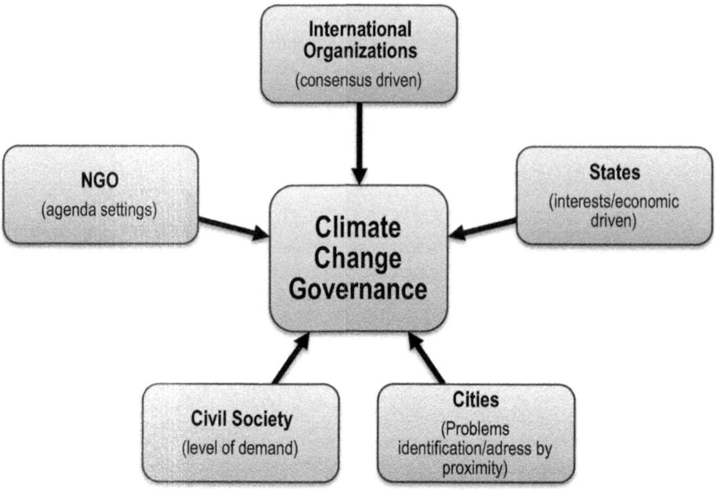


**Fig. 3** Climate change governance

performance of the latter. Also, they identify that taking account of the production stages, both engines have highly detrimental environmental effects in the use phase. This highlights again the importance of consumers' choices towards more environment-friendly vehicles. This chapter also highlights that a technological change that incorporates more electric power potential within hybrid vehicles would have important climate change mitigation effects. In addition to the automobile industry (one of the greatest contributors to climate change), tourism as a growing industry is highlighted by Luiz Pinto Machado and António Almeida. This is important because climate change can be a trigger of changes in tourist demand. As an important consequence of this dependence of the tourism demand due to climate change-driven events, tourism can also be a sector that contributes to climate change mitigation, if good governance policies are implemented. For example, tourist tax revenues may be directed to climate change mitigation measures.

## 2 Topics for Further Discussion and Research

This book has sought to bring together a number of interdisciplinary contributions to the discussion of climate change and global development, but does not close the discussion. In fact, there are a number of issues that need to be monitored in the coming years, including:

- The response of different agents (consumers, families, and the states) to climate change, accounting for economic incentives, political arrangements, and intergovernmental frameworks

- The post-Paris Agreement and the assessment of its effects on the mitigation of the effects of climate change, taking into account the heterogeneous climate change effects that have been identified for high-income and low-income countries
- Necessity of continuing interdisciplinary social, economic, and environmental research on monitoring climate change and the danger of a climate change catastrophe, looking closely at both global and local effects
- Necessity of communicating and influencing public opinion and thus preferences of agents concerning the scientific results on climate change

Many positive trends seem unstoppable particularly at the level of the international "green agenda". In effect, the foreign policy of the main state actors has welcomed the environment and mitigation to climate change as an undeniable resource of soft power. It also seems consensual that most states remain committed to environmental advocacy and sustainable development initiated at the Rio Conference in 1992. Still, the editors of this book consider it to be of crucial importance that we continue to follow the international agenda of all international actors (state and non-state actors) and the evaluation of the results of the Paris Agreement.

As stated by Wyligała (2012, pp. 2–3), "Environmental foreign policy shows how environmental changes, as climate change influence foreign policy processes and how actors and institutions—both domestic and international—recognize national activity towards global problems at local, regional and international level. Green diplomacy is associated more with negotiation-based approach and focused on international struggle within the frameworks of environmental regimes or conferences". In fact, greening of foreign policy emphasizes integrative approach in which domestic agencies incorporate environmental concerns into different aspects of policy, such as investments, development cooperation, defence, and diplomacy. Using "greener missions" improves economic prospects of development.

## Bibliography


Keohane, R., & Nye, J. (1974). Transgovernmental relations and international organizations. *World Politics, 27*, 39–62.

Vaz, D. M., & Reis, L. (2017). From city-states to global cities: The role of cities in global governance. *JANUS.NET e-journal of International Relations, 8*(2), 13–28.

World Economic Forum. (2017). *Global risks report 2017*. Available at http://reports.weforum.org/global-risks-2017/. Accessed 15 May 2017.

Wyligała, H. (2012). *Environmental foreign policy as an element of soft power strategy: The German example*. Paper prepared for the IPSA XXIInd World Congress of Political Science, Madrid, IPSA.

Printed by Printforce, the Netherlands